Born to be Riled

The collected writings of

JEREMY CLARKSON

PENGUIN BOOKS

PENGUIN BOOKS

Published by the Penguin Group
Penguin Books Ltd, 80 Strand, London WC2R 0RL, England
Penguin Group (USA), Inc., 375 Hudson Street, New York, New York 10014, USA
Penguin Group (Canada), 90 Eglinton Avenue East, Suite 700, Toronto, Ontario, Canada M4P 2Y3
(a division of Pearson Penguin Canada Inc.)
Penguin Ireland, 25 St Stephen's Green, Dublin 2, Ireland (a division of Penguin Books Ltd)
Penguin Group (Australia), 250 Camberwell Road, Camberwell, Victoria 3124, Australia
(a division of Pearson Australia Group Pty Ltd)
Penguin Books India Pvt Ltd, 11 Community Centre, Panchsheel Park,
New Delhi – 110 017, India
Penguin Group (NZ), cnr Airborne and Rosedale Roads, Albany, Auckland 1310, New Zealand
(a division of Pearson New Zealand Ltd)
Penguin Books (South Africa) (Pty) Ltd, 24 Sturdee Avenue,
Rosebank, Johannesburg 2196, South Africa

Penguin Books Ltd, Registered Offices: 80 Strand, London WC2R 0RL, England

www.penguin.com

First published by BBC Worldwide Limited 1999
Published in Penguin Books 2006
5

Copyright © Jeremy Clarkson, 1999
All rights reserved

The moral right of the author has been asserted

Typeset by Rowland Phototypesetting Ltd, Bury St Edmunds, Suffolk
Printed in England by Clays Ltd, St Ives plc

Except in the United States of America, this book is sold subject
to the condition that it shall not, by way of trade or otherwise, be lent,
re-sold, hired out, or otherwise circulated without the publisher's
prior consent in any form of binding or cover other than that in
which it is published and without a similar condition including this
condition being imposed on the subsequent purchaser

ISBN-13: 978–0–141–02899–6

This book is dedicated to —
all those people who have bought it.

Contents

Foreword   xv

Norfolk, twinned with Norfolk   1
GT90 in a flat spin   3
Blackpool Rock   6
Gordon Gekko back in the driving seat   9
All aboard the veal calf express   12
Speedy Swede   15
Drink driving do-gooders are over the limit   18
Car of the Century   20
The Sunny sets   23
Who's getting their noses in the trough?   27
Ferrari's desert storm   29
Killjoys out culling   32
Flogging a sawn-off Cosworth   36
Weather retort   38
Burning your fingers on hot metal   41
Speeding towards a pact with the devil   44
Road rage – you know it makes sense   47
911 takes on Sega Rally   50
A laugh a minute with Schumacher in the Mustang   54
Girlpower   56
Nissan leads from the rear   59
Cable TVs and JCBs   62
Mystic Clarkson's hopeless F1 predictions   65

Commercial cobblers   68

Struck down by a silver bullet in Detroit   71

You can't park there – or there   74

Sermon on Sunday drivers   77

A riveting book about GM's quality pussy   81

Aston Martin V8 – rocket-powered rhino   83

Caravans – A few liberal thoughts   87

Blind leading the blind: Clarkson feels the heat
    in Madras   90

Norfolk's finest can't hit the high notes   92

Car interiors in desperate need of some Handy
    Andy work   96

New MG is a maestro   99

Darth Blair against the rebel forces   102

Riviera riff-raff   105

Objectivity is a fine thing unless the objective is
    to be first   109

Kids in cars   111

Brummie cuisine is not very good   115

Last bus to Clarksonville   118

Land of the Brave, Home of the Dim   121

Only tyrants build good cars   124

The principality of toilets   128

Clarkson the rentboy finally picks up a Ferrari   130

Hate mail and wheeler-dealers   134

No room for dreamers in the GT40   137

A rolling Moss gathers up Clarkson   140

Can't sleep? Look at a Camry   143

Big foot down for a ten gallon blat   147

Car chase in cuckoo-land   149

Frost-bite and cocktail sausages up the nose   153

Bursting bladders on Boxing Day   156

Lies, damn lies and statistics   158

Radio Ga Ga   162

Spooked by a Polish spectre   165

Boxster on the ropes   168

Concept or reality?   171

Top Landing Gear – Clarkson in full flight   174

A fast car is the only life assurance   177

Rav4 lacks Kiwi polish   180

Cuddle the cat and battle the Boche   183

Secret crash testing revealed   186

Diesel man on the couch   189

Stuck on the charisma bypass   193

Travel tips with Jezza Chalmers   195

Capsized in Capri   199

Noel's Le Mans party blows a fuse   202

The Skyline's the limit for gameboys on steroids   205

Henry Ford in stockings and suspenders   208

NSX – the invisible supercar   211

Corvette lacks the Right Stuff   214

Footballers check in to Room 101   217

Big fun at Top Gun   220

Traction control loses grip on reality   223

Driving at the limit   226

Global Posting systems   229

Fight for your right to party   233

Gravy train hits the old buffers   235

Weird world of Saab Man   238

Freemasons need coning off   242

The curse of the Swedish smogasbord   244

Pin-prick for the Welsh windbag   248

Showdown at the G6 summit   251

Spelling out the danger from Brussels   253

Dog's dinner from Korea   257

New Labour, new Jezza   260

Sad old Surrey   263

A frightening discovery   266

Hannibal Hector the Vector   269

F1 running rings round the viewers   272

Big cat needs its tummy tickled   275

Elk test makes monkeys of us   278

At the core of the Cuore   281

Last 911 is full of hot air   284

False economies of scale   287

Blowing the whistle on Ford and Vauxhall   290

Hell below decks – Clarkson puts das boot in   293

Country Life   296

Beetle mania   302

Football is an A Class drug   304

Yank tank flattens Prestbury   307

Supercar suicide   311

Bedtime stories with Hans Christian Prescott   313

Clarkson soils his jeans   317

Burning rubber with Tara Palmer-Tailslide   321

Jag sinks its teeth in   324

Kraut carnage in an Arnage   327

Absorbing the shock of European Union   330

Minicabs: the full monty   333

Supercar crash in Stock Exchange   336

The school run   339

Voyage to the bottom of the heap   343

Van the Man   346

'What I actually meant was . . .'   349

Mrs Clarkson runs off with a German   352

Un-cool Britannia   355

Move over Maureen   358

Toyota gets its just deserts   361

Kristin Scott Thomas in bed with the
   Highway Code   364

Time to change Gear   368

Even soya implants can't make a great car   370

Lock up your Jags, the Germans are coming   373

Well carved up by the kindergarten coupé   376

Fruit or poison?   379

Left speechless by the car that cuddled me   382

One car the god of design wants to forget   385

Can a people carrier be a real car? Can it hell   388

Hell is the overtaking lane in a 1-litre   391

Forty motors and buttock fans   394

Audi's finest motor just can't make up its mind   397

Keep the sports car, drive the price tag   400

Out of the snake pit, a car with real venom   403

The Swiss army motor with blunted blades   406

Perfection is no match for Brian and his shed   409

Waging war with the motoring rule book   412

Evo's a vulgar girl, but I love her little sister   415

At last, a car even I can't put in a ditch   418

Trendy cars? They're not really my bag   421

Why life on the open road is a real stinker   424

Cotswold villages and baby seals   427

Shopping for a car? Just ask Rod Stewart   430

Gruesome revenge of the beast I tried to kill   433

Out of control on the political motorway   436

Old sex machine still beats young fatboy   439

Whatever happened to the lame ducks?   442

Bikers are going right round the bend – slowly   445

Freedom is the right to live fast and die young   448

A shooting star that takes you to heaven   451

Congratulations to the Cliff Richard of cars   454

David Beckham? More like Dave from Peckham   457

A prancing horse with a double chin   460

£54,000 for a Honda? That's out of this world   463

It's Mika Hakkinen in a Marks & Spencer suit   466

Like classic literature, it's slow and dreary   469

Prescott's preposterous bus fixation   472

Take your filthy, dirty hands off that Alfa   475

Yes, you can cringe in comfort in a Rover 75   478

Don't you hate it when everything works?   480

The kind of pressure we can do without   483

Three points and prime time TV   486

Every small boy needs to dream of hot stuff   489

Footless and fancy-free? Then buy a Fiat Punto   492

Now my career has really started to slide   495

The best £100,000 you'll ever waste   498

Styled by Morphy Richards   501

The terrifying thrill of driving with dinosaurs   504

Perfect camouflage for Birmingham by night   507

Another good reason to keep out of London   510

My favourite cars   513

Need a winter sun break? Buy a Bora   516

Driving fast on borrowed time   519

I've seen the future and it looks a mess   522

Nice motor; shame it can't turn corners   525

Stop! All this racket is doing my head in   528

Looks don't matter; it's winning that counts   531

It's a simple choice: get a life, or get a diesel   534

Insecure server?   537

Ahoy, shipmates, that's a cheap car ahead   540

So modern it's been left behind already   543

Something to shout about   546

Appendix   551

Foreword

As a motoring journalist, you spend much of your life on exotic car launches, feeding from the bottomless pit of automotive corporate hospitality. And then you come home to tailor a story that perfectly meets the needs of the public relations department that funded it. For sure, you dislike the new 'xyz' but what the hell. Say it's fabulous and you're sure to be invited on the next exotic press launch. And so what if some poor sucker reads what you say and buys this hateful car? You're never going to meet him because by then, you'll be on another press launch, in Africa maybe, trying out the 'zxy'.

I used to live like this, and it was great. But sadly, when I climbed into *Top Gear*, I had to climb off the gravy train. This is because, all of a sudden, people in petrol station forecourts and in supermarket checkout queues started to recognise me. These people had bought a car because I'd said they'd like it. And they didn't like it because it kept breaking down. So now, they were going to fill my trousers with four star. And set me alight.

I learned, therefore, pretty quickly that the single most important feature of motoring journalism - or any kind of journalism for that matter - is speaking your mind. You mustn't become Orville with a PR man's hand up your bottom. I know that over the years, these columns from the *Sunday Times* and *Top Gear* magazine have caused PR men to choke on their canteen coffee, and that makes me happy. I have been banned from driving Toyotas, I've had death threats, and my postman once had to deliver letters from what seemed like the entire population of

Luton. But at least I can sit back now and know that every single opinion on these pages was mine. I just borrowed a car, and told you what I thought. No sauce. No PR garnish.

I never said you had to agree with my opinions but I can say that in the last 10 years, I've only been on maybe five press launches and I've sat through all of them with my fingers in my ears, singing old Who songs at the top of my voice.

Sure there are some things I wish I'd never written. I wish, for instance, that I'd learn to stop predicting the outcome of a Grand Prix championship and I wish I'd never been so rude about horses. But most of all, I wish I wasn't growing up quite so quickly. Just seven years ago, I had an Escort Cosworth and wanted a minimum speed limit of 130mph on motorways. Babies, I thought, were only any good if served with a baked potato and some horseradish sauce. And here I am now with an automatic Jaguar, three children and a fondness for the new 20mph inner city speed limit.

So, as you read through the book, you might find what you think are contradictions, some evidence perhaps that I told the truth one day and some bull the next.

Not so, I've just got a bit older.

I expect soon that I shall start to favour cars that have wipe down seats, denture holders in the dash and a bi-focal windscreen - but don't worry. Even when my nose has exploded and all my fingers are bent, I still won't like diesel, or people carriers or Nissans, and I shall still be happy to point out the weirdness of America. 250 million wankers living in a country with no word for wanker.

And be assured that when I'm dead, they'll find a note at my solicitors' saying that I want to be driven to my grave at 100mph in a something with a V8.

Jeremy Clarkson, 1999

Norfolk, twinned with Norfolk

In a previous life I spent a couple of years selling Paddington Bears to toy and gift shops all over Britain. Commercial travelling was a career that didn't really suit – because I had to wear one – but I have ended up with an intimate knowledge of Britain's highways and byways. I know how to get from Cropredy to Burghwallis and from London Apprentice to Marchington Woodlands. I know where you can park in Basingstoke and that you can't in Oxford. However, I have absolutely no recollection of Norfolk. I must have been there because I can picture, absolutely, the shops I used to call on in, er, one town in this flat and featureless county.

And there's another thing, I can't remember the name of one town. The other day I had to go to a wedding in one little town in Norfolk. It's not near anywhere you've heard of, there are no motorways that go anywhere near it, and God help you if you run out of petrol.

For 30 miles, the Cosworth ran on fumes until I encountered what would have passed for a garage 40 years ago. The man referred to unleaded petrol as 'that new-fangled stuff' and then, when I presented him with a credit card, looked like I'd given him a piece of myrrh. Nevertheless, he tottered off into his shed and put it in the till, thus proving that no part of the twentieth century has caught up with Norfolk yet.

This is not surprising because it's nearly impossible to get there. From London, you have to go through places such as Hornsey and Tottenham before you find the M11, which sets off in the right direction, but then, perhaps sensibly, veers off to Cambridge. And from everywhere else you need a Camel Trophy Land Rover.

Then, when you get there and you're sitting around in the hotel lobby waiting for the local man to stop being a window cleaner, gynaecologist and town crier and be a receptionist for a while, you pick up a copy of *Norfolk Life*. It is the world's smallest magazine.

In the bar that night, when we said we had been to a wedding in Thorndon, everyone stopped talking. A dart hit the ceiling and the man behind the counter dropped a glass. 'No one,' he said, 'has been to Thorndon since it burned down 40 years back.' Then he went off, muttering about the 'widow woman'.

Moving about Norfolk, however, can be fun. I am used to having people point as I go by. Most shout, 'Hey, look, it's a Cosworth!' but in Norfolk they shout, 'Hey, look, it's a car!' Everywhere else people want to know how fast it goes, but in Norfolk they asked how good it was at ploughing. The spoiler fascinated them because they reckoned it might be some sort of crop sprayer.

I'm sure witchcraft has something to do with it. The government should stop promoting the Broads as a tourist attraction and they should advise visitors that 'here be witches'. They spend millions telling us that it is foolish to smoke, but not a penny telling us not to go to Norfolk – unless you like orgies and the ritual slaying of farmyard animals.

The next time some friends get married in Norfolk, I'll send a telegram. Except it won't get there because they haven't heard of the telephone yet. Or paper. Or ink.

GT90 in a flat spin

Earls Court becomes the fashion capital of the western world this week as the London Anorak Show opens its doors to members of the public.

Better known as the Motor Show, families will be donning their finest acrylic fibres and braving the Piccadilly Line so that they may gawp at all that's new and shiny.

However, if you want to see all that's really new and shiny, you need to stay on the Piccadilly Line until you arrive at Terminal Four. And then you should catch a plane to Japan.

The trouble is that the London Motor Show clashes with the Tokyo Motor Show, and there's no surprises for guessing which one is rated most highly by the exhibitors.

So, if a car manufacturer has spent all year developing a new concept to wow the crowds at an exhibition, it goes to Japan, leaving London with the mainstream stuff, the kind of cars that are parked in your street anyway.

That said, it will be your first chance to see the Ferrari F50 (which makes the show worthwhile all on its own) and the TVR Cerbera, but as its astonishing engine will be off, onlookers will be deprived of its USP.

Other notable debutantes include the MGF, the Renault Megane, the really rather nice Fiat Bravo and, of course, the fascinating and interesting Vauxhall Vectra which, in case you can't find it, is the one that looks pretty much the same as a Cavalier.

However, pretty well all the one-off concept cars will be in Tokyo, and in case you're wondering why we don't move the dates of our show, I should remind you that we once did. But because it no longer straddled the half-term break, no one came. And anyway, the new dates meant we were competing with Paris.

And all the manufacturers thought France more important than London anyway. We could, of course, move our event to June but I've just checked and there's a show on then in Pune, a small town 120 miles from Bombay. And I'm pretty damn sure that's where the car makers would concentrate their resources.

The upshot of all this is that you won't be able to see the Ford GT90, and that's a pity because it's America's first attempt at a supercar.

At this point, I'm sure, Wilbur and Myrtle will be running around waving their arms in the air and pointing to the Corvette ZR-1 and the Dodge Viper, saying that these are supercars. But they're not.

And nor is that absurd Vector which is made in agonizingly small numbers in California, and nor was the Pontiac Fiero.

Supercars are what the Europeans do. We are the only ones who know how to make a car go quickly . . . round corners.

People at Ford in Detroit say the old GT40 was a supercar and that they made it, but again, they're wrong. It may have had an American engine but the rest of it, the important stuff, was as American as Elgar.

The GT90 is their first attempt and it seems to work rather well, because it is capable of 235mph, making it the fastest road car in the world. It does 0 to 60 in 3.1 seconds so it is pretty sprightly on that front too. And because it is mid-engined, light and sits on a modified Jaguar XJ220 chassis, it should be pretty nifty through the bends too.

Under the engine cover, you will find a 6000cc V12 which has four turbos. Total output is a staggering 720bhp, making it not only more powerful than the McLaren F1 but, significantly, more powerful than the McLaren F1 driven by Mika Hakkinen.

It's a looker too. They say there are hints of the GT40 but I couldn't find any. For a kick-off, there isn't a single curve on the car – every line is straight, except the roof which is a glass dome.

And that's why, when I drove this monster, the turbo's wastegates were jammed open, limiting me to just 440bhp. They say that if the engine were working at full noise, and the chassis could handle the onslaught, the glass would crack, splinter and break.

They were very, very worried about this $4 million one-off car as I set off for a couple of laps at Le Mans because the very next day it was off to the Tokyo show.

That's why I went so fast. To punish Ford for sucking up to the Japanese and ignoring just about the only market in the world that truly loves their ordinary cars, I had fun with the GT90.

Until one of its wheels came off. I'd been enjoying the bark of that V12 and experimenting with the radar sensors which ignite a red light in the door mirrors when you're being overtaken, when it all went horribly wrong.

The tail stepped out of line and even the downforce from its truly huge tail spoiler failed to prevent a spin.

Unfortunately, there was no damage and I didn't hit anything, which means the fastest car in the world is sitting right now under the rising sun.

At the London Motor Show, Ford is hoping the new Fiesta will be enough to draw the crowds. And though it's a nice little car, they're as muddleheaded as that fine band, REM, who once said, 'I forget my shirt at the water's edge. The moon is low tonight.'

Blackpool Rock

The Alfa Romeo GTV6 had the worst gearbox I've ever encountered, the worst driving position and the worst record for reliability. Nevertheless, I bought one.

I knew it was a hopeless basket case but I'd become smitten by the noise its engine made: a rumble in the jungle at low revs and an almost eerie howl as it neared the red line.

I would put up with the massive bouts of truculence, the deep discomfort and the absurdly heavy steering because no car before, or since, has ever made such a glorious sound. It was music to the enthusiast's ears, like a cross between 'Ode to Joy' and 'Nessun Dorma'.

However, its title as the best sounding road car of them all is under threat from the wheeled equivalent of Aerosmith. You really could call the new TVR Cerbera heavy metal were it not fashioned from plastic.

The best way to first experience this car is to be about seven miles away. As it comes towards you, it's like being in a horror movie. The monster is getting closer. The Thing. The Blob. Terror has no shape. But God, what a noise.

Amplify the sound of someone ripping calico a thousandfold and you're getting near the mark, but you'll miss out on the gear changes, each one accompanied by a frantic popping and spitting as unburned fuel crackles and fizzes its way down those two Matrix Church superguns that TVR calls exhaust pipes.

As the car tears past, at its top speed of 160mph or so, your eardrums will burst. There's no music here, just volume. Woodstock just went by. It was even pink.

Now for 15 years or so, Blackpool-based TVR has used the Land Rover V8 in its cars and they've sounded good in a beefy, brutal sort of way, but this Cerbera is on another level altogether. So what's the story?

Well it would seem that the company's charismatic boss, Peter Wheeler, was not very pleased when Rover was bought by BMW. It's reported that he said, 'I'll not have anything bloody German in my cars.'

And so he set out to build his own V8. It's designed to be just like the unit you'll find in a Formula One car except it has just two valves per cylinder and displaces 4.2 litres instead of 3.5.

It only produces 360bhp but that, in a car which weighs about a ton is enough, believe me. It's enough to get you from 0 to 60 in four seconds for a kick off. Six seconds after that, with blood pouring from your ears, you'll be past a hundred.

Now in a normal car, the more responsible motorist can trundle around, knowing the power is there but only using it when necessary. This is not an option in the Cerbera.

If this car was a drug, it would be crack cocaine. Its power is viciously addictive and you find yourself holding the throttle wide open just to hear what the motor will sound like at 6000rpm. You take it to the limiter every time, not caring that there's a corner coming up and that, really, you should be standing on the brakes.

A lot of people are going to lose their licences with the Cerbera; that much is for sure. Everything else is less clear because I was driving a prototype with wonky brakes and a suspension set-up that was not finished.

I therefore don't really know how the finished car will handle but if it's anything like other TVRs, it'll be average. However, though it might not be fast through the bends, it will be like lightning between them.

And comfortable too. Thanks to the long wheelbase, it rides with a dexterity and suppleness I wasn't expecting.

It also has film star looks. Though it's essentially a lengthened, hardtop version of the droptop, two-seater Chimera, it manages to look completely different: like a chopped 1950s Mercury in many ways.

Inside, it's even more wild. To get there, you hit the remote control plipper once, to unlock the doors, and then again, to open them. There are no handles.

Once inside, there's a boot and two back seats which could handle anyone up to about 5ft 5in, but you tend not to notice because the dash is straight from the pages of Isaac Asimov. The cream-coloured dials are grouped above and below the steering wheel, which in turn is festooned with buttons.

With all the controls taken care of, the designer has been allowed to let his imagination run wild. And what makes it so good is that unlike Aston Martin and Lotus, TVR doesn't use switches from Metros and Vauxhalls and Sierras. They make their own.

Despite this, the Cerbera will cost less than £40,000. So does this mean TVR is running a social service, providing cars at a loss?

This is unlikely when you know the boss. A few years

ago, when the Labour Party was holding its conference in Blackpool, Paul Boateng rang to ask if he could borrow a car while he was in town. Peter Wheeler was heard to mutter: 'If he wins the election, he can have the bloody company.'

It seems TVR can afford to sell its cars cheaply because it makes so few. Instead of having to buy robots to make thousands of parts a year, the designers can wander into the factory and simply ask the line workers to 'do it this way from now on'.

And that means we have the chance of buying, for half the price of a Ferrari, a hand-built, all British supercar.

That, all on its own, would be enough to swing it for most people but what makes this car so desperately appealing to me is not the power or the speed or even that wonderful dash. Yes, I love the looks too, but they're not the issue either.

I would buy this car because it's the living embodiment of counter culture rock and roll. Today, when most cars are packaged like Michael Bolton, or rely on past glories like the Stones, the new TVR gets back to basics. You would not want it to marry your daughter.

Plus, to use a word the Cerbera would undoubtedly choose, it's as loud as a bastard.

Gordon Gekko back in the driving seat

As the 1980s drew to a close, Britain was gripped by a recession which would see car sales fall from 2.2 million a year to just over 1.5 million. Hundreds of thousands of people lost their jobs. Factories closed. House prices plummeted. So did hemlines. It was all horrid. Through-out those dark and gloomy days, gurus told us that the glorious times of easy credit, greed and avarice were over and that in the 1990s we would all be busy gathering wood for pensioners and helping to set up community service projects. Cars would have catalytic converters and airbags. Films where everyone got shot would be replaced by films where women wandered around meadows in beekeeper hats, making daisy chains and falling in love with gallant and good men on eco-friendly white horses. It sounded like the worst nightmare I could possibly imagine and it all looked like coming true when, in *Terminator 2*, Arnie refused to kill anyone.

But, thankfully, the British recession has ended and those old values are back on line. Girls who had been forced into long and tedious skirts now insist on huge slits up to their ladies' areas, estate agents are selling houses in Chelsea for £25 million, the stock market is up above the ionosphere. Greed is good. And greed is back. Phew.

And nowhere is this phenomenon more apparent than in cardom. At the Motor Show, I talked with thousands of visitors and not one asked about safety, or economy, or value for money. They wanted to talk power.

In ten days no one suggested that the new Golf Estate was a good car because of the space in the back for meals-on-wheels deliveries. No one talked about how BMW's recycling programme might conserve the earth's resources.

No one noticed that there wasn't a single electric car in Earls Court, but the aisles were full to overflowing with people lying on the floor having paddy fits because the McLaren F1 was an absentee. When they came round, they talked about the Aston Martin Vantage, the 7 litre, twin-supercharged Lister Storm, and the Lamborghini Diablo VT. Suggest that we should rip out all the cats, fit six downdraught Webers and prime them with five-star fuel and they wet themselves. And so did I.

Outside, ladies in Puffas and corduroy trouserwear handed out leaflets demanding that cars be banned from city centres. If they could have had a pound for every time someone told them to get back to Greenham, they could have afforded nicer leaflets, and a Lear jet to drop them from. Inside, you couldn't get near the TVR stand. All the other manufacturers with their airbags and their safety videos and their girls in ankle-length skirts were watching tumbleweed blow by, while the boys from Blackpool had to fight off the crowds with sticks. Their Cerbera no doubt meets the letter of the environmental law but as regards the spirit it's a V8-sized joke, a 5 litre two-fingered salute to the world's whales and all who love them.

The safety lobby with their meat-free fridges and their green-tinted specs had their 15 minutes of fame in 1991, but they must now realize that Gordon Gekko is back in the driving seat, with his foot flat down in a tyre-squealing slide back to 1986. And even though the insurance companies are doing their best to ensure we can't afford cars that will squeal tyres, we, like all clever capitalists, still have an answer.

We are buying more and more off-road cars so that we can drive through the countryside. Literally.

All aboard the veal calf express

It is a fact that most people in the major financial institutions go to work by train, which means they harbour a deep-seated hatred of British Rail.

You can see them all piling out of the station at 11 a.m., six hours after they left home in Kent, clutching their Customer Charter form and muttering to one another about how it was leaves the last time. And the wrong-shaped snow before that.

This is a big problem for British Rail as it heads towards privatization. Without the support of the City boys – and they're hardly likely to get it, having wrecked their lives for so long – the flotation will be a disaster.

So they've come up with a cunning plan, which involves demonstrating to their customers that the alternative to rail travel is even worse.

Ever since I was thrown off a train by the police for arguing with a guard – who should have been drowned at birth – I've made it my business to avoid British Rail's pitifully inadequate, overpriced, badly run, slow, sick-making service, but last weekend I had to go to Harwich.

And for all sorts of complicated reasons I couldn't drive, so with bad grace I set off for Liverpool Street station, where the man at the ticket desk said, without looking up, that Aborigines have their fish in the laundry.

Seriously, this guy had not mastered the art of speech and if he'd been on one of those customer care programmes that BR is always harping on about, I can only deduce that the lecturer was on holiday that week. Or Goebbels.

I explained that I couldn't understand a word he was on about and that it might be better if he looked up so I could see his lips. This helped a lot and I was able to work out that the train to Harwich had been cancelled. No sorry. No nothing. His head just flopped down again like his neck had suddenly broken.

Which it would have done had there not been a piece of glass separating us. Why do they have that glass anyway? Who'd want to rob those dunderheads?

Instead of a train, there was to be a coach and this sent shivers down my spine. A coach. I'd rather have gone to the dentist. I don't go on coaches. Coaches are for old people on tours of North Wales. Coaches are for students who want to go from Northampton to Sheffield for 10p. Coaches don't have seatbelts and they roll down embankments, killing everyone on board.

To make sure we boarded it, there were some small Chinese women marshalling the crowds, shouting at shufflers. It was like a scene from *Schindler's List*. Exactly where was the bus's exhaust outlet?

Now, British Airways have proved that it is possible to fit a human being into a space 30 per cent smaller than his body but the coach operators have gone much further.

What you do on a coach is get yourself roughly near the seat and then a Chinese woman comes along with a mallet and hammers you into position. Then some Scandinavians pile rucksacks on your head.

I began to wonder why on earth anyone needs a seatbelt on a coach. The driver could have driven into a wall at 100mph, and I wouldn't have felt a thing.

There is no smoking on board but that's OK because not even Harry Houdini could have got into my pockets.

And anyway, the packet had taken a direct hit from the mallet so all the Marlboros were bent and broken.

Now I've just come back from Cuba and I remember staring in open-mouthed wonderment as the buses there, huge 300-seaters, trundled by with 500 dismal faces pressed to the glass. What I hadn't realized is that I was staring in the face of sheer luxury.

Bus travel in Britain is far worse, and the pain is doubly bad here because we know what we're missing. We know that it isn't beyond the wit of man to fix up a buffet bar or a lavatory or, indeed, to space the seats in such a way that I could breathe properly. They didn't even have women coming down the aisles offering to empty a pot of coffee into your lap in exchange for £1.20.

I swear as we went past one field in Essex, a herd of veals were pointing at us and waving placards.

On a coach, you pay your money and, crashes permitting, it takes you there. That's it. This is frill-free travel, and at the other end of your journey more people come with spatulas to ease you out of your seat.

It was, without a doubt, the worst two hours of my entire life and when we emerged at Harwich docks I found myself staring wistfully at the trains there. They looked so big, and so fast, and the staff all looked like angels – a bit fatter perhaps – but with their sandwiches and their teas they were definitely God's children.

And they are working for a bunch of people who are, very obviously, brilliant. To have thought up the idea of putting disgruntled rail customers on a coach once in a while to shut them up is inspired.

And anyone who can think like that gets my vote. When the flotation comes, I'll take 400 shares please.

Speedy Swede

If the makers of Blue Nun were to convince the entire nation, within the space of two years, that their sickly interpretation of wine is better than Chablis, I wouldn't be surprised.

Given enough money, it can be done. I know this because in just 18 months Volvo has turned itself from a music hall joke into a serious and credible BMW rival.

It all began, gently at first, with the introduction of the 850, a surprisingly nice car to drive but you'd never know it, what with those Etch-a-Sketch lines and that badge.

Never mind that the top models had a creamy 2.5 litre, five-cylinder 20-valve engine and a truly sophisticated rear suspension, it was still bought by old people who indicated left a lot and turned on their high-intensity rear fog lights during June, in readiness for autumn.

The rest of us were still safe. We could still spot a Volvo coming the other way and get out of its way. Motorcyclists could see one approaching a main road and know, for an absolute certainty, it wasn't going to stop.

Then Volvo gave us the T5. And before everyone woke up to the fact that this was a really nice car to drive, and seriously fast too, Volvo had entered the world's most prestigious racing series for saloon cars. Only they'd entered an estate.

There was turmoil in the motoring world. 'No,' we said to them politely, 'It's not the British Towing Car Championships. Please pull out. It will be embarrassing.'

And it was. In the first year, they lost spectacularly. And as the Swedish tidal wave cruised round in the middle of the field, the crowd pointed and made bovine lowing noises.

The trouble is that in the car parks at these events you started to notice, among the Sierra Cosworths and BMWs, a growing number of the aforementioned T5s, finished in black, and lowered, and sitting on 17 inch gunmetal grey alloys.

They looked very good and the cognoscenti were impressed, in a confused, what's happening here, sort of way.

Then all hell broke loose. Volvo started its second year in the Touring Car Championship with a brace of saloons that actually won some races. And they bought every single advertising slot on ITV so that we could see stunt men and photographers and meteorologists whizzing around in their Volvos.

Never slow to leap onto a bandwagon, I got hold of a T5 for my wife, and pretty soon everyone who'd ever had a Volvo was saying that they'd been right all along and that they knew I'd come round to their way of thinking in the end.

Things by now were out of control because to run alongside the T5, Volvo brought out an even meaner T5R. And now there is, simply, the 850R. Or as my wife calls it, the R2D2. Or as I call it, Terminator 2.

You'll have to think up your own name because it just says Volvo on the back. However, no one is fooled for long, thanks to the rear wing, the vivid red finish, the six-spoke gunmetal grey alloys and the chin spoiler that grazes the road.

Inside, there's powered, heated Suedette seats and all the fruit. There is also a wooden dashboard, the likes of which I have never seen before. You see, it's made from what looks like polished pine and it is absurd.

It's useful, though, because there's no way you'd climb inside and think of the car as ordinary in any way. OK, so it starts with a key and the clutch pedal is on the left, but once you've let it up a bit you're at the controls of a wheeled neutron bomb.

The huge turbo means the 2.3 litre motor now develops 250bhp, and that, translated into bald figures, equals a 0 to 60 time of six point something seconds and a top speed of 160mph. In a Volvo.

It will cost about £32,000 whether you have the saloon or the estate, manual or automatic transmission, and while that's a lot, I have to say, you do get a lot of car for the money.

What surprises is the sophistication. Instead of being bad and loud, it's all quite subdued. You even get traction control which does its best to mediate as the explosion of power fights with the front-wheel drive.

Saab once said you can't put more than 170bhp through the front wheels. But Volvo has anyway and they've ended up with a car that you drive like your trousers are on fire.

There's a Terminator 2 outside my house right now, and as it's three in the morning I'm sorely tempted to take it for another drive. I love looking at the body language of those in front as they struggle to see what on earth is behind. It's a Volvo Jim, but not as we know it.

And it isn't either. It flows through the bends and while the ride is firm on those unbelievably low-profile tyres, it's never jarring. It's just like a BMW really, only faster.

And there you have it: an entire piece about Volvo where the word 'safety' didn't crop up once. Mine's a Neirsteiner.

Drink driving do-gooders are over the limit

How very heartening it is to see that the government is to step up its fight against the bubonic plague. Even though they admit that it was wiped out by the Great Fire in 1666, they still feel that more funds and more hospitals are needed to combat this dreadful disease. It's also good to note that, at last, they are to prevent the Royal Navy from using press gangs to recruit new sailors. 'They have offices in most town's high streets and I don't know why they won't use them,' said a spokesman last week.

Other recent announcements from Whitehall couldn't have come a moment too soon. Kings will no longer be allowed to behead people they don't like very much, Wessex is to get its own legislature and the campaign against drink driving is to be moved up a gear. What?

There is now lamb chop all over my television because there I was, eating supper, when the Roads Minister, Robert Key – who looks like he's seen rather too many lamb chops in his time – came on the news to talk about his war on people who drink and drive. All seven of them.

In 1982, 43,341 people were breathalysed and 31.1 per cent of them were over the limit. Something needed to be done, and something was. In 1992, 108,856 people were breathalysed and under 8 per cent were found to be positive. In other words, the government has won its battle.

But Mr Key says 610 people died in drink-related accidents last year and that his fight goes on. Well, my dear chap, most of those were wobbly pedestrians who fell in front of sober drivers and, short of adopting a Muslim attitude to drink, you aren't going to do much about that sort of thing, are you?

Apparently, yes. In America, dinner party hosts are being sued by their friends for failing to provide soft drinks and, while that is unlikely to catch on here, Mr Key does ask that we encourage sobriety when we have people round. Now look, I spend most of my time these days sitting around dinner tables not being allowed to smoke or eat meat – and now Key says that I can't have a glass of wine either. Bet he never bans food.

His next point is that young people often find it difficult to say no to a drink because of social pressure. The last time someone was this wrong, he was called Neville Chamberlain and he had a piece of paper in his hand.

It is, in fact, old people who are far and away the worst offenders. And the reason they get away with it is because, at night, the police tend to stop youngsters in hot hatchbacks rather than rosy-cheeked farmers in Jags.

Key has proved that he is not in the real world and that he should be fed to the lions. But he has yet more to say. It seems he wants to lower the legal limit, arguing that one pint affects a person's ability to drive. Sure does, fatty, but so does being old. A 17-year-old with one pint in his triangular torso has faster reactions than a sober pensioner, so why not ban old people from driving? Or people with a cold, or those who need to go to the loo, because I sure as hell can't concentrate when I'm bursting for a pee and you haven't provided any service stations. And anyway, what do you lower the legal limit to? Nought? And when does someone have no alcohol in their blood? Five hours after a pint? Five days? No one would ever dare drive again.

We've had a long line of idiots in the Transport Ministry but this one tops the lot. And weighs the most as well.

Car of the Century

As the motor car edges towards its 100th birthday in Britain next year, it seems like a good time to ask the question: what is the best car ever?

Top Gear magazine recently surveyed everyone in the know and found opinions varied somewhat. The Mini, the Model T Ford, and various old Mercs, Alfa Romeos and Ferraris were popular choices. Gareth Hunt even suggested the winner should be a Humber. Damon Hill went for the Renault Laguna.

But actually he's probably nearer the mark, because new cars are bound to be better than old ones. The Renault Laguna, though dull and tedious, is faster, safer, kinder to the environment and blessed with better cornering prowess than a Bugatti Royale.

Bernd Pischetsreider, BMW head honcho, said the best car ever was the old BMW 507 Roadster. Well hey, if it's so good matey, how come you don't start making it again?

Here's the thing. Every month, one car firm or another invents a new way of fitting a new piece of techno wizardry into one of its products. And usually, it makes the car better.

Sure, there have been some daft ideas like rear-facing video cameras instead of mirrors. Er, what happens when the lens gets dirty? And I still don't know why the old Nissan Bluebird had two trip switches, despite a number of letters from various people in cardigans.

Look at aerodynamics for instance. Only 12 years ago, Audi gave us the 100 which had a drag co-efficient of just .30. It had flush-fitting glass and was rounded like a blancmange. Today, we have the E Class Mercedes which slips through the air even more neatly, despite its wavy and weird front end.

This means that even the larger-engined versions can sip petrol through their fuel injectors at the rate of one gallon every 33 miles. That would have been impossible even 10 years ago.

Then there's power. There was a time when people cooed over Ferraris that developed 200 horsepower, whereas today 2.0 litre Escorts can manage that. It's almost impossible to buy a car that won't do a hundred. (If you really want one, various Mercedes diesels make a pretty good stab at it.)

Then there's the environment. The Volkswagen Beetle could kill a rain forest at 400 paces whereas today's Golf trundles around with tulips coming out of its exhaust. The gas coming out of a Saab is actually cleaner than the air that went in. That's true, that is.

And we mustn't forget safety. If Marc Bolan had hit his tree in a modern car, T-Rex would be at the top of the charts today with 'Fax Sam'. If James Dean had been in a 928 Porsche, we'd all be at the cinema this evening watching *Rebel With a Pension*.

I can remember being hawked around motor shows when I was a boy, clutching my crotch with excitement because my father was thinking of buying a Peugeot 604 that had electric windows.

Today, you can buy a Ford Fiesta for £10,000 that has air conditioning, a CD player, a heated front windscreen and anti-lock brakes. Traction control is commonplace and BMW is even fitting televisions now. Mercedes – that name keeps cropping up – will sell you windscreen wipers that come on automatically when it rains.

Cars today are quieter and more comfortable too, but more importantly, they're cheaper. In the 1960s, only the very upper echelons of the middle classes could afford a medium-sized Vauxhall, whereas today the Astra is yours for half what a petrol pump attendant makes in an hour.

I therefore mock and taunt anyone who says that the best car ever is some hopeless old classic with drum brakes.

The best car ever absolutely must have been introduced within the last year or so, because then it will incorporate all the advances we've seen recently.

This means the best car ever is out there now, in a showroom. But which one is it?

I'm tempted to say the new Ford Fiesta because here we have a car that does everything you could reasonably expect, and a whole lot more than you could have expected in 1972.

But I think it's more likely to be a Mercedes. These cars are built like no others, with an integrity that would leave the people at Sellafield gasping. I think you could buy an E Class today and never tire of it. Everything that is sensible is on that car. There's no waste, no silly frills, no nonsense. It just gets out there and does the job, exquisitely well.

And therein lies the problem. The greatest car ever should get out there and do the job, but it should do more besides, which is why I have to say it's the Ferrari 355.

This car is as much a piece of sculpture as a lump of engineering. You could derive as much pleasure from putting it in your sitting room, where the piano used to be, and looking at it as you could from going for a drive.

But if you get out there, you will have a V8 with five valves per cylinder. You will rev it to 9000rpm between gear changes. And all the time you will know you have airbags and catalytic converters and anti-lock brakes and all the other stuff that a great car should and must have.

There's one other thing too. No car can truly be great *unless* it's a Ferrari.

The Sunny sets

The question I am asked most frequently is this: what's the worst car you've ever driven?

The FSO Polonez, a Polish built Fiat cast-off with styling from the pupils at Form IVb at High Wycombe primary school, is an obvious contender.

But then there's the Mahindra Jeep, an Indian-built four-wheel drive vehicle, and the Vauxhall Nova. Oh heavens, I nearly forgot the Lada Samara. And the Volvo

343, which was only safe because it could never achieve a high enough speed to cause injury. And the Morris Marina, which usually did the decent thing and disintegrated before leaving the dealership.

Wait . . . The worst car I've ever driven is the Nissan Sunny. Pick any one of the countless different models from the seven generations of Sunnies which have come and gone in the last 29 years, and it will be worse than anything you have ever driven before.

My colleague on *Top Gear*, Quentin Willson – the second-hand car dealer – once part exchanged a Sunny 120Y for a packet of Benson and Hedges, and still maintains he was ripped off.

Here's the problem. In the 1970s Red Robbo was running things in the Midlands, and on the rare days when anyone turned up for work at a car factory, the machines which left the factory gates were outstandingly unreliable.

Then, all of a sudden Datsun brought out the Sunny 120Y and it was the answer. It may have been ugly beyond the ken of mortal man and it may have handled like Bambi but it didn't break down.

This, to the overheated, stranded British motorist was like an epidural to a pregnant woman. You know it's a bit dangerous. You know it's not good for you. But when you're lying there bathed in sweat and shouting a lot, you don't give a damn. You would sell your soul to the devil to have that needle rammed in your spine.

It's a commonly held belief that Japan is not an innovative country, and that they can only copy the USA and Europe, but that is not so. Japan taught the world that it is possible to make a reliable car.

And pretty soon, everyone else was making reliable

cars too, which made the Sunny look a little bit hopeless.

However, people kept on buying them. Ford and Rover stood on Ben Nevis telling everyone that their cars were now clever, and practical and well equipped and good-looking . . . and reliable, but no one was listening.

Mr Sunny Driver had been late for an appointment in 1974 when his Allegro broke down and there was no way he'd EVER buy a European car again.

Nissan, as Datsun is now called, did everything in its power to make him change his mind with an endless succession of Sunnies that just got worse and worse. When I started testing cars in 1984, I absolutely couldn't believe how bad the Sunny was.

On a roundabout under the A3 in Surrey, it just careered into the kerb for no real reason. And when I got out to inspect the damage I remember feeling dumbstruck at just how ugly the car was.

Nissan was not to be deterred. They teamed up with Alfa Romeo – who at the time were still nailing their cars together with spit and Kleenex – to create the ARNA, a Nissan Sunny which was built in Italy, so you got the worst of both worlds.

Then came the ZX Coupé. Ah, now here was a car that perhaps echoed the old 240Z, a stylish two-door fastback that might just cut some ice outside Shitters Disco on a Friday night.

Er, no. Here was the most angular piece of design since Etch-a-Sketch went west. It had a feeble engine too, and to make sure you knew it would be a slow, evil-handling piece of junk it wore a Sunny badge.

Now Nissan has come up with some spectacular names in its history – Cedric tops the list – but you just can't call a two-door coupé a Sunny. You can call it a Thunder-storm or a Lightning or even a Rainbow, if you like a meteorological theme, but Sunny means the car has no cred. You might as well call it the Drizzle. You should have done actually – it would have been more honest.

By this time, I'd stopped testing the Sunny lest any late-night revellers in South London should mistake me for a minicab driver, but I had one last stab, about a year ago.

I was fooled into thinking that the 100NX was not a Sunny at all, but a new little sports coupé to rival the Honda CRX and the Toyota MR2. It wasn't and it didn't.

It looked like a buttock but the worst thing was its engine – a miserable little 1600cc unit that developed just 101 horsepower. The top speed of 121mph wasn't so bad, but it took a startling 11.2 seconds to lumber itself from 0 to 60mph.

I've been in faster lawnmowers. This was, to the world of cars, what Sloggis are to the world of underwear. My daughter corners better and she's only 14 months old.

It was the worst car I've ever driven and that is why I shed no tears at the news that the Sunny is no more. It is dead. Nissan have done the decent thing and pulled the plug. No flowers please.

STOP PRESS I've just seen the new Almera. Can we have the Sunny back please?

Who's getting their noses in the trough?

There's been a debate for some time now about whether films mirror real life or whether it's the other way round. For the vast majority of the cinema-going public, films are all about escapism. They're like drugs – fine when you're in there, but sooner or later the honeymoon ends and you have to go back to the real world.

Now, strong-willed individuals can handle that. They can separate fact from fantasy. Michael Howard, however, obviously cannot. This wouldn't be so bad if he was just a regular guy, but unfortunately he is the Home Secretary.

Now, at some point he has been to see *Robocop* – a film set some time in the future, when the private sector runs everything and employs some morally questionable tactics to police cities. After I'd been to see it, I came out of the cinema marvelling at the director's vision, and all the way home talked excitedly about that bit when *Robocop* shot a rapist in the testicles.

Unfortunately, Mr Howard took a different view. While the rest of the audience was urging *Robocop* on in his war with the greed-is-good merchants of 1999, Michael was hatching a plan. Over the years, his government has privatized pretty much everything it can lay its hands on, so why not go further? Here was a film about a privatized police force – maybe it would work here. So last month, Mr Howard declared that he wanted to see the police service freed from non-essential tasks.

Naturally, the Police Federation and every motoring organization from the Tufty Club up is furious because they know what's coming. Everyone knows what will motivate the traffic police if they are privatized. Share-holders. People in suits. The faceless minority that was actually happy to see Rover become German.

At the moment, on a clear sunny day, Britain's traffic police can just sit on their pig perches or back at base, playing cards. But when you introduce shareholders into the equation, they're going to have to make a profit.

Someone will have to pay for their wages, so the money will have to be raised from people caught doing 71mph up the M4. We've seen how the privatized parking enforcement agencies operate, so can you imagine what a privatized traffic police force will be like?

Behind every tree, there'll be a man with a moustache and a hairdryer. And you can forget all about a caution, because someone will have to pay for the fuel he used to chase you and the time you've spent chatting.

The only good thing is that to keep costs down, the suits will buy their officers cheap, slow cars. Tool around at 120 and there's no way that Roboplod in his Proton will get close. Well, not until you get to the next toll booth, anyway.

Yes, speeding is a minor misdemeanour and Howard is right to call the enforcement of it a 'non-essential task', but what about drinking and driving? Is that a minor offence which the police shouldn't bother with? And if it is, why do they spend so much money every Christmas on television commercials telling us that it isn't? And do they really think a profit-led police force will tell people to stop doing something that feeds its coffers?

No, they'll want us to speed and jump lights when we're pissed, or they won't keep their bloody shareholders happy.

Now, when BT, British Gas and all the rest were floated, I was standing in line waiting for a slice of the action, but when we're offered a chance to buy some of the police, I'll be at home watching television. And I urge you to do the same.

That will keep the share price way down. And then, for a month or so, let's all stick absolutely to the letter of the law. Let's not put a single foot wrong. We'll go straight and starve them to death.

Think about that, and smile.

Ferrari's desert storm

We don't have roads like this in Britain. Coming off the roundabout it plunges, arrow straight, into the heart of the Arabian desert for mile after unrelenting mile.

It doesn't really go anywhere special and so there's not much traffic, but even so the oil-rich local authority has made it super smooth, and with two lanes in each direction.

Nature has helped out the safety because on either side there is soft sand which will firmly and sternly bring any errant car to a halt.

That's the hardware, but the software is even more impressive. Up front, there's a Ferrari F40 which the owner has modified so that its turbocharged engine now gives a colossal 600bhp.

As it rockets away from the roundabout, each gear change is signalled by a huge sheet of flame from the exhaust system. The Ferrari F40 is desperately fast in standard tune, but this one, in the half light of evening, is something really very special indeed. Fast is too small a word. Unbebloodybefrigginglieveable is nearer the mark.

However, it is in my way. As its driver shifts from third to fourth, I can see him glancing in his rear-view mirror at the monster that will not be shaken off. As he hits fifth, I'm still there, a little closer now, and flashing my lights to warn him I'm coming through.

This was my first ever drive in a Jaguar XJ200 and I simply couldn't believe the size of its power. I knew this was a car built to be the fastest in the world – a record it held for a short time – but it had never really excited me before.

It was just so damn big, and while I respected Jaguar for making it look odd and unusual, I always thought the overhangs looked wrong, like the wind had eroded some vital part of the car away.

Plus, its miserable little six-cylinder engine – reputed to be the same unit that had been used in the Metro rally car – just didn't send out the right aural signals.

I still think all those things but there's no getting away from the fact that it is unbelievably fast. I mean, it just loped past that F40 and kept on going to 195mph – the fastest I've ever driven.

My previous record – 186mph – was achieved in a Lamborghini Countach and I vowed I'd never go that quickly again. It was terrifying, as the car bucked and writhed and shook and rattled. It was patently obvious that while the engine was capable of such things, the dear old car was not.

But in the Jaguar, all was serene. Sure, the wind noise had built up to double hurricane speed, but all the seals stayed intact, the steering stayed calm and the suspension never felt like a pebble would put me in the desert, upside down and on fire.

In fact, truth be told, the reason I slowed down was boredom, and also because I wanted to have a go in the other cars – the aforementioned F40, a Porsche 959 and the new Ferrari F50.

They all belong to the same guy, who uses them infrequently and insures them even less. The Jag wasn't even registered, while the 959's service history suggested, wrongly, that it was still the property of the Sultan of Oman.

Whatever, it was a mouthwatering selection of machinery to find in a lay-by, especially when I could choose what to drive next.

I drove all of them in turn and found myself marvelling at how national characteristics shine through even in the quest for speed. The Italian duo were lively and vocally active. If they'd had arms, they would have waved them about as they pootled along.

The Brit was a heavyweight; leather-lined and rather calm. In town, it was as unruffled as an embassy dinner. This was a car that would know its fish fork from its dessert wine.

And then there was the German, the 959, which tried to achieve world domination by a dazzling array of computerized goodies. Ordinarily, I have no trouble figuring out even the most bizarre dashboards but in the 959 there are dials and knobs that are meaningless.

One light was winking away, warning me that the pressure in the rear off-side tyre was a little low. It was all very impressive, but so is a dentist's drill.

And therein lies the problem. I actually had the time and the inclination to think about these things as I roared around in the desert in four of the most exciting cars the world has ever seen.

The trouble is that there were no passers-by to enjoy the looks and the sense of occasion. If there had been, then it would have been a built-up area and I'd have been doing 30.

And that's the other hassle. Speed in itself is not exciting. As you sit in a Boeing, are you thrilled that it's ripping up the sky with a 500mph orgy of big numbers? No, and it's the same deal in a straight line on a straight road in a car. Two hundred mph. So what.

What matters is acceleration and handling, an ability to take corners as though they're not there, and this is why the Ferrari F50 has been so well received by those who know. It's light, and simple, like a choux pastry in a world full of suet pudding.

Small boys may argue that its top speed of 206mph means it is slower, and therefore less good somehow, than the 237mph McLaren F1. But that is nonsense.

The Ferrari is built to take corners and it only uses that ex-Formula-One V12 engine for dealing with the rather tedious straight bits.

It seems that, once again, Ferrari has got it right.

Killjoys out culling

1995 saw the start of a new trend which began like a cancer, undetected and not remarkable.

Nissan announced that its really rather good 300ZX

would not be imported any more because the cost of making it comply with new noise and exhaust regulations meant it was no longer viable.

So what? It was a nice car sure, but it was like losing a distant acquaintance, a bloke who you used to see in the pub from time to time. There are lots of other blokes in the pub.

Then word crept out of Italy that Bugatti had financial difficulties, and that production of the turbocharged V12 EB110 had ceased. This was a bit more serious.

But let me assure petrolheads everywhere that in 1996 things are going to get a whole lot worse.

By far the most significant casualty is the Ford Escort Cosworth. Launched in 1992 amid reports that there were now more car thieves in Britain than motorists, it was burdened with whopping insurance premiums, and Ford didn't help matters by making it rather expensive.

So far, they've sold 7000 and that, in Ford's book, is just not enough. It's an interesting point this, but if you pick up a dictionary in Ford's headquarters, you'll find the word 'sentimentality' has been deleted.

No really, it's true. Ford owns 25 per cent of Mazda and it too has started 1996 with an announcement that by far its best car is off to God's recycling bin.

The RX7 was a Japanese car that didn't look Japanese. It's what the E type would have looked like today and, with its turbocharged Wankel engine, it could go like stink too.

There was no room inside for anyone over 2ft 6in but that didn't really matter. I used to enjoy my rare sightings of this oriental Batmobile – only 150 were sold in Britain – but not any more.

And it's the same deal with the Volkswagen Corrado. Here was a handsome coupé which could provide a driving experience that was almost sexual. No car costing less than £20,000 could even get close, but last I heard it was coughing up blood and looking a bit green around the air intakes.

I am also concerned about the health of the Dodge Viper. There's talk of new hardtop versions but I've heard that the roadster is about to be axed. It seems that each one sold costs Chrysler a few quid, and that in business this is a 'bad thing'.

Other cars causing concern include the Lamborghini Diablo, the Honda NSX and the Lotus Esprit.

You're starting to get the picture. Fast, truly exciting cars are being killed off so that pretty soon the officers will all be gone, leaving us with a field full of enlisted men.

The people with green hair are winning, and this is more baffling than an early Genesis lyric.

When I was 18, I had many big ideas which neither my housemaster nor parents would take seriously. I was a stupid and spotty person with a predilection for cap-sleeved T-shirts and ELP, and I was therefore daft. Almost daily, I was told, 'No Clarkson, you are a fool and you may not wear training shoes for school.'

But today, people with army jackets and odd hair decide to live in trees, and yet we think they have something intelligent to say. If I was a reporter charged with the task of talking to these morons, my first question would be, 'If you're so bright, how come you aren't running Unilever?'

Newbury is being strangled by traffic jams. Local traders are watching their businesses go to the wall. The environment suffers as trucks and diesel cars sit belching their carcinogenic fumes into primary school classrooms.

Common sense dictates that a bypass should be built, but no . . . a few not-in-my-backyard locals have teamed up with a bunch of women who like to breast-feed sheep in public, and there's going to have to be a re-run of the Somme to get the bulldozers through.

The trouble is that a bypass means that cars and trucks will be able to travel more quickly, and 'quickly' in 1996 is a dirty word, like 'profit' was in 1979. We are forever being told by imbeciles that they want people to think of speed in the same terms as drunken driving.

And they're winning. According to a report a reader sent in, I see that Buckinghamshire County Council is to recruit thousands of volunteers to drive around at or below the speed limit to slow everyone down.

More amazing still, the local police force is backing the scheme where interfering busy bodies and pensioners will tootle around country lanes in their Austin Maxis at 7mph.

To these people and their green-haired brethren in Berkshire the fast car is a symbol of all that's wrong today, but what they just can't understand is that fast cars don't have to be driven quickly. I have been known to go past schools in my 145mph Jaguar at 15mph.

The trouble is that this is a common-sense argument and that do-gooders with their politically correct shoes don't have any common sense. Arguing with people as stupid as this is like arguing with a mug of tea.

Flogging a sawn-off Cosworth

For three reasons, I've always had my doubts about Prince Charles. First, it's said he talks to plants – a singularly unrewarding experience, because all they do in return is die. Mind you, my plants also die when I don't talk to them, when I water them and when I don't water them. My garden is a herbaceous version of Fred West's cellar.

Second, he believes in all sorts of alternative medicine which, from personal experience, I know to be nonsense. I have had a man stick needles in my ears to cure hay fever, but I sneezed so much they all shot out. And when I have a hangover, I could eat dock leaves till the cows come home but you just can't beat a Disprin.

And third, he is separated from the world's second most beautiful woman.

Against this sort of background, it is easy to see why I was on Urquhart's side in *To Play the King*.

But now Prince Charles has shown himself not only to be a decent cove, but also the kind of leader that this country needs. Unlike John Major, he obviously has rather more than two pubic hairs in his underpants.

In fact, he must be hung like a horse because he has actually dared to take on the tiny, tiny minority of homosexuals who sit around in Camden telling everyone else not only what they can and can't say, but also what they can and can't think.

Time and again in recent years I've been cornered by these wishy-washy liberals who think that the motor car is the seven-headed beast from Revelation, and there's been no way out. My passion is killing the planet and they're on the moral high ground, wagging their fingers.

Well, now the future King of England has given me a royal seal of approval to fight back.

So here goes. Women can't drive. Immigrants need to take a driving test before being allowed on the roads in Britain. Old people should have to hand in their licences at the age of 65.

And when I meet someone I know to be homosexual I can't help staring at their bottom, wondering.

But most of all, that American boy who went round Singapore spray-painting cars got what was coming to him. Apparently he'll be scarred for life. Diddums. My heart bleeds.

In pubs throughout the land and on every golf course, people are whispering to one another about how they think flogging is a good thing.

When I left the house the other morning to find that someone had sawn the rear spoiler off my Escort Cosworth during the night, I was speechless with rage. To the casual observer, my body language suggested that I had inadvertently spilled some sulphuric acid in my lap.

Now, the politically correct say that the thief, when apprehended, should be taken to court, addressed by his Christian name and let off with a stern warning. I want him beaten to within an inch of his life, and everyone I know who has had a car vandalized thinks the same way.

Apparently, a radio station did some kind of straw poll about that American vandal and found that 97 per cent of its listeners think the Singapore authorities treated him fairly. The other 3 per cent, by the way, had beards. Yet if anyone says 'bring back the birch' on television or in the papers, they're labelled as far right extremists. So no one dares do it. Political correctness has stifled discussion and debate in this country to the point where 97 per cent of the population daren't voice an opinion in case someone from the Camden thought police is listening.

But now, thanks to Prince Charles, you can go home tonight, beat your kids, eat a rare steak, go out in your V8-powered car, sleep with a woman without using a condom and then have a Capstan Full Strength. In other words, you should do what you've always done.

Only now you've no need to be embarrassed about it.

Weather retort

Someone once suggested that Britain is the only country in the world to have weather. Everywhere else has a climate.

But here in the Cotswolds, in the last two weeks, we've had an ice age, a plague of frogs, mice and fog so thick you've needed a chain saw to get out of the drive in the morning. Today, the locals are expecting locusts.

For 15 years I lived in London and every day it was the same: 57 degrees and a bit of drizzle round tea time. I'd forgotten what extremes can do to a man, and his car.

Yes, we had burst pipes too, and it was a nuisance, but this was a flea bite compared to what was happening in the drive.

On the day when it dipped to minus 20 in Glasgow, it was minus 68 here and the weather people were talking about freezing rain. Well what we got in Oxfordshire was a sort of see-through gravel which encased my Jaguar in an ice shell.

The doors should unlock themselves when commanded to do so by the remote control device, but they were as good as welded shut by the ice. I was out there with most of what Saxa made in 1995, and all sorts of sprays, for nearly an hour before I could get in.

As you'd expect, the car started, but all was not well. Both the anti-lock brake and traction-control warning lights stayed on. In exactly the sort of weather I needed these things, they had decided to stay in bed.

I couldn't even get out of the drive. It may be gravel but the see-through ice pellets had encased it in a sort of ultra-slippery aspic, so that each touch of the throttle pedal was interpreted by the rear wheels as an instruction to impersonate a washing machine on its final spin cycle.

I'd been missing London badly and this was icing on the cake. The traffic may only move at 13mph in the capital but at least it moves. After ten minutes of trying to go forwards, I'd slithered backwards about 10 feet, and bumped into the stable block.

So what. We had the Volvo, a car built for such extremes. Er, no. I chiselled my way in, fired it up, hit the heater button and directed the warm air flow at the windscreen . . . which split clean in two.

What really annoyed me was that my mother-in-law's Y-registered Fiesta worked perfectly. It would have rubbed salt in the wounds, but we didn't have any left.

I considered firing up the lawnmower, but as it's a convertible I resorted to the Volvo, complete with its bifocal windscreen.

Largely, the roads were passable provided you never felt

the need to exceed 20mph, but every 15 minutes someone came on the radio to say that we should all stay at
home unless our journey was a matter of life and death.
Well we'd run out of lavatory paper, so does that count?

I wouldn't mind, but they're still at it. Since the arctic
weather moved over to New York we've had fog, high
winds, torrential rain, and on each occasion the radio
traffic people have told us to stay at home.

Well listen here guys, if I stay at home, you get a cat
playing with some wool on a Thursday instead of *Motorworld*. If a doctor stays at home, people die. If snowplough
drivers stay at home, the roads get even worse. If shop
assistants stay at home, we can't buy loo roll.

The trouble is that people listen to these radio idiots
and overreact. They still go out, but they set their mental
cruise controls at one.

Last Saturday, I pulled out of the drive right behind
a B-registered Maestro which was being driven by a
man who had turned his high-intensity rear lights on in
November.

The fog was just bad enough to make overtaking
dangerous so I was forced to do the 16 mile drive to
Banbury, and the blessed relief of the M40, at 1mph.

When I came home at night, the fog had been replaced
by astonishing rain and a wind that was moving people's
bungalows around. But I could see, and I was going to
overtake people . . . except for one thing. I've forgotten
how to do it.

A recent survey said that the average driver only uses
full beam headlights for 2 per cent of the time at night,
which seems about right. In town you never use them, on
a motorway you never use them and on country roads

these days, something is always coming the other way.

Think about it. Only as recently as 15 years ago, you would brush aside slower-moving traffic like you dismiss bits of mince pie that have dropped onto your new Christmas jumper. But hand on heart, when did you last overtake someone?

It's no better during the day either. On the rare occasions you find yourself on a normal road you can see a stream of cars heading off into the distance, so even when it's safe to go past the car in front you don't bother, because you'll only have to do the same thing over and over and over again.

Plus, there is a similarly train-like concoction coming the other way. The British A road today has become like a railway line. The carriages are the cars, and the engine is that B-registered Maestro.

Overtaking has become a forgotten and pointless art for people in this country, as Damon Hill seems happy to prove every other weekend.

Burning your fingers on hot metal

I had three economics teachers at school. One was a Ugandan who'd let me go round to his house at night to practise smoking. Another never shook himself properly after a trip to the lavatory. And the third was a communist.

I learned very little, but I do recall being taught that the human being was greedy because of the anytime, anyplace, anywhere Martini advert.

I didn't bother finding out why because I was in the middle of the *Melody Maker* crossword, which I'd cut out earlier and pasted in a copy of *The Economist*.

But now, 20 years on, I've discovered the Martini advertisement is not to blame for our acquisitive streak. It's magazines.

Back in the days when *Melody Maker* and the *NME* were my bibles, I'd spend all my money on albums and ever more sophisticated hi-fi equipment. I really believed that 'Snow Goose' sounded better on my Garrard 86SB than it did on Andy Byrne's miserable SP25.

I was out of the traps like a greyhound with chilli up its backside when CD hit the scene, but since Neil Young told *Q* magazine that analogue is better I've dusted down all my all LPs again.

I have a voracious appetite for magazines, even though I know the cover price is a tiny fraction of the resultant costs.

Last year I lived in London surrounded by friends and restaurants but, having picked up a copy of *Country Life* at the dentist's, I now live in the countryside, where there are wasps and murderers and low-flying Tornadoes. The cinema is showing *Mad Max*, and everyone at the pub is saying there should be a sequel.

Naturally, we've started to take *Homes and Gardens*, and now the kitchen floor is being replaced with stone flags, a company called Smallbone is being asked to check out the units and Harrods have just delivered a bed so large that it encompasses three time zones.

When you pick up a magazine, you'd better have nerves of titanium or you'll go broke.

But I challenge anyone to stay out of the bankruptcy

courts if they even casually browse through an organ called *Auto Trader*. This, I just know, is published by Lucifer himself. This is bad news on bog roll.

It's a chunky 350 pages and it's stuffed full of advertisements for secondhand cars, each one usually accompanied by a poor-quality black and white picture.

And yet it is one of the most compulsive reads in the entire universe. When I take a normal, glossy magazine to the loo in a morning I get pins and needles, but with this tome you develop gangrene.

I think the basic problem is money. *Auto Trader* concentrates its efforts on stuff we can afford, stuff we drooled over in the glossies five years ago, which is now being sold for beer money.

Let me give you a few examples. *Mercedes 500SEC. B reg. full spec includes air con, electric seats, cruise control etc. New 16 inch wheels and tyres. Very clean car. £6,795.*

Think about that. What we have here is the classic football manager's car with a 5.0 litre V8 engine, all the fruit and three-pointed star reliability for less money, after you've haggled a bit, than a Mini. Go on, admit it: you're tempted.

Well what about this one then? *BMW 750iL. 1988. Diamond black with black hide upholstery, electric everything, cruise control, full service history, a truly stunning car. £8,345.*

So there you have it. For the price of a downmarket Ford Fiesta, you can have a V12 BMW.

Every single page throws up a fistful of bargains which make a sale at DFS look like some kind of rip-off.

The book has just fallen open on the bargain basement section and there, at the bottom of the page, is a Mitsubishi Starion, which is a sort of Japanese Capri.

I remember testing this car back in 1985 and I thought it was wonderful, a real hooligan's special with its 2.0 litre turbo motor, its simple rear-wheel drive layout and 170 horsepower. Nice seats too.

Well now you can have one for £995. Or, if the Starion is a bit too garish, how about a V12 Jaguar XJS for the same price? Or for a tiny bit more, a Porsche 944 or a Range Rover?

I don't doubt for a moment that these cars have been clocked, stolen, pushed in boating lakes, crashed and welded together in a school project, but for nine hundred quid we're not talking about the BCCI are we?

Yes, they will cost heaps to insure and, sure, a big V12 will eat fuel, but let's be honest: the biggest single cost with any new car is depreciation, and you won't lose much sleep over that.

Cars like this are best used as funsters, at weekends, so you can consider your purchase as a sort of gamble, a punt on an outsider in the 3.30 at Lingfield.

Its doors may fall off the first time you take it out, or it may sail through its MOT six months down the line. But either way, you'll be able to stand around at parties telling everyone who'll listen that you have a Jaguar XJS.

Speeding towards a pact with the devil

In recent months there have been several distressing moments on television. We were all moved by the scenes of poverty and deprivation from Rwanda, and my mother was shocked by the language and violence in *GoodFellas*.

But according to an obscure government quango, the most irresponsible and dangerous programme on television is *Top Gear*. The quango in question is called PACTS (Parliamentary Advizzzzzzzzory Council for Something or Other) and it says that when *Top Gear* refers to a car's ability to 'knock on the door of 150mph', we are guilty of 'glamourising' speed. Funny that – I never knew 'glamorising' had a 'u' in it.

PACTS also says that speeding costs 1200 lives a year. Well, they've obviously researched this subject with the same diligence that they spell their words, because if speed really does kill, Concorde would be the most dangerous means of travel. I've just done a quick calculation and reckon the number of people killed by Concorde so far is zero. And that makes it pretty damn safe in my book. When will people learn that speed cannot kill someone? It needs to be mixed with something else first, like the sort of bad driving you see in Whitehall at half-past five when all the quangos are shutting down for the night.

Besides, if speed is so lethal, how come motorways, which carry 15 per cent of all the traffic in this country, account for only 3 per cent of the casualty accidents? And if you do crash on a motorway, you are three times less likely to die than if you crash in a built-up area.

PACTS is undeterred by facts, though, and backs up its claims by saying that about one-third of all fatally injured vehicle occupants are involved in a speed-related accident. What speed? 90mph? 40mph? 0.002mph? It doesn't say.

If the people who make up PACTS are typical, I know exactly what we're dealing with here – wizened old has-beens in Hondas who suffer from the upper-class disease of too much money and not much brain. Unable to get a proper job, but duty-bound to do something constructive, they sit on endless committees doing good things. And just because the patron is a marquis or a baroness or a marquee, everyone they write to is supposed to fall on their sword and promise never to stray again.

When I test a car, I don't leave out the price just because some viewers can't afford it, and I won't leave out the top speed either. It's a salient point. And if I described it in a drab monotone everyone would throw chairs at the television.

If there is one character trait I despise even more than reasonableness and socialism, it's idealism. Yes, it would be lovely if no one was killed on the roads and there was no war, but they are and there is and that's tough titties. It's like the NHS. It would be ideal if I had a nurse, a GP and a selection of specialists in attendance 24 hours a day, but this cannot happen. We have to be realistic, but you can bet that someone, somewhere, is prancing about on a bloody quango telling anyone who will listen that Stow-on-the-Wold needs nine new hospitals. Yes, it does, but it can't have them and that's an end to it.

Do you know that there are a bunch of wimmin outside Greenham Common even today. Though the base is now only used for *Top Gear* photo shoots and police driver training, they say they won't move until the last nuclear weapon has been removed from the face of the earth. But if the entire American Pacific fleet can't persuade North Korea to stop making its atom bombs, I really don't think a bunch of hippies in Berkshire has much of a chance.

There's bound to be a quango up in Whitehall where people with gout meet once a week to decide how best to deal with these grubby New Age campers. The odd thing is that both groups of people are as daft as each other.

Road rage – you know it makes sense

Like the rest of Britain, I was saddened to see that Britain's schoolchildren cannot read or write.

It seems that teenagers are leaving school these days well versed in the dangers of ecstasy, but with no real idea how to spell it.

Worryingly, these people are driving around in cars, peering at road signs and wondering why they always end up in Colchester when they were trying to get to Weston-Super-Mare.

Presumably, they cannot understand any of the information being provided by their dashboards either. 'Why', they will wail as they splutter to a halt on the hard shoulder, 'have I run out of petrol?'. And how do they know whether they're doing 40 or 90mph?

What concerns me most, though, is that these people are just as likely to be stopped in the street and counselled for their opinions as clever people like Stephen Fry or Jonathan Miller.

That is why I am always deeply suspicious of market research. I mean, if it were so good at predicting things, we'd have a Welsh prime minister.

Nevertheless, I've been completely absorbed this past week by the *Lex Report on Motoring*, a huge tome that's been compiled by one of Britain's foremost car retail and leasing operations.

It says here that six out of ten people supported the road protesters' cause, which is an extraordinary finding when you learn that 72 per cent of drivers say traffic congestion is a 'major' problem.

So, what we have here is a majority of people wanting fewer jams, and a majority of people saying there should be no new roads. Hmm.

How about this one? Sixty-one per cent of the British public – the people who brought the world jet engines, hovercrafts, communism, optical fibres, television and the telephone – say that cars are only a 'little' more environmentally friendly than they were 10 years ago.

Nine per cent – the real dimwits – say that cars have become more damaging to the environment in the last decade.

Unbelievable. Ford has just announced that a new Fiesta produces the same amount of toxic gases as 20 Fiestas did a few years ago which, in my book, means there's been a twentyfold improvement.

And who had heard of recycling centres in 1986? Car firms are making huge efforts to shape up, but obviously the message is not getting across.

Ah, I see now why that should be so. The report says that only 19 per cent of people trust car advertisements, and that friends and acquaintances are considered to be a great deal more knowledgeable than newspaper journalists.

I may as well give up now because *Top Gear* gets a special mention. Only 34 per cent of private buyers trust us. Right: now it's personal.

So now I shall switch my attention to the huge section on so-called road rage.

This is the bit that's been picked up by radio stations

and television networks all over the country but, again, I find myself wondering . . .

In 1995, 1.8 million people were forced to pull over or off the road, 800,000 were physically threatened, 500,000 had their cars deliberately rammed, 250,000 were attacked and another 250,000 had their cars damaged.

Add the figures up and you'll find that 3.6 million people were abused, threatened or hit on the roads last year . . . which isn't enough.

You see, I have a great deal of sympathy with people who become angry and frustrated while in their cars, because losing your temper is part of the human psyche, as natural as smiling or having sex.

Wetties ask why we don't lose our rag quite so readily while walking down the pavement, but that's a stupid question. If someone inadvertently brushes past you in a shop doorway, it's no big deal.

If, however, by not paying attention, their car brushes against yours, you will be without wheels for a week or so, there will be a fight with the insurance company and you will almost certainly end up poorer as a result.

And that's if you are lucky. If you're on foot, even the biggest Mickey Skinner-type impact won't cause much damage, but on the road, it's different. You could wind up dead or paralysed, and that's certainly a good enough reason to get out of your car and smash the other guy's teeth in.

A few years ago, I was desperately late for a wedding and, while overtaking a Volvo, found another car coming the other way. I dived back to my side of the road and very nearly caused a huge shunt.

At the next set of lights, a huge Irish person heaved

himself out of the Volvo and spent a couple of minutes trying to throttle me. That was road rage.

But it was my fault. I deserved it. I nearly killed the poor bloke and I consider myself rather fortunate to have escaped from the encounter with mild bruising. I deserved more.

Frankly, if more people behaved as responsibly as that large Irishman the standard of driving would improve. You'd think twice about cutting someone up if there was even the remotest possibility that you'd end up impaled on your gear lever.

When I see that there have been 3.6 million examples of road rage in the last year, I say to myself that there must have been 3.6 million examples of bad, inattentive or selfish driving.

911 takes on Sega Rally

If you were to enlarge Birmingham a thousandfold, you would end up with Australia. Sydney is like a bigger version of Edgbaston. Perth is the National Exhibition Centre. Alice Springs is Handsworth and the rest is Canon Hill Park.

I think it's fair to say that you can judge a city by whether or not you feel the need to go to an amusement arcade. If it's sunny and warm and the bars are full of lively and interesting people, you won't give 'Space Invaders' a thought.

But if it's dull and the people are awful, then the idea of pouring hundreds of pounds into an arcade game becomes quite tempting.

In Perth, we went to an amusement palace every night and I discovered the Sega Rally machine.

With this computer game, you choose what sort of car you want and whether you need manual or automatic transmission, and then you're among the make-believe mountains in an increasingly difficult game of pure skill.

When the car slides, the wheel fights in your hand. When you crest a brow the seat moves, and all the time a computerized co-driver is warning you of unseen hazards ahead.

Of course, the other cars are driven by silicon chips, but in our arcade four machines were linked so we could race each other.

Now this was something else, because what we have here is racing without the fear. It is driving quickly and irresponsibly with no risk of death or injury. Even Steven Norris would be forced to admit it's safe and environmentally friendly.

Obviously, you can't go shopping in a Sega, but people don't go shopping in supercars either. Supercars are designed to be fun, to put a big grin on your face. And so is the Sega.

When you come flying over the crest of a hill in a Porsche, only to find there's a 90 degree right-hander ahead, you will probably wind up dead. At best, it'll be written off, your insurance premiums will go nuclear and you'll be off work for a week while they mend your nose.

Do the same thing in a Sega and you will spin, your seat will rock about a bit and the cars you've just overtaken will get back in front. You will lose the race, of course, but it will only cost you a pound to have another go.

At a stroke, therefore, the Japanese seemed to have delivered a hammer blow to Europe's sports car industry.

Porsche is obviously aware of what they perceive to be a very real threat. They know that no one will spend £50,000 on a fast car if they can spend £15,000 on a machine that lets you go even faster in complete safety . . . and with no Gatsos in every village.

Thus, they've tamed the 911.

In the past, these three little numbers have been the automotive equivalent of 666. The rear-engined Beetle on steroids took no prisoners and would punish any mistake by hurling itself into a hedge upside down, and on fire.

I've always bowed my head slightly whenever a 911 burbled by because the driver, very obviously, was a reincarnation of Sir Galahad – brave beyond the ken of mortal men and hung like a baboon too.

To drive a 911 quickly required more than talent. It needed bravery on a scale that would leave even Michael Buerk breathless.

All these thoughts, and more, were crossing my mind as I trudged across the drive last week for my appointment with the Reaper. There, under a foot of snow, was a 911 targa Tiptronic.

As I fired up the beast, my Adam's apple ricocheted twixt chin and sternum like a pinball. A thin sheen of sweat formed on my brow even though it was minus four out there.

Here was a car with tyres like garden rollers and a top speed of several hundred miles per hour. And here on the radio were a bunch of stupid DJs telling everyone to stay at home. Me? I was off to Telford.

Since then, I've been to Doncaster, Lincoln, Birmingham (twice), London and Oxford and I haven't crashed. I've driven with the roof tucked away under the rear window. I've had the gearbox in manual mode, swapping cogs by pressing little buttons on the steering wheel, but not once did the Porsche deviate from my chosen line.

Yesterday, I had built up enough confidence to really fly on a road I'm learning to love, and it was a sensation. When the understeer built up, I just backed off and the car shuffled back into line, with no fuss and no drama.

I didn't much care for the Tiptronic because four gears aren't enough, and I still think the dash is worse than Winalot, but this was a car you drove through the seat of your pants, a car that talked to you, a car that was alive.

It was also a supercar that I had dared to drive when the weather suggested I should have been on a bus. No way would I have taken a Ferrari out in conditions like those.

The Porsche has that Beetle-like unburstability so that no matter what is coming out of the sky, it will always be fun.

But you can't race your friends in a Porsche. You can't tear down mountain roads with your hair on fire. And if you go through a village at 120, the residents won't be lining the streets to cheer you on.

And that's where it loses out to the Sega. As cars go, the 911 is right up there with the best but, when it comes to a battle between European flair and Japanese technology, Japan wins.

A laugh a minute with Schumacher in the Mustang

Michael Schumacher is a German. Which means that he should, by rights, be fat, loud, vulgar and in possession of some ridiculous clothes to go with his absurd facial hair. Yet his torso is the shape of Dairylea cheese, and his face is unburdened with any form of topiary. At post-race press conferences, he is intelligent and modest when he wins, and quick to congratulate when he doesn't.

So when I met him at Silverstone this month I was rather disappointed to note that he was surly, impatient and about as communicative as that Red Indian chappie in *One Flew Over the Cuckoo's Nest*. I have had more inspiring conversations with my pot plants. And they're dead. I told him my wife hoped he would be world champion, and he gave me a look which made me think that I'd inadvertently said, 'You are the most disgusting human being I have ever had the misfortune to encounter.'

Later I tried again, asking him what he thought of the Mustang. Which, judging by his reaction, translates into German as 'I know that you like little boys and I'm going to tell your team manager unless you give me some money.' Had he driven a Mustang before, I asked, fully expecting another withering glance from the driver's seat. 'Yes,' came the reply. 'Where?' I asked, not realizing that 'where', in German, means 'I hope you fall into a combine harvester, you maggot-faced creep.'

So I gave up with the conversation and settled back to watch the fastest man in Formula One deal with the slowest sports car in the world. On lap one there were other cars on the track, so we pottered round. Then on lap two, instead of giving me the ride of my life, Mr Schumacher chose to demonstrate the driving positions. On lap three, we were following *Top Gear*'s camera car so I asked if we could see some wild and leery tailslides. We did, but sadly each one ended up with a spin. I couldn't help wondering if these gyrations might have been avoided if Mr Schumacher had kept both hands on the wheel. But who am I to question the ability of the greatest driver Germany has ever produced? And apart from muttering about how the Mustang had plenty of grip and wasn't bad for an American car, he told me nothing about what it was like to drive. So I set off on my own, and fell head over heels in love.

The new Mustang's body is not particularly pretty or brutal but it is big and eye-catching. Everyone turns to look and everyone knew what it was, even though this was the first in Britain.

To drive, it's American and rather good in a cheesy grin, firm handshake, hi, howya doin' sort of way. It's a big, open, honest sort of car which despite the air conditioning, cruise control, power seats, power windows, power roof and 5.0 litre V8 engine, costs just $22,000 in the USA. It's not very fast – ask it to go beyond 130 and it gives you a look of pure incredulity – and it treats corners with the same disdain I reserve for vegetarians. It will do everything in its power to go straight on, but there's never a moment when you think it might go round a bend, so there are no surprises. You know where you are with this car. It also makes a good noise, unless you take it past 3500rpm when it sounds strangled. But hey, have you ever heard Stallone hit a high C?

No, the Mustang is musclebound, dimwitted and slow,

but it's a good guy to have around town at night, looking mean and threatening.

It's the automotive equivalent of Carlsberg Special, which is probably the reason why Mr Schumacher was so underwhelmed. He, after all, is sponsored by Mild Seven, which are the most limp and pathetic cigarettes I have ever encountered. They have about as much to do with hairy-armed Mustangs as fish.

Girlpower

It's been a bad winter for everyone, especially out here in the Cotswolds, where even the mud has frozen. Each morning, I've needed to strap tennis racquets to my feet to retrieve the milk.

But today, the sun is out, the snowdrops have poked their way through the ice and there's a definite hint of spring in the air, so it seems like a good time to talk about convertibles.

It's an especially good time, in fact, because last week I went to Germany to test the MGF VVC, the BMW Z3 and the drop-dead gorgeous Alfa Spyder.

I wanted all of them, and the Renault Spider, and the Mazda MX5, and the forthcoming Mercedes SLK. They're all damn good. They're all my kinda cars.

I was still agonizing over a final verdict when some friends rolled up for the weekend, one of whom was driving a new MG. Now she's a thirty-something stock-broker, single, pretty and fully conversant with all that's hip in the City.

I normally make a point of never talking cars with

friends but I needed to know what she thought of the MG. She reckons it's far too noisy – no surprises there; that it's fun – yawn; and that it's a girly car. WHAT?

'Oh yes,' she said, 'everyone at work came down for a look when I bought it, but they don't want one. They think it's too girly.' By now, my countenance had adopted the look of a goldfish. 'Well it *is* girly. That's why I chose it,' she added.

This was a major shock because I have made it my business, in 12 years of writing about cars, never to make such a point. Talking about cars being aimed specifically at one sex or the other is about as dangerous as French-kissing a shark.

I mean, there's a girl on this newspaper who, though heavily pregnant, commuted to and from work every day in that most hirsute of things – a TVR Chimera. Dawn French has a 150mph Mazda RX7. My wife refuses to drive anything unless it has 'at least 200 horsepower'. And have you ever seen a woman in a Robin Reliant?

I vaguely recall my sister asking once why car interiors had to look like men's lavatory bags, but that was a trivial thing. I'd got it into my head that women are just as likely to be interested in cars as men. And yet, apparently, the MGF is 'girly'.

I've spent the last couple of days trying to figure out why this should be so, and now I'm going to attempt an answer. And yes, I know, the resultant postbag will be the size of Wakefield.

The MG is a girly car, first of all because it's small. It only has two seats, which says to other road-users: 'Look at me. I don't have children. I'm single and have no baggage.'

I know of one girl who used to drive around in an equally tiny Honda CRX. Whenever she saw an attractive man alongside, she'd make sure her ringless wedding finger was clearly in view. This was annoying because, at the time, she was my girlfriend.

By way of comparison my wife drives a Volvo 850R, which is equally clear-cut. 'I am married and I'm on my way home from the supermarket, in a hurry, because the girl-child needs feeding.'

The MG is also cheap. Perhaps women feel that they have no real chance of being a main board director or a guitar wizard and are therefore more ready to settle for second best. I, for instance, want a Ferrari and am half-heartedly striving for that goal.

If there were no chance of achieving it, maybe I'd quit saving and settle for the £17,000 MG, which, let's face it, is a mid-engined, droptop two-seater. And in the real world, it's just as fast as a 355 anyway.

Or perhaps women are more sensible and practical than men. Maybe, even if they could afford a Ferrari, they'd still buy an MG.

Then there are the lines. You could say it's small and curved and almost dainty, which makes it a sort of designer handbag in metal, a pink mackintosh which is to be worn rather than driven.

I think people will say the same of the Alfa Spyder when it comes out later this year, but the BMW Z3 is different. It has an aggressive, out-of-my-way stance. Ideal for the City boys but less appealing in the Peter Jones curtain department.

The trouble is that I'm floundering around here in uncharted waters. Having given it some thought, and having spoken to a number of girls, I now believe there is such a thing as a 'girly' car.

And I do think the MG may be such a machine. Men I've talked to don't like it much and, when pressed to explain why, most say they don't know. I, on the other hand, do like it, which calls my whole sexuality into question. I can't play football either. This is worrying.

A Rover spokesman told me to relax. 'You're not gay,' he said, 'men are buying the MGF in British racing green and red. Girls are buying it in bronze and that horrible lilac. It's not a girly car, but there are girly colours.'

Phew. Mine's an MGF then. In red.

Nissan leads from the rear

If you're the sort of person who casually scans this column on a Sunday looking for a bit of controversy and some silly metaphors, then I have bad news.

This week, I have my sensible trousers on and I'm only talking to petrolheads; people who see the car not as an art form, nor as a conversation piece, but as a machine.

When I learned to drive in the late 1970s, just about everything on the road had rear-wheel drive. Front-wheel drive was something in a Mini, something that had no place in the hubbub of saloon bar, saloon car chatter.

But today, of the 200 different cars on sale in Britain, over half send their power to the road via the front wheels. As far as the manufacturer is concerned, such technology makes the car cheap and light.

And in a traffic jam in Newbury, the average motorist

doesn't really care. Just so long as the damn thing moves when the traffic shuffles forwards, who cares whether there's a prop shaft or not?

Well me, for a kick-off.

In a small-town car, front-wheel drive is eminently sensible. Apart from the lightness, the simplicity and the vital cost factors, there are fewer bulky components taking up space that, more properly, should be used for people and stuff.

It's the same deal with ordinary saloon cars and, especially, people carriers. No one buys a Ford Galaxy so they can give it the full moo on the A40.

But in a thrusting, hirsute sports car with bulging pecs and a rippling, washboard torso, front-wheel drive is rather suspect. Saab once said that it is 'undesirable' to feed more than 170 horsepower through the front wheels.

Odd then, that its top models are now putting out more than 220 horsepower. This means the front wheels have to deliver that to the road, while dealing with the burden of steering too.

Anyone who's accelerated hard in a powerful front-wheel drive car will have felt what's called torque steer, as the steering wheel squirms from left to right. It's horrid.

It seems perfectly obvious to me that in a performance car the front wheels should steer and the rear wheels should be the propellant. I really do believe that BMW continues to dominate the sports saloon market because all their cars are rear-wheel drive and consequently feel more . . . together.

And it isn't just the 'feel' either. Rear-wheel drive cars are faster, otherwise they wouldn't have to carry a weight penalty in the British Touring Car Championship.

Yes, Golf GTis and similarly powerful Peugeots and Fords were a laugh but they were a triumph of engineering over design. If these companies had been serious about making performance cars, their high-speed models would have had rear-wheel drive.

And it's the same story today with coupés. The Fiat is lovely and fast and it really does grip and go but, as far as the enthusiast is concerned, drive goes to the wrong end of the car. And it's the same story with the VW Corrado, the Ford Probe, the Vauxhall Calibra, Honda Prelude – even the new Alfa GTV.

Try to hustle any one of these cars through a bend and the front wheels will run wide in a nasty bout of bowel-loosening understeer.

However, help is at hand from an unexpected source. The Nissan 200SX coupé has rear-wheel drive. The turbocharged engine's 200 horses are fed down a prop shaft through a limited slip differential to the tyres at the back.

Hustle this car through a corner and the tail will slide, giving a fun-filled few moments of laugh-a-minute oversteer. Accelerate hard and the rear will squat, forcing more pressure on the driven wheels, giving even more grip.

Yes, cry the poseurs, this is all very well but the Nissan looks like the dinner of a dog, and have you seen those Bri-nylon carpets? I don't like the dash either, because it's too shiny, and why can't it be a hatchback like all the others?

And I can hardly go down to the golf club and tell everyone I've bought a Nissan. I mean, a Nissan for Chrissakes. They might think I have an Almera.

Shut up. I agree the Nissan badge is to cars what Findus

is to *haute cuisine*. I know the 200 is no styling *tour de force* and I agree that the interior was designed by someone who should really be out in the factory flowerbeds, digging up weeds. But who gives a damn?

The aristocracy is known for concealing its enormous wealth. Lord Fotherington-Sorbet drives a crap car and wears rags because he is accustomed to money and feels no need to play the peacock every time he goes out.

And it's the same deal with the Nissan 200. It doesn't need a Rolex on one wrist and a chunky ID bracelet on the other. It has no need for medallions or satellite dishes. It's a serious player and can therefore sit back with its light under the thickest bushel on the road. Hell, I wouldn't be surprised if Nissan didn't deliberately get a gardener to do the interior.

If you truly like driving, by which I mean you derive pleasure from how a car feels, then this is the one for you. You may look good in a Fiat and you may be able to snap knicker elastic if you have a Probe, but if you want a 'real' coupé, it has to be a Nissan.

Cable TVs and JCBs

Roads wear out and every so often we must expect the Cavaliers in the outside lane to be replaced by men whose trousers fail to cover all their bottoms. Cones will go up and the traffic will stop.

It may well be irritating to sit there, being gently marinated in your own sweat. But the fact is that road-works are the inevitable result of a thriving society in which 42 ton trucks thunder up and down the highways and byways, bringing fresh produce to your corner shop. However, there's a worrying trend. For the last four weeks, London's South Circular Road has been closed due to an entirely new sort of roadwork. I have been marooned by gridlocked traffic for more than a month. And it's not because the road had worn out or because some vital underground maintenance needed to be carried out. No, they have dug up the main artery between south-west England and the City because Cableguyz, our local cable TV company, decided to drive one of their JCBs through a water main.

You'll know when the cable people are about to come round, because you'll wake up one morning to find the pavement outside your house looks like one of Joseph's more vivid overcoats. All the electricity, gas and water routes are individually marked out in different coloured chalk so that they know exactly where to dig when the time comes.

When the time does come, your street begins to look like the Somme. If they don't park a JCB on your car, they'll encase it with mud. And then, when all their care-fully laid chalk marks are covered with more mud, they'll wait for you to step into the shower before they drill through the water pipe. You get out and are half-way through writing a book on your computer and they'll cut the power. Then, in the evening, when you have eight people coming for dinner, they'll sever the gas. Outside, there will be troughs both at the top and bottom of your road, so even if you could get to your car, there's no way you'll be able to drive it anywhere.

A day or two after they finish, a man with a bad suit

and a cheesy grin will knock on your door asking if you'd like the cable service which, in case you hadn't noticed, is now available in your street. If this happens, there's only one course of action – you must punch him straight in the mouth. What you must not do is invite him in and sign all the various forms which spew out of his plastic briefcase.

If you do, more men will come round to drill great big holes in your walls, just so that your television can show exactly what was coming in anyway, via the big council house wok on the roof.

I now have cable television and it is a disaster. It tells me what is happening in Lewisham, and at night it shows me a bunch of overweight German blondes with black pubic hair having simulated sex. There are two 24-hour-a-day news services, both presented by people whose teeth are so white I can't look at them, and reruns of programmes which weren't funny 25 years ago – and which are very not funny now.

I can see French game shows and, if I tune into QVC, I can buy a video recorder from Tony Blackburn. Yesterday, a woman spent one hour trying to sell me a necklace, so I tuned to MTV, where Prince was singing a song called 'My Name Is Prince'. Blimey.

Most of the 36 channels on offer are scrambled, and if I want better porn or big bucks films or, perish the thought, football, I need to dig even deeper into my pockets. And I refuse to let my money be used to dig up your street. It's not sociable.

To be fair, I do get a great deal of motorsport on my television these days, but car racing without the Murray Walker soundtrack is like holidaying in a caravan – it's not really a holiday at all. The only advantage Eurosport has is that it covers post-race press conferences, whereas *Grandstand* switches immediately to cricket as the chequered flag falls.

But is this worth £168 a year, when you get the BBC for half that? Plus, the BBC doesn't dig up your road, sever all your essential services, cut off your telephone for two days or send cheesy salesmen round wearing awful clothes.

Mystic Clarkson's hopeless F1 predictions

Before giving the result of a football match which is to be televised later, news readers usually invite us to put our fingers in our ears and hum.

But this morning, as you lay in the bath listening to the radio, I bet it went something like this. 'In Northern Ireland today, Sinn Fein leader, Gerry Adams, likened the situation to . . . Hill won . . . the conflict in Israel . . .'

Bang. There was no warning and those two little words took all the suspense from the subsequent televisual feast. Plus, with Grand Prix, knowing who won the first race means you have a pretty good idea of who's going to win the world championship.

Furthermore, when you know who's won, there is little to be gained from finding out how he did it. He simply drove faster than everyone else.

But if I take my cynical trousers off for a moment, and slip into a nice pair of sensible slacks, in beige, from Marks & Spencer, it's worth having a little look at what might happen in 1996.

The experts are suggesting that Michael Schumacher stands no chance in his all-new Ferrari. They point to the winter testing programme, saying that the car arrived too late to be shaken down properly, and that first indications suggest its new V10 engine is too gutless and too unreliable.

Well I've met enough racing drivers to know they don't choose to lose. Michael Schumacher could have stayed with Benetton, a team he knows and enjoys, and very probably won the crown for the third year in succession.

No one with a 'need to win' like his is going to throw the chance of another trophy away because he feels like a change. He's gone to Ferrari because he knows something we don't. I have no idea how the car performed in Australia because I wrote this before the event but, mark my words, Schumacher – a man I hate more than butter beans and Jeffrey Archer – is my tip for 1996.

Damon Hill, we are told, has spent the winter psyching himself up for the battles that lie ahead. He is now a lean, mean fighting machine who will slice through the field in what everyone says is the best car.

Well Damon's a nice chap and that's where his problems start. Nice chaps with wives and children do not go wheel to wheel at 160mph in a fight to the death. To do that, you must be a berk, and Damon is not at all berkish, which is why he is destined to be the runner-up. Again.

Some are saying his new team mate, Jacques Villeneuve, is a more realistic bet. He, after all, is the son of possibly the greatest entertainer of them all – Gilles Villeneuve. Yes, well my dad understood how to do his VAT returns but that doesn't make me a chartered accountant.

Damon's fans hit back, saying he has trounced all-comers in the American Indycar series. Oh for heaven's sake, that's like saying you can be a Red Arrows pilot because you're good at Monopoly.

We've seen these Indycar boys come over to F1 before – Michael Andretti was the last – and they make complete and utter fools of themselves. Look at Nigel Mansell. In America he became used to duelling with fat has-beens like Mario Andretti, so when he came back to F1 last year he looked as stupid as his facial topiary.

But back to F1 and Benetton. My sources suggest they do have some reliability problems and that Berger and Alesi are finding the car's twitchiness a nightmare. And anyway, the likeable Gerhard Berger seems more interested these days in putting a plastic dog turd under your pillow than actually winning a race.

I hear that McLaren is now back as a force to be reckoned with. David Coulthard has promised not to spin off on the warm-up lap anymore, or run into the pit wall when coming in for tyres, and Mika Hakkinen is fit once more after his awful crash in Adelaide. Indeed, he smashed the lap record while testing at Estoril only last week.

This means he is more deranged than ever. He has an awesome reputation in Grand Prix as a madman and there is now talk that his head injury has made him even nuttier. I like the guy hugely, but don't think he'll win.

First, he will continue to crash a lot as he ekes out levels of grip which are not available; and second, while the new Mercedes engine goes like a bomb, it will also go off like one fairly often.

That will be mildly entertaining but it won't really compensate for the tedium that will result from a new rule in 1996. No car is allowed to qualify unless it can get within 107 per cent of the poleman's qualifying time. Thus, there will be no Fortis and Minardis cruising round to get in everyone's way.

Being held up by a dawdling backmarker added some spice to the race, and gave Murray Walker something to shout about. But now it has gone, and next year Murray will go too.

There is some good news though, because when ITV takes over the reins in 1997, the BBC will have to concentrate its resources on the British Touring Car Championship.

This is 26,000 times better than Formula One, with more overtaking in one lap than you get in the whole Grand Prix Championship.

The Touring Car season begins on Easter Monday. You want to know who'll win? Haven't a clue. You want to know who'll crash? Most of them. Can't wait.

Commercial cobblers

Have you seen that hideous man in the Boots commercial on television? The one who spices up his tedious life by choosing a designer pair of spectacles. So that I can't poke him in the eye should we ever meet in a lift, or on a railway station. 'He'll take care of that. And it's good to know . . .'

Oh for God's sake man, please shut up. We've got the message. Boots do designer glasses. If things get so bad that I can't read a newspaper without being in another room, I'll feel my way straight down there in my blazer and slacks.

This is the point, surely, of television advertising. In the tiny timeframe available it's only possible to give the audience one little nugget. The product may be a dodecahedron, but in the ad slots, we only get an atom.

Unless the subject matter is cars, in which case the trick is to hand over absolutely no information whatsoever.

In a Volvo, it is possible to drive across the Corinth canal on railway lines should the more conventional bridge be blocked for some reason.

How much does a T5 cost? How fast does it go? Can you get a chest of drawers in the boot? Dunno, but if anyone ever starts to throw packing cases at me out of a DC3, I'll wish I had one.

The point, of course, is that the advertising agency is trying to create an image. If you have a Volvo T5, you are the sort of person who is likely to be chasing Dakotas. And while your next-door neighbours are doing the garden, you're out in the eye of the hurricane.

Buy a T5 and you'll be at every dinner party in town, being anecdotal and getting laid.

Unless someone turns up with a Peugeot 406. This guy gets raped in a restaurant, just after he's pulled a little girl from under the wheels of a truck. He plays rugby, is a mercenary and wears a sharp suit.

There's no such thing as an average person. Absolutely. But there is such a thing as an average car, and the 406 is it. I'd rather have a Mondeo, but in the knicker-elastic snapping stakes, the 406 is streets ahead.

Today, the most important man in the car design process is the advertising copywriter. All cars in the mid-ranges are basically the same, so the only way people can choose is by selecting an image.

The 406 is an endearing and well-priced family saloon with the usual features, the usual economy and the usual performance. There are the usual mistakes too in the shape of poor seats, a lousy gearbox and rather too much noise from the 2.0 litre engine.

I am sure it will be a massive sales success for Peugeot though, and that is entirely down to the admen.

Look at the Vauxhall Vectra. Here is another dull and tedious family saloon car with all the usual features and all the usual mistakes. It is being annihilated in the sales charts. Why? Because the advertisements are crap.

In an attempt to encapsulate the essence of New Age imagery, the film director responsible has obviously studied every special effect in the book and, for the reputed cost of £1 million, has apparently ended up shooting the commercial through marmalade.

To date, Lowe Howard Spinach, the advertising agency responsible, has spent another £6 million to ensure that a whopping 96 per cent of the UK population will see the commercial 17.8 times. Ten per cent will see it 30 to 40 times and 1 per cent will see it over 60 times.

You will be able to recognize this 1 per cent in the streets. They will be gently banging their heads against brick walls.

As far as I'm concerned, we should bring the millennium forward by 45 months so that this stupid commercial can be taken off the air and out of the magazines, and put into the wastepaper basket.

Where it will nestle alongside the tripe without onions that Rover has served up just recently.

'An Englishman in New York' cost Rover £1.3 million to make. And the joke is that they don't even sell their cars in the USA because Americans grew tired of the endless mechanical maladies.

When it was announced that BMW were to take over Rover, two senior executives were apparently heard discussing their futures. One said to the other, 'Do nothing. That way you can't be blamed for anything.'

If only he'd listened. But no. He sanctioned that ridiculous advert which served only to line the already bulging pockets of Sting's accountant.

Where's the image association? I can't afford a parking space, so I watch television from inside my car. I'm a berk. Cross the road if you see me coming the other way.

Dear Rover. Large warehouse flats are out. The couple in the Findus advert had one of those back in 1987. *Wall Street* is in the discount bucket at my video rental shop. It's 1996, boys, and Peugeot are making mincemeat of you.

Your engineers did a good job with the Rover 200, and your stylists were wide awake too, but your admen must have been having a large and luxurious lunch with plenty of wine that day.

Probably with the guys from Nissan. The Car They Don't Want You To Drive. Good. I wasn't going to drive it anyway.

Struck down by a silver bullet in Detroit

Last night, in one of the world's five great cities, I shared an alligator with Bob Seger. Ever since that long hot summer of 1976, when I ricocheted around Staffordshire desperately trying to shake off those awkward teenage blues, I have worshipped the ground on which old Bob has walked. I know that it is desperately train-spotterish to have heroes, but here we have a man whose lyrics are pure poetry, whose melodies are a match for anything dreamed up by Elgar or Chopin and whose live act is, quite simply, the best in the world.

After a gig at the Hammersmith Odeon in London in 1977, the manager wrote to *Melody Maker* to say that in all his years he had never seen a better concert. I was there, and it was even better than that. And there I was, 18 years later, in a restaurant in downtown Detroit, sharing a piece of battered alligator with the man himself. My tongue wasn't just tied – it looked like a corkscrew. I wanted to talk music but Bob's a chatterbox with the laugh of a cement mixer, and he wanted to talk cars. He was born in Detroit and apart from a brief spell in Los Angeles, which he hated, he's lived there all his life.

He argued, quite forcefully, that if you're a Detroiter you are bound to be part man and part V8. The only jobs are in car factories, all your neighbours work there, and the only way to escape the production line is music. It's no coincidence that Motown began in the Motor City.

The buses move around empty, as does the hopeless monorail. The train station is derelict. Everyone drives a car in Detroit because cars are everyone's soul. And Bob Seger is no exception.

A point that's hammered home by the GMC Typhoon in which the great man had arrived. He has a brace of Suzuki motorcycles on which he tears around the States, getting inspiration for songs like 'Roll Me Away', but for family trips to Safeway he uses the 285bhp, four-wheel drive truck – you may remember that we took its pick-up sister, the Syclone, to a drag race on *Top Gear* last year.

Bob's mate, Dennis Quaid, has one too apparently, which made me itch to ask what Meg Ryan was like – they're married to one another – but Bob was off again, telling us between mouthfuls of reptile how things used to be in Detroit, how he used to go and race tuned-up musclecars between the lights, how a side exhaust gave an extra 15bhp and how they posted lookouts for the cops.

This was heaven. The man I've most wanted to meet for nearly 20 years is a car freak, but the best was still to come. When we'd finished dinner, he sat back and pulled a pack of Marlboro from his pocket. He smokes, too! And so, he added, does Whitney Houston. By this stage, I had regressed to the point where I could easily have been mistaken for a four-year-old boy – I may have even wet myself slightly – but the full flood was saved until later that night.

Do they, I enquired gingerly, still race their cars on the streets. 'Oh sure,' came the reply. 'Most Friday and Saturday nights up on Woodward you can find some races going down.'And this, I'm happy to tell you, was not just some rock-star-close-to-your-roots-SOB. Because they do. Big money changes hands as a hundred or more guys turn up in Chargers and Road Runners and God knows what else. And then, from midnight until dawn, they simply line up at the lights, wait for the green and go. We watched it all, and happily, from your point of view, we filmed it too, for a new series called *Motorworld*.

We learned, too, that in days gone by the big three American manufacturers used to take their new, hot cars to these races to see just how quick they were. And that, even today, engineers may sneak a new development engine out of the factory and down to Woodward to see if it can cut the mustard.

And all this is set to a backdrop of Martha Reeves, Marvin Gaye, Smokey Robinson, Don Henley, Ted Nugent and Bob Seger – plus the thousand or so other stars that were born and raised in the Motor City.

And we have Longbridge and Take That. Which makes me want to throw up.

You can't park there – or there

Having tried to go shopping in Oxford last week, I now know why Inspector Morse needed to be a two-hour televisual feast. It took him that long to get across town.

On that inner ring road you get to be very good indeed at crosswords.

On paper, I'm sure the traffic management system looked like a brilliant idea. Remove all the on-street parking. Keep cars out of the town centre. And encourage vibrant bus companies to run shuttle services.

On the town planners' maps, the pedestrianized streets would have been full of carefree shoppers and dainty trees, but in fact they're choked with buses from endless different companies who, to stay competitive, run older vehicles that belch out blue smoke.

The end result is a cauldron of chaos and lost opportunities.

My wife was 8.9 months pregnant so, as a bus was out

of the question, we were forced to park near Abingdon and walk. The Japanese tourists were hard enough to circumnavigate, but on the High Street there was a wall of buses jammed nose to tail for as far as the eye could see. The air was unbreathable and the distances were simply too great for someone who, in fact, gave birth that afternoon. We trudged back to the car, empty-handed, and headed for home.

On the way, we found a Toys 'Я' Us on the outskirts, where I burdened my credit card to the tune of £275. That's £275 which, thanks to the council's idiotic transport policy, has been kept from the town centre shopkeepers.

I only live 17 miles from Oxford but I will never ever go back. I will never eat in an Oxford restaurant. I will never go to an Oxford pub. I will never buy anything from an Oxford shop.

I will go instead to towns where they are wise enough to welcome me and my car. Banbury. Cheltenham. London even.

This town planning business is becoming a triumph of vegetarianism over common sense.

Council people are obsessed with commuters, people who work in town centre offices, people who create the rush hours. But in their blinkered drive to solve this problem they're forgetting that towns should be a hub for the outlying villages, centres where people go to shop and eat and be entertained. And these people, whether councils like it or not, need to come by car. You can't take a fridge-freezer on a bus.

If the car is banned and out-of-town superstores are encouraged, town centres will die. Already, privately run bakers and haberdashers have been replaced with estate agents and building society offices. Oxford is spoiled because the Japanese tourists will keep on coming, but other towns whose spires are not quite so dreamy should be very, very careful.

If you hammer the commuters, shoppers will go elsewhere, and if you keep on hammering, business leaders will pack up and go too. What's left?

I wouldn't mind, but the solution is so desperately simple. Instead of removing parking spaces, councils should provide as many as is humanly possible. They should analyse every last yard of yellow line and wonder whether it's absolutely necessary.

Stick up pay-and-display units. Charge us a pound an hour. It's OK. We don't mind.

If you make parking easy, you will automatically reduce congestion because you will not have cars going round and round the block any more. Seriously, I cannot think of a more idiotic use for a car than looking for somewhere to stop it.

But, of course, a town planner who admits this is talking himself out of a job, so over the coming years we're going to be treated to a series of schemes which are, quite simply, bonkers.

Last week, a bunch of European ministers met to discuss the issue and heard that in Turin there is now an advanced booking service for parking spots, to prevent motorists from driving into the city on the off chance of finding a space.

Yes, but when I need some cigarettes, I want them now and not after Mrs Miggins has finished buying her cat food.

In the Netherlands, the city of Groningen is divided

into four quadrants. Traffic can whiz round the ring road and enter one sector, but if you then want to go to another sector you must get back on the ring road again.

Why? I lay awake all last night trying to figure that one out and I can't think of a single advantage.

In Zurich, sophisticated bus priority signals keep cars sitting at junctions for hours.

This is galactically stupid. A fat Swiss banker is not going to leave his Mercedes 600S at home and take the bus. He is just going to set off from home a little earlier to compensate, and then he's going to sit in the jam with that huge, 6.0 litre V12 chewing up the world's resources like Pac-Man.

And there's the rub. In Britain, according to the RAC, 80 per cent of all journeys are dependent on the car. It doesn't matter how much the government taxes motorists or how miserable life is made for them by councils, there is no alternative.

Yes, cry the dissenters, but what about the 20 per cent of journeys where there is an option. When I go into my local town I could easily walk, or use a bicycle.

But here's the thing. I never ever will.

Sermon on Sunday drivers

There's a bar in Austin, Texas where the locals gather on a Thursday to dance the night away . . . country style.

Strangely, even though some of the chaps are the size of a double garage and their womenfolk are even larger, it's a graceful sight to behold.

And so, during the week, are Britain's motorways.

Stand on a service station footbridge and you are treated to what can only be described as automotive ballet.

Get yourself on to the M40 on a Tuesday and you will see a display of driving that would leave Damon Hill breathless. Sure, he can control a car at 180mph, but unlike the reps in their Mondeos and Vectras he doesn't have a phone in one hand and a sausage roll in the other.

Formula One chiefs are concerned about the speed differential between the top cars and the Fortis, but guys, guys, guys. Mondeo Man is out there every day on the motorway, doing 90, juggling with juggernauts that can barely crack 50. And he doesn't whinge.

He can't because he has no grounds. British truckers are in a class of their own. Think. When was the last time one of these seven-axled giants caused you even so much as a moment of concern on the motorway network? It never happens.

They get on with their lives, getting lettuces to the shops before brownness sets in, and you get on with yours.

Of course, you're good too. I drove up from London to Oxford last Tuesday and I have never seen such fine driving. These guys were harassed and bored but they made deft, precise and well-signalled moves. They kept up when it was right, hung back when it was necessary and, as a result, never gave any cause for concern.

Like anything, practice makes perfect. The more you drive, the better you'll get. If you're out there every day, be it in a truck or a Mondeo, clocking up 50,000 or more miles in a year, you will be damned good.

You'll learn to recognize the danger signals. That's a Datsun. He is likely to do the unexpected. That's a T5 doing 50. It must be Plod. It's starting to rain. I'm easing off now.

I use Britain's motorway network a great deal during the week and it is like being part of a huge, perfectly synchronized, well-oiled dance routine. Everything is fluid. Everything is inch perfect.

Unfortunately, I also use Britain's motorway network on a Sunday and it is an experience that takes me to the outer edges of fear and trepidation. Two miles and my colly is well and truly wobbled.

Last week, I found a Vauxhall Nova with a lone woman on board trundling down the outside lane of a dual carriageway at 15mph. What in God's name was she doing in possession of a driving licence? Enid Nun 004. Licence to kill . . . and be killed.

Now I don't believe there are many people left who, just for the hell of it, go out for a drive on a Sunday. And even if there are, I doubt they'd chance their arm on a motorway.

But there are undoubtedly a great many people who, after they've done the *Mail on Sunday*, head off to see Auntie Flo, via the garden centre. These people probably never drive during the week, and in a whole year probably clock up fewer than 2000 miles.

They've never been trained to drive on a motorway and they have had no practice. Allowing them out there is like letting me take the role of principal violinist the next time the London Symphony Orchestra is in town. I'd be crap and they'd sack me.

These people get into the lane they'll want 20 miles early and will do everything in their mealy-mouthed little minds to ensure you and I do likewise. If they can get their wheezing asthmatic old crocks up to 70, they'll sit in the outside lane making sure no one gets past. It's against the law, you know.

They clutter up the petrol stations with their awful cardigans, putting unleaded in their diesels and Wendy's Panties on their hideous, hateful children.

And then they crawl down the slip road at 4mph, joining the motorway when they're up to 6.

Suddenly, the professional, talented, regular driver finds his space is full of no-hopers. The trucks may not be out to play on a Sunday but you still have an odd and dangerous cocktail.

On a Tuesday, 99 per cent of all the cars out there will do exactly what you expect them to do. But on a Sunday, half will do exactly the reverse.

I have discussed various solutions with all sorts of clever people but there don't seem to be any that are practical. You can't include motorways in the driving test because the good people of Norfolk and Cornwall would be stuck.

You can't post lookouts at the top of every slip road to pull over people they suspect may be a nuisance when they get down there.

And we can't encourage people in bad cardigans to drive around with a huge sign in their rear windows saying 'I'm really no good at this.'

Or can we?

A riveting book about GM's quality pussy

Quentin Willson has read a great many books and is prone to inserting large and complicated pieces of Shakespeare into normal conversation. My wife's bedside book table, on the other hand, is filled entirely with those orange-spined Penguin Classics, all of which are about women in beekeeper hats who walk around fields full of poppies, doing nothing. These make for good bed-time reading, only on the basis that you need to go to sleep. 'A Saturday afternoon in November was approaching the time of twilight and the vast, unenclozzzzzzz . . .'

With Quentin's books I'd have to spend the whole time buried in a dictionary, finding out what all the words meant. The guy reads Chaucer for fun, for Chrissakes! All my books have either a submarine or a jet fighter on the front and they're full of goodies who seem like they're going to lose but who, on the last page, do in fact win. I like plots, and Hardy wouldn't recognize a plot if one jumped out of a hedge and ate his foot.

A book is no good, as far as I'm concerned, unless I just cannot put it down. I missed a plane once – on purpose – because I was still sitting at home finishing *Red Storm Rising*. If Princess Diana had walked into my bedroom naked as a jaybird just as I was three-quarters of the way through *The Devil's Advocate*, I wouldn't have looked up long enough even to tell her to get lost. My wife, however, has just taken two years – yes, years – to read *Wild Swans*, which is about a woman in China who has a daughter who goes to live somewhere else.

But I have just read a book which has no plot, no F-16

on the cover, no goodies, no baddies, and I absolutely loved it. Which is a bit of a worry. It's called *Rivethead* and it's by an American person called Ben Hamper who, in the review section, describes it as 'an enormously enjoyable read. I laughed. I cried. I learned. I got naked and performed cartwheels for my repulsed neighbours'. My kinda guy.

Basically, *Rivethead* is the story of one man; a man who gets up every morning and goes to work at the General Motors truck and bus plant in Flint, Michigan. Really, it should have an orange spine, but mercifully it doesn't. Because if it did, I never would have heard about GM's answer to the Japanese threat. You see, when American cars were being sold with tuna sandwiches under the driver's seat and Coke bottles rattling in the doors, GM decided it must impress on its workforce the need for better standards. The workforce, largely, was a doped-up bunch of ne'er-do-wells who thought only of their weekly pay cheques and how much beer they could cram in at lunch time, which is why GM's decision to have a man dress up as a cat and prowl the aisles, spurring people on, is a trifle odd. That they called him Howie Makem is stranger still.

Equally peculiar was the later scheme, which involved the erection of several sizeable electronic notice boards all over the plant. These kept the people informed of sales, production figures and such, but could also be used for messages. One day it would say, 'Quality is the backbone of good workmanship' and on another, 'Safety is safe', but Hamper saves his vitriol for the day when he looked up from underneath a suburban pick-up to see the sign: 'Squeezing rivets is fun!' He goes on to wonder whether, in the local sewage works, there are boards telling the guys that 'Shovelling turds is fun'. And asks why, if the 'demented pimps' who had dreamed up this message thought riveting was so much fun, they weren't all down on the line every lunch time, having the time of their lives.

Hamper also lays into the likes of Springsteen and John Cougar Mellonfarm, asking what they know about the daily grind. He says they should be forced to write about things they understand, like cocaine orgies, beluga caviar and tax shelters. I made an exception and read this book because I am interested in the car industry, but I can recommend it to you even if you have never been in a car plant, and don't ever intend to.

I tried to get Quentin to read it, but as the first word is 'Dead' and not 'Sibilance', he said he couldn't be bothered . . . and asked how Janet and John were these days.

Aston Martin V8 – rocket-powered rhino

From time to time I peer through *Esquire* and *GQ* to see what I should be wearing, but it's hopeless. You can't go shopping in a red plastic vest if you have a belly like a Space Hopper.

And I'm sorry, I just don't like those jackets which have lapels like a butterfly's wings. Nor will I ever do my top button up unless I'm wearing a tie.

Last weekend a footballer called Paul Gascoigne was in the *News of the World* wearing what can only be described as a dogtooth dog's dinner. It was a suit, in that the top and bottom matched, but the jacket was down to his knees.

I have never seen such a ludicrous garment, and can only assume his mother had knitted it.

But then again, this Gascoigne person probably looks at my Lee Cooper jeans and Toggi shirts and thinks he's been through a time warp. Away man, it's 1976 all over again.

And that's the point. Each to his own. Those of us with a penchant for chunky gold jewellery will go for a Toyota Supra. Paul Gascoigne would be bewitched by the Honda NSX whereas Clement Freud, obviously, has a Lexus. I have no idea what David Attenborough drives, but would hope it's a Jaguar. A Bentley, these days, is a bit too Paul Daniels. Know what I mean?

So what about me? Well from a fashion point of view, it would have to be the new Aston Martin V8 coupé.

This is a brute of a car. It weighs 2.2 tons. It's 17 feet long. It's wider than an ocean liner and it has a monstrous, hand-built V8 which can propel it to 60mph in less than six seconds. It's a rocket-propelled rhino.

Basically, what we have here is a Vantage without the artificial lungs. Aston has removed the superchargers but kept the high-performance pistons, camshafts and valves to create a replacement for the unloved Virage.

In terms of styling, it does without the Vantage's hugely flared wheel arches and massive tyres, but the rear end is identical. To follow this car is to be in the presence of evil.

When you see it in your rear-view mirror, be afraid. Be very afraid.

Get out of its way or be prepared to look like a

waxwork dummy at gas mark six as each of its eight lights begins to flash.

If you still choose to block its path, you should know that its driver could swat you out of his way and not even know. A big Aston could head butt a tower block and the tower block would lose.

Some say it's nothing more than a bespoke Corvette, a big American-style tank with leather innards, and I say yes to all that. I can't think of anything better than a V8-powered gentleman's club.

I can, however, think of a great many cars which are nicer to drive. A Ferrari 355 will run rings round it and a Mercedes is not only more nimble but undoubtedly more reliable too. Round a race track, I doubt the big Brit could hang onto a Golf VR6.

But for all the reasons already outlined, the Golf had better hope the Aston didn't catch up on the straight bits. Which it would.

None of this matters though. The point is that when I looked at my reflection in a shop window, I felt good. It is my automotive Lee Cooper and Toggi combo. The interior of the V8 may be surprisingly cramped but, despite that, this is not a car for small people. You'd look stupid driving this unless you were at least 6ft 3in and 14 stone.

Other people who would look stupid in it include Liberal Democrats, Freemasons, folk singers, nancy-boy footballers, vicars, scoutmasters, people who like DIY or Michael Bolton, women, environmentalists and anyone who has ever been to a poetry reading.

You can't even think about driving this car if you like salad.

Socialists are right out. So are people who use the words 'toilet', 'nourishing' or 'settee'. If you read the *Daily Mail*, talk about tasty square meals and country fayre then, along with ramblers and people with limp wrists, lisps, or sticky out ears, you must buy a Datsun instead.

Are you a new man? Do you like to help around the house? Are you proficient at changing nappies and running up a set of curtains? Have you ever read a Barbara Taylor Bradford novel? Well go and buy a Honda then, because the Aston will break your kneecaps.

The V8 is for those of us who like our beer brown and our fags to be high on tar and low on lentils.

What I love about this car is that while it does nothing to hide its immense power, it comes trimmed in the finest leather. The carpets are so expensive you wouldn't fit them in your house, and the wood is lustrous enough to cause a mass fainting on *The Antiques Roadshow*.

You mustn't be fooled though. If you slide a Phil Collins CD into its stereo, the airbag will spring forth to punch you in the face.

It likes Elgar and its favourite rock track is Bruce Springsteen's 'Born to Run', though Led Zep's 'Black Dog' will do. If you treat it like a hard-drinking, hard-playing soul mate, it will reward you with a spine-tingling range of growls, and the power to knock down copper beeches as you fly by. The only trouble is that it costs £140,000 which is an awful lot of money. I have a suggestion though. To raise the funds, rob a bank. It would like that.

Caravans – A few liberal thoughts

After much careful thought in the bath this morning, I have decided that we don't really need an elected parliament.

These 650 guys are concerned not with what's good for the country or the environment, but with power. Every decision they make is based on a quest for votes.

I remain absolutely convinced that the Labour Party's apparent shift to the right has nothing whatsoever to do with the elected members' beliefs. They're just saying what they think the middle classes want them to say.

And the Conservatives are no better. Here are a bunch of people who'd done all that was necessary by 1989. They could have just sat back and let things tick over, but no: half of them now want to privatize my shoes.

We should replace them all with a bloke who has a bit of common sense. Every Thursday, he would pop down to Westminster so that civil servants could ask for advice.

Should the Spanish be allowed to fish in our waters? No.

Should Peter Blake be allowed to keep his ninety grand? No.

Should we ban scoutmasters from keeping guns? Yes.

Should we shoot people who let their dogs crap in the street? Yes.

It's all so simple. We don't need 650 people making noises like farmyard animals five days a week, when most of the burning issues could be settled over a cup of coffee by a bloke in a cardigan.

Certainly, if we were to introduce this new system, and

I really think it's one of my better ideas, the roads would become free from caravans.

Should this question ever be brought before the Commons, the member for Devon North would argue forcefully that caravans form part of his constituency's life blood, and that if they were to be banned so soon after all the cows were burned there'd be anarchy and looting on the streets of Minehead. And then someone else would rise to their feet and point out that some of his voters work in a caravan factory and that they'd be out of work, claiming benefit.

And that would be it. Caravans would stay.

Whereas under my system the bloke in a cardy would weigh up the issues over a slurp of Kenco and say, 'No, they must go.'

In twelve years of writing about motoring I have only touched on this issue once because it did not seem important. I lived in London, and on the rare days when I sallied forth to the Provinces I was on a motorway.

But now I live in the Cotswolds and it's unbelievable. I've just taken delivery of a new supercharged Jaguar, and so far I haven't had it past 20 because round every corner the road is blocked by a Sprite Alpine.

I was stuck behind one called Sprint the other day. How can you call a caravan a 'Sprint'?

And when they're parked in a field they hardly blend into the environment. As Mark Wallington says in his magnificent book, *500 Mile Walkies*, 'Why can't they be painted black and white, and given udders?'

As a child I went on a few caravan holidays and I remember wondering what we were doing there. I mean, we lived in a large farmhouse in the countryside and now, here we were decamped in a small box in the countryside – feet away from a fat family whose daughter, Janet, had woeful diarrhoea.

This, however, is not the issue. If people want to spend their precious vacation in a metal container, in a field full of other metal containers, eating shabby food and defecating in a bucket, fine.

The problem with caravans is that you can't simply beam them to a site, *Star Trek* style. You must hook them up to the back of your wheezing, asthmatic car and, with absolutely no training whatsoever, tow the damn thing into some of Britain's greener parts . . . like here.

People. As you look in your rear-view mirror and see a trail of cars stretching back to the horizon, do you not feel even the smallest pang of guilt? Do you not feel that it might be a good idea to pull over and let everyone by once in a while?

Do you not vow that next year you will undertake the journey at night, when you would be less of a bother?

Or do you secretly relish having the power of being part of a tiny, tiny minority who, for a few hours a year, can control something huge like traffic speed. Did you dream as a child of being a councillor? Or joining the parks police? Go on, admit it, you did.

You are a mealy-mouthed little twerp with no regard for others. In the last few weeks you've made me late for every single appointment, and you don't give a damn.

If caravans can't be outlawed, and without my new system of government they never will be, there should at least be some new rules.

Anyone wishing to tow one should be forced to take a complicated driving test. They cannot be towed by any car with less than 300 ft/lbs of torque. They can only be taken on the roads between 2 and 6 a.m. on a Wednesday morning. And they should incur road tax of £200 a foot.

Blind leading the blind:
Clarkson feels the heat in Madras

This is what it said on the first page of my joining pack for the world's weirdest motorsport event. 'Rallying has never featured very significantly in the lives of blind people.' No, and neither will it. Men can't have babies. Fish can't design submarines. BBC producers can't make up their minds. And blind people don't make very good rally drivers. However, they can navigate. More than that, in the last six years there have been 25 rallies in India where the co-drivers have had more in common with a bat than Tony Mason.

Now, to be perfectly honest, I'm not talking about the sort of rally where the car's wheels only ever touch the ground in service halts. No, this sort is best described as a treasure hunt. Even so, disappointingly, there are rules, the worst of which is that all cars must be fitted with seatbelts. This meant that when I took part there were only 66 competitors, which isn't good enough in a country with nine million blind people. But hey, I'm used to rules, and the best way round them is to indulge in a bit of Boss Hoggery. I figured that if I nicked the notes from the navigator, he'd never know and we'd win. But the organizers had that one covered; all the directions were in Braille, a language which means as much to me as Swahili or German. Like everyone else, we had to use the force. But unlike everyone else, we went wrong at the very first turn.

Let me explain. The Braille was in English and this was not a language that featured on my co-driver's CV. So he spelled out each instruction, letter by agonizingly slow letter. Thus we left the base and headed off towards the centre of Madras in our Maruti Gypsy, with Mr Padmanabhan muttering t-y-r-d-i-n-a-k-l-m-t-e-y-r-l-e-f-f. Which, if you have a pen and a piece of paper, and a fortnight, you could work out meant turn left in a kilometre. Trouble was it took me nearly five miles to figure it out, by which time we were completely and hopelessly lost. Not only do I not speak Braille but my Tamil's not that good either. And there I was, with a blind man, in a city that I've never been to before (and never want to go to again, incidentally), on the same land mass, worryingly, as Portugal and Yemen. Things could go wrong here.

We'd be drifting down a road and, all of a sudden, Mr Padmanabhan would look up from his notes to ask: 'What is l-k-j-r-i-j-l-s-s-a-e-q-j-t?' And to be honest, there isn't really much of an answer.

But somehow, and I guess quite by chance, we did happen upon a checkpoint. Relieved, I wound down the window and asked just how far behind we were. But here's a funny thing; they said we were the first to come through, which was strange as we'd been the last to leave. However, it all became crystal clear when they told us that we were at checkpoint six and that we had somehow missed one to five. I knew damn well how we'd missed them. We'd been in Tibet. Nevertheless, we ploughed on until suddenly I was told to stop. 'We are now at checkpoint seven,' Mr Padmanabhan said. But we weren't. We were in the middle of an industrial estate, and it's hard to point out to a blind man that he's gone wrong. Again. 'I'm sorry,' I said, 'but we're not.' 'Yes we are,' he insisted. And to avoid hurting his feelings, I had to leap out of the car to get my card stamped by a non-existent official at a checkpoint which wasn't there. 'Told you so,' he said when I got back in the car.

Back at base, the event over, we learned that we'd been scrubbed from the running order altogether, on the basis that we'd only found one of the checkpoints. They all figured we'd given up and gone home. We didn't even get any lunch, which was no bad thing because it seemed to consist of stillborn blackbirds which had been trodden on then coated with curry powder, bay leaves and ginger.

Oh how we all laughed as the navigators tried to pick bits of beak out of their teeth. And oh how they all laughed as they reminisced about how hopeless all their drivers were. We must see this sport in Britain. All you Round Table, Rotarian types, stop pushing beds up the high street, jack in the three-legged pub crawls and give the RNIB a call. And then call me to say where and when.

Norfolk's finest can't hit the high notes

In my early twenties I filled my days by spending money. Then, in the evening I went out and spent even more, often in a casino.

Inevitably, my bank manager wrote regularly. Now he could have called me all the names under the sun, he could have used language blue enough to make Quentin Tarantino blush. He could have made threats of violence and it would have had no effect.

Instead, he used the strongest word in the English language: 'disappointed'.

'I am disappointed to note . . .' he would begin and my Adam's apple would swell to three times its normal size. By saying he was disappointed, he was suggesting that he had had high hopes but that I, personally, had let him down.

Since those days I've been very careful about using the 'D' word. When reviewing cars, I've said that they were foul, or dull, or that they couldn't pull a greased stick out of a pig's backside. But I have never used the word 'disappointing'. Until now.

The Lotus Esprit V8 is disappointing. The cruellest six words you will ever read. I had high hopes of this car, and *it* let *me* down.

On paper, the car looked good. Here was a machine with OZ racing wheels and Brembo racing brakes. Here was a car with the latest generation of ABS and, to top it all, a twin-turbo V8 with 350 horsepower. Here, in short, was my kinda car.

When it arrived there was nothing to suggest anything was amiss. The Esprit may have been around since the Bay City Rollers, but with its curvier corners and its eyebrow wheel-arch extensions, it looked really rather good sitting in my yard.

I wasn't entirely convinced by the leather and wood interior, which seemed out of place in a mid-engined supercar, but that's like saying you won't eat pizza because you don't like olives.

I admit it: for a few brief moments, I actually thought about buying one.

Then I went for a drive and I was reminded of *Top Gun*, a film that, like the Esprit, promised so much. I'd seen the trailers and noted that there was unprecedented aerial footage of state-of-the-art American fighter planes.

I'd read reports which spoke of unparalleled access given to the producers by the Pentagon. And there was Val Kilmer and Tom Cruise to boot.

It all opened well too, with those slow-mo shots on board the aircraft carrier. There was purple haze and yes, from seat H16, a few excitable squeals. This was going to be two hours of kapow, whiz bang action.

But, in fact, it was two hours of sheer, unadulterated drivel during which time grown men punched one another on the shoulder instead of shaking hands, and talked nonsense. Furthermore, they were all called improbable names like Ice Man and Maverick and Goose.

When Maverick refused to engage the enemy because he'd killed Goose while trying to show off to Ice Man, I was nearly sick. And when he returned to the ship, and was given a hero's welcome, I vowed that if I ever met Tom Cruise I'd hit him.

Then the little pipsqueak went and married Nicole Kidman, making the need to wallop him even more urgent.

I digress. The point is that the Lotus was as much of a disappointment as that film.

The reason why its 3.5 litre, blown engine only develops 350 horsepower is because the Renault gearbox would blow up if it were asked to deal with any more.

As it is, it feels like the lever is set in concrete. To

get from third to second requires a two-week course on anabolic steroids. If you want reverse, you need the negotiating skills of an anti-terrorist policeman.

And when you get the beast rolling, another problem rears its ugly head: vibration.

The new engine has the same basic design as a Formula One unit, where refinement is not really an issue. In a road car though, the constant buzziness does get to be a bit of a bore.

When you take the motor past 5000rpm, the gear lever vibrates so much it feels like you've just grabbed a high-voltage cable. It's a brave man who reaches for it, especially when he knows he can't move it anyway.

Then there's the noise. I was expecting one of two things: a V8 bellow or a crackly strum. In fact, you get almost nothing to write home about, a result, say Lotus, of forthcoming noise regulations.

Well listen here guys. When I hum gently, it's an awful noise that makes the children cry. When Pavarotti hums quietly, people will pay £200 to listen. You can make a car sound nice *and* quiet.

You can see now, I'm sure, why the Esprit was such a terrible letdown. But unlike the Maserati Quattroporte I drove recently, there are at least some up sides.

First, it handles even better than any Esprit before, which is to say it handles absolutely beautifully. There is real class in the chassis here.

And second, it is truly fast, not daft like the F50, but quicker than a Ferrari 355, and that's saying something. At less than £60,000, it is cheaper too.

But the Esprit has always been good value, fast and blessed with great, supercar handling. Its real problems in the past have been a poor gear change and not much aural excitement.

Meet the new boss; same as the old boss.

Car interiors in desperate need of some Handy Andy work

Televisions are grotesquely ugly but I have never even thought, for one minute, of hiding ours away in a reproduction Georgian cabinet.

We have no Dralon either, and button-backed leather furniture is a bit thin on the ground too.

Should you drop round at sixish, in need of a drink, it's in a cupboard in the kitchen. I'm sorry, but we don't have a globe which opens to reveal the bottles inside.

So why, then, do I rave about the interior of a Rolls-Royce? Only a footballer would ever dream of fitting inch-thick, royal blue shag pile in his drawing room so why is it acceptable in a car?

And look at the wood on that dashboard. Perfect. Unblemished. Polished like a guardsman's shoes. Now look at the wood on your refectory table. Knackered. Riddled with sixteenth-century woodworm. Ill-fitting pieces. And worth ten grand of anybody's money.

Then there's the Roller's seats. There's no doubt that the contrasting piping lifts the magnolia trim, but would you buy a chair for your hall which had cream hide and pale blue piping?

Things get even worse down the automotive scale too because once you're in Roverland the wood becomes plastic. You wouldn't dream of having anything made out of plastic fake wood in your house and yet you'll pay more for it in your car.

It's madness and I don't have an answer. I'm as guilty as the next man. I love the innards of my Jag but there isn't a single square inch of it that would be allowed over the threshold of my house.

And as cars go, the Jaguar is good; tasteful, refined and, rarely for a car these days, fitted with a radio that anyone over 50 could operate.

But there are plenty of cars out there which not only have foul trim but which have been laid out by idiots as well.

Take Ford. It's all very well designing a swooping dash-board which rises and falls like the distant hills in a child's painting, but what if you want to put a can of Coke down somewhere? You wouldn't accept curvy worksurfaces in your kitchen, would you? All your Brussels sprouts would roll on to the floor.

And sticking with the kitchen theme, there's the Ford Galaxy people carrier. Whoever designed that upholstery had a cauliflower fixation.

It is a Dralon type fabric, textured like the walls in an Indian restaurant, but instead of fleur-de-lis sculptures, which might have been acceptable, they've ended up with something that looks like a fuzzy vegetable basket.

Renault, though, still holds the title of 'worst ever fabric design'. Check out the seats on a Clio Williams to see what I mean. You could invite an entire rugby team to be sick on them and the owner would never know. Apart from the smell, perhaps.

One of the latest trends is to litter the interior with

leather that's not only grey – unacceptable on shoes and worse in a car – but perfect: smooth like Formica, and odourless too. Leather is a natural thing, so kindly leave the blemishes in place. We like them.

We do not, however, like your passenger-side airbags. These prevent owners from fitting baby seats in the front, and reduce the size of the glove box to a point where its description becomes literal. A glove box. Not a gloves box, you'll note. And what, pray, is the point of an airbag for the passenger? A driver needs one so that in a frontal impact his head will not hit the steering wheel, but what is the passenger going to hit? Nothing. Two airbags are as unnecessary as two slippers by Douglas Bader's fire.

Car interior designers would be better employed finding space for little cubbyholes where we may keep our telephones, cassettes and fags. Thank you for the fuzzy felt coin holders, but as most parking meters take pound coins these days it means leaving a fiver's-worth of metal in plain view all the time. Which means you need a new side window every day.

Thank you too for making sun roofs such a common fitment these days. They don't allow any breeze to get into the car, but if you open them at anything above 3mph your eardrums implode.

And as a by-product of their uselessness, sun roofs also rob up to two inches of headroom, which renders the car useless to anyone who's registered at the doctors as a human being.

I'm 6ft 5in, so you might say I must pay the price for blocking your view in a cinema, but my wife is a 5ft 1in midget who has never blocked anyone's view of anything.

Yet despite her public-spiritedness, there are some cars – yes you, TVR – where she can't reach the pedals.

Surely to God if a car firm can use platinum to extract poisonous gases from the exhaust, they can design an interior which can accommodate all forms of human life, and not just those that are average.

Yes, we've recently had the Fiat Coupé, which represented a significant step in the right direction but I really do think it's time for car interiors to be radically altered. Why can't we have raffia seats and straw matting on the floor? Maybe a real fire or a wood-burning stove instead of a heater. Why not?

Land Rover employed Terence Conran to design the interior of the Discovery, which showed spirit, but I was crestfallen at the result. I'd expected something really radical with new fabrics, new shapes, new ideas.

Instead, I got some map holders above the sun visors and a zip-up centre console bag. Wow.

New MG is a maestro

At a major league party, there are certain rules you won't find in any book of etiquette. And the most important one is this: when called upon to move into the dining room for dinner, never arrive at the table first because you will have no control over who sits next to you. And don't get in there last either, because when there's only one space left you can be assured that the people on either side of it will be ghastly.

Unless you pay attention to these simple rules you could

find yourself sandwiched between a footballer and a vegetarian. Or a homosexual and a lay preacher. Or a caravan-
ner and a socialist. There are any number of shiversome
combinations, but the absolute worst is finding yourself
between two members of the MG Owners Club.

For a kick-off they will have beards, bits of which will
fall in your soup. And because they like fresh air, they
are likely to be vegetarians. This means you'll be told, at
length, about the plight of dewy-eyed veal calves and
baby foxes with pointy ears and snuggly tails . . . and
chicken feathers stuck to their rabid fangs. By the time
their nut cutlet is served, the subject will have turned to
their horrid cars.

Now you and I know the old MG was a gutless bucket
of rust which leaked every time it rained, broke down
every time it was cold and overheated every time the sun
put his hat on. It turned with the agility of a charging
rhino, stopped with the panache of a supertanker and
drank leaded fuel as though it had a Chevy V8 under the
bonnet. However, our bearded friends don't see it quite
like this. These people actually enjoy the frequent break-
downs because it gives them an excuse to get under the
damn thing.

And then, in the pub that night, they can talk liberally
about exactly what went wrong and precisely how they
fixed it. To you and I a track rod end is very probably
the dullest thing in the world but to MG Man it is a steel
deity, an almost religious icon, an automotive Fabergé
egg. MG Man can talk about a track rod end for two hours
without repetition or hesitation. And the only reason he
stops after two hours is because you've shot him. MG
fanatics are the people that give all car enthusiasts a bad

name. These days you only need mention that you like cars – meaning that you'd buy a Ferrari if you won the lottery – and the person you're talking to will run away screaming. They'll recall a conversation they once had about track rod ends and they will assume that you're about to do the same, that you are a member of CAMRA and that you only drink beer if it has some mud in it.

For this reason, I am concerned about the new MG. If you can be labelled an anorak for simply liking cars, can you begin to imagine how you will be spurned if you walk into the pub brandishing an MG key ring?

Other people at the bar will conclude that you have a 1970s Midget in the car park and that you're about to regale them with the interesting tale of how you adjusted the timing that morning. They will all feign illness or urgent appointments so they can get out.

Except, of course, for the landlord, who'll be stuck. His only escape is suicide. He may even impale himself on his hand-pump levers and die horribly without even realizing that, in fact, you have a new MG. I don't doubt that this is a wonderful car, what with its clever engine, cleverly arranged between the axles. It is lovely to look at too, and those white dials make what's an ordinary interior look a bit special. I feel sure that the hood won't leak and that, mechanically, the MGF will be as bulletproof as your fridge. And though no journalist has driven it yet – contrary to what many would have you believe – I don't doubt that it will handle tidily and be fast. And it's British – which automatically makes it better than the Barchetta and the Speeder and the MX-5 and the SLK and the Z3 and all the other roadsters that are due to be launched in the coming months.

The trouble is, though, that if you do buy one of the new foreign convertibles you will be perceived as someone whose feet are loose and whose fancy is free. But if you go, instead, for anything with an MG badge on the bonnet, people will think you are a git.

Darth Blair against the rebel forces

Anyone who wants to be a politician is very obviously unfit to actually be one.

The would-be politician is weak and craves power so that he may impose his will on the people who bullied him at school.

Of course, when he gets elected, he finds it doesn't really work like that. Whether his boss is Mr Major or the Joker, he is told to sit at the back and shut up.

'Your views are irrelevant. You do as we say. You agree with us publicly and we shall be elected. We shall have the power.'

He no more wants a single European currency than he wants his children to catch typhoid, but he knows that if he votes with his heart, he'll go home that night with a cattle prod up his bottom.

At the next election there will be 650 Labour candidates, and if the Joker is to be believed, every single one of them agrees with his new transport policy. Of course they agree – it's hard not to when the alternative is having a strimmer put down your underpants.

Well frankly, I'd rather feed my toes to a lawnmower than live in a country where the roads are run by Mr Blair who, it seems, wants to stabilize traffic levels by 2010 and reduce them to 1990s levels by 2020.

There would be taxes on the car parks at out-of-town superstores and anyone who takes their car to work would be forced to pay £8 per week, in tax, for the privilege of parking it on company premises. The extra cash will pay for the extra bureaucrats.

There'd be road tolls and local authorities would be given the power to introduce charges to manage traffic in their area. Well that's brilliant. In my experience, most local authorities can't even decide whether to put the lavatory seat up or leave it down.

Got a company car? Well you're in it right up to your neck because the car police will be round to empty your pockets on the hour, every hour.

Oh and you can forget about harrying the fleet manager for a better set of wheels next time round because he'll be under orders to buy cleaner, greener cars that run on manure or potato peelings or some such nonsense.

I'm damn sure there are a great many prospective Labour candidates who would agree that this is idealistic claptrap, but such is their fear they won't dare speak out. Remember, the woman who thought it up was sacked for leaving an interview too early.

I loathe out-of-town superstores too, but they do make life easy for shoppers, and they keep traffic out of ancient town centres, so surely they're a good thing?

Statistics show that a family's weekly bag of groceries weighs a whopping 66 lb, so I wonder how Mr Blair thinks a woman with two children and a pushchair can get a load like that home on one of his infernal buses.

I guess the solution is to follow Harriet Harman's

example and cut the weight down – perhaps by sending your children away to a private school.

So, does anyone know where the Society of Motor Manufacturers and Traders is secreted? Because the very industry it is supposed to represent is under attack.

If someone threatened to burn my house down I would do everything in my power to stop them, but when Labour says it wants to damage the car industry the SMMT doesn't even chirp.

It doesn't even climb on a soapbox when stupid environmentalists go on the news to spout a lot of nonsense about pollution. The report tells us that cars are killing everyone, a man with a beard backs this up and then it's back to the studio with Michael Cheerful Buerk.

Where's the bloke from the SMMT, pointing out that cars do less damage to the environment these days than lawnmowers, or that houses produce more greenhouse gases than anything Ford has ever built? He's in an office somewhere having a meeting. Or he's taking a sympathetic MP out for lunch to indulge in a spot of gentle and pointless lobbying.

Meanwhile, every motorist in the land is on a massive environmental guilt trip, soaking up Labour's new plans and accepting them as inevitable. If you commit murder, you pay the price.

The defence is left to a tiny little organization called the Association of British Drivers, who put out a scrappy little newsletter every so often. However, scrappy though it may be, it's the best read since Alistair MacLean finished *HMS Ulysses*.

In the most recent issue it tells of an accident that was caused by a new speed camera, and of a speed trap near Dover which netted £7500 in fines in one hour.

They talk about how a Honda Accord costs £14,000 in the UK and less than £10,000 in America, and on the letters page a Mr Bishop argues that breaking the speed limit can be either a heinous crime or of no consequence at all, depending on conditions.

Labour's plans are torn apart and the chaps mock Railtrack for urging its employees to use their cars.

And just in case you were thinking it's a right-wing propaganda machine, I should tell you the government fares no better. Believe the ABD and you'd believe that the Tories' plans for toll roads represent a bigger threat to the future of mankind than Aids.

They don't of course. The biggest threat facing mankind right now is Tony Blair and his new Transport Division which, this week, is headed up by a man from Oxford East.

Riviera riff-raff

We've all been there. The stewardesses have taken your coat and you're thumbing through the in-flight magazine to see which Godawful John Grisham film you'll be watching this time round.

The bloke sitting next to you has already started to pick bits of fluff from his navel, and you've already spotted the Disque Bleu in his shirt pocket, but that's OK.

What is definitely not OK is the family that's just coming down the aisle. The family with the baby. The baby with the lungs like Zeppelins.

You don't hear the pilot's welcoming speech, and you would only have understood the safety briefing if they'd done it in semaphore.

As you pass through 15,000 feet and the screaming reaches a fever pitch, you feel like organizing a collection among fellow passengers, so that the child can be upgraded to Business Class.

You regularly trip over no-smoking flights these days, but I have never heard of an airline that runs a guaranteed baby-free service.

So I have taken the bull by the horns and vowed that I will never take any child of mine on a long-haul jet until he or she is 32.

Which is why I've just come back from a holiday in France – a pretty country spoiled, like Wales, by the people who live there.

With the money we'd saved by not going to the arse end of Chile, we decided that we'd gorge ourselves stupid, only eating in the very best restaurants.

So, for ten days, we became veritable Michael Winners, lurching from rum ba ba to sauce Siam in an orgy of four-figure bills and two-rosette excellence.

I know this is a motoring column but in case you're interested, the Château Eze does the best view, and l'Oasis in La Napoule does the most wondrous food.

However, our enjoyment most nights was tempered by the *maître d's*.

I learned, over time, that a jacket and tie teamed with chinos improved the welcome somewhat; in that they stopped looking at me like I'd just urinated all over their trousers. But we were still made to feel about as welcome as plague-carrying rats.

Then I worked it out. It's the damn car. It's the bloody diesel-powered Renault Espace that Hertz had rented to us – after we'd queued for nearly an hour.

These *maître d'* chappies figure we've been saving for this meal for our whole lives and that we're going to choose the cheapest things on the menu, drink tap water and not tip.

I began to form a hitherto unseen hatred for the van with electric windows. Not only could it not climb the hill to our villa, but using it to trumpet our arrival at a flash restaurant was like being introduced at a party by the master of ceremonies as Mr Syphilis Trousers.

I was still considering this as we arrived at the Domain de Saint Martin near Vence. The electric gates swung open and we parked in a car park far away . . . which meant no one knew whether we'd come by Bentley or Raleigh Wayfarer.

The *maître d'*, maybe coincidentally but I doubt it, was brilliant, effusive, obsequious, welcoming and efficient. He was the best *maître d'* in the world.

Flushed with success, we tried the same thing again the next night at the Eden Roc on Cap d'Antibes. We parked outside and went down the drive on foot.

Well the welcome we got couldn't have been more cold if it had been in the deep freeze for a week. The last time I was greeted like that, it was by my headmaster just before he expelled me. We were shown to the worst table, and sneered at.

I wouldn't mind, but most of the customers in these places – not the Eden Roc specifically – look like Mafia hitmen and murderers. I know exactly where the Brinks Matt gold is. All of it is round one bloke's wrist at l'Oasis.

But these overtanned, fat boys with trophy wives and big suits turn up in Ferraris and Dodge Vipers.

The car is the first thing the *maître d'* sees, and way before he has a chance to clock the contents he must already have decided what table to give them, what face to pull as he opens the door and how big the tip will be.

A Ferrari gets you the sea view. A diesel-powered Renault Espace puts you in the broom cupboard with a lettuce leaf and a glass of Blue Nun.

And this is all very disappointing because it turns my view of France upside down. I've always figured that the French had cars all sussed.

Even in the leafier bits of Paris, people who could well afford a ship are happy to run around town in a battered Peugeot diesel, while a lowly secretary might have a BMW. A car, out there, I always figured, was not a measure of your wealth, only of your interest in motoring.

But the South is very definitely different. It's the cousin that's done rather well for itself. It's the family rock star, the orphan Annie that became a Hollywood celebrity. It's part of France in the same way that Elton is part of the Dwight dynasty.

For all that though, I simply love it down there. The food, the weather, the light and my starter at l'Oasis make it all worthwhile.

But to enjoy those restaurants properly, you need a real car. I saw on the way home that Europcar can do a BMW Z3 for 700 francs a day.

With a car like that you could go to the Eden Roc, park on the *maître d's* foot, and still get a kir royale on the house.

Objectivity is a fine thing unless the objective is to be first

No journalist has driven the new MGF yet, but already I know that it throttle steers very neatly, that it grips like a limpet and that there's a whiff of initial understeer on turn-in.

Wow, sounds like quite a car. And there's more. The MGF combines the best handling features of the Mazda MX-5 and the Toyota MR2, gripping well but offering adjustability at the same time. And it rides far better than its two main rivals.

I know this because I read it in *Autocar* magazine, who, in turn, were enlightened by that most balanced and unbiased of sources: Rover.

'How do we know all this?' they ask, in print. 'Because Rover's engineers told us and, in our experience, engineers never lie.' Dammit. All those years I've spent on frozen hillsides trying to work out why the car behaves the way it does have been wasted.

Instead of agonizing over a verdict, I could have simply telephoned the manufacturer and asked for its impressions. Lada, undoubtedly, would have told me that the Samara was a modern, front-wheel drive equivalent to the Escort, and that it offers unrivalled value for money. Volkswagen would have claimed that their new diesel Golf was fast, and instead of calling the new Scorpio ugly, I'd have said it was bold and imaginative. The Saab convertible was conceived as such from the very early stages of the model's design and suffers no scuttle shake whatsoever. And the best car in the world is a Ferrari, an Aston

Martin, a Mercedes, a Bentley, a BMW, a Lexus, a Cadillac and a Jaguar.

Engineers never lie, my arse. They're hardly likely to spend the best part of eight years working on a new car and then present it to the press as 'a bit of a duffer'. When I was at the launch of the Escort a few years ago, I never heard anyone on the podium say that it 'handles like a dog'. Not once did the people at McLaren say the F1 was 'a bit pricey'.

The worrying thing is that *Autocar* may be on to something here. I mean, every popular newspaper in the world relies on gossip, most of which is untrue. Divorces and affairs can happen entirely in the imagination of the writer, in the same way that handling problems and steering stodginess can happen entirely in the imagination of a car journalist.

Look, if you are a gossip writer on the *Sun* and you see Tom Cruise and Nicole Kidman having a row over lunch in San Lorenzo, you've got a story – even if they were merely arguing over what colour to paint the west wing's sitting room.

Same goes with cars. You feel a bit of a wobble over a particularly nasty pothole and as far as you're concerned, the car is crap. I've said it before; car testing is an inexact science, same as writing gossip stories.

But then along came *Hello!* and all of a sudden celebs were queuing up to open their hearts. Here, at last, was an outlet where they knew they could put their side of the story without fear of contradiction. As a result, *Hello!* gets into everyone's lovely homes while the rest of the paparazzi are camped outside looking at the action through a Nikon F2. *Hello!* buys up all those photographs of Diana with her breasts out to ensure the world never sees them – the bastards – but you can be damn sure that Diana now owes them one. When she's ready to talk about the new man in her life, *Hello!* will get the story first.

And it's basically the same with cars. Every magazine fights to be first with the road test of a new car – and I shouldn't be at all surprised if *Autocar* beats everyone to it with the new MGF. The trouble is, how do we know that what they'll write about it isn't total bollocks? We don't.

Take Paula Yates. I suspect she's a silly two-timing bird who ditched her husband and children for a fling with a hirsute Australian who looks like he needs a good bath. This is a line most newspapers are free to take. But in *Hello!* we get her side of the story, which isn't quite the same. And nor, frankly, do I find it rings very true. Bob Geldof deserves better.

And you, the reader, deserve better than what *Autocar* has in store. They may get the stories first, but if you want opinions rather than public relations puff, stick with the Beeb.

Kids in cars

So what's the daftest lyric you ever heard? I always go for Mink Deville's immortal 'He caught a plane and he got on it.'

Or what about McCartney's magnificent 'In this ever changing world in which we live in'?

Ten years ago, some would undoubtedly have cited

Mungo Jerry's 'Have a drink. Have a drive. Go out and see what you can find.'

But not any more. The war has been won. Nobody in their right mind even thinks about drinking and driving any more.

Oh sure, we need the occasional prod and at Christmas time victims are wheeled out to get the message across a bit more. The police step up their vigilance but the hit rate is miserable.

They pull over anything that moves, and in some regions only 8 per cent of drivers are found to be watching the world go by through haze-coloured spectacles.

Britain's drivers are about the safest in the world. Well done. Let's hop on a bus, go down the pub and get rat-faced.

But no. The thought police decided that a new menace must be dreamed up. No one is drinking and driving any more so let's point our big guns at . . . eenie, meenie, minie, mo . . . people who drive around talking into mobile phones.

Unfortunately, Nokia and Ericsson and all the other mobile phone manufacturers were too quick. Before the government could get into its stride on this one, the boffins came up with the new digital phone . . . which doesn't work.

Today, my airbag is better at communicating messages than my phone so I simply talk into that. This may look funny but it's not against the law.

So the eenie meenie game began all over again and settled on people with bad tempers. Yes, you. You keep losing your rag while behind the wheel and you are therefore suffering from road rage.

Then it was E and then it was joy-riders and then it was youngsters who'd just passed their test and were driving at 80 on motorways. Then it was old people whose reaction times were measured in light years.

In recent years, the thought police have had a go at just about everyone. No one is safe. But, astonishingly, they've missed what is easily the biggest menace the roads have ever seen. It tends to affect sensible, mature people in their early thirties. Law-abiding citizens who read the *Daily Mail* and vote Conservative.

Never mind drink driving. Never mind E. Never mind mobile phones or speeding or road rage. I'm talking about . . . children.

According to the RAC, 91 per cent of parents admit that they have been distracted by children while driving, and 7 per cent have crashed as result.

They list the top five distractions as children crying, kicking the back of the seat, fighting, throwing toys and pulling hair.

And they give us case studies to contemplate. Rebecca, aged three, threw a toy which jammed under the brake pedal. Jake, aged five, kept climbing into the front seat and changing gear. Antonia, aged four, had a mint imperial stuck up her nose. Let me add some of my own observations. Emily, my two-year-old, can produce such vigorous and sustained bouts of vomiting that the entire car is full of sick in three minutes.

Finlo, who's my boy child, can cry so loudly that the front windscreen regularly shatters. Only last week he perforated his nanny's ear-drum.

I freely admit that his 400-decibel chants drive me to distraction. In Antibes the other day I leapt from the car while it was still moving and buried my head in the sea, telling my wife that I wouldn't come out again until he'd shut up.

Let me tell you this. In a country with no drink driving laws, I have driven a car while so drunk I couldn't talk without dribbling. I have driven while bursting for a pee. I have done 90 while attempting to talk on the phone. And after 40 aborted attempts to get through, I suffered from road rage so badly that I pulled the steering wheel out of the dash.

But on each occasion I was a lily-white angel compared to how I am when driving around with the children.

I know of one woman who turned round to slap one of her kids while driving down a motorway in a Range Rover. She veered off course and, in trying to straighten up again, rolled the car into a bridge parapet.

So what can be done? It's no good giving them toys because in a car the most harmless Fisher Price drawing kit becomes more deadly than a thermonuclear missile.

It's no good giving them nothing either, because they then scream with boredom, and don't try taping up their mouths with duct tape. This doesn't work. I've tried it.

Noddy cassettes shut them up for a bit, but how many times can you hear that infernal signature tune before you start to foam at the mouth? Frankly, I'd rather let them scream. I'd rather listen to Radio One even.

It's taken a couple of years to work it out but my wife and I now use heroin. Before we go anywhere we slip a little smack into their peanut butter sandwiches and they're good as gold.

You might think us a little irresponsible but the goverment doesn't. Drug smugglers now get let out of prison after just one year, leaving more cells free for people who drive without due care and attention.

Brummie cuisine is not very good

This week, I shall herald the arrival of the British Motor Show with an even bigger sigh than usual.

Now don't get me wrong. I love the whole glitzy shebang. I love the old cars. I love the new cars. I love the dancing girls. I love the kids running round collecting brochures. It's a billion-dollar party thrown by a multi-billion-dollar industry.

And this year, the girls should be even prettier and the metal even more gleamy because 1996 is the hundredth year of car production in Britain.

But unfortunately, on a global scale, Britain's annual showcase is as highly regarded as an Albuquerque tractor pull.

Today, the major shows are Frankfurt, Tokyo and Geneva. That's where the important cars are launched. That's where you bump into the mandarins and the moguls. That's where you'll find on-stand special effects which make EuroDisney look like a garden shed.

At this point, I guess you're sighing too. Yes, yes, yes, you'll be saying, but that's the price we must pay for our industrial unrest in the Seventies. If you want a strong motor show, you must have a strong motor industry.

To which I say pah, and then pah again. Geneva doesn't have any motor industry at all, and Detroit –

Motown itself – has a motor show which feels like it was put together by *Blue Peter*.

No, if you want to know why the British Motor Show is so widely ignored by the world's motor industry, you need look no further than the *Michelin Guide*.

Turn to the section marked 'Birmingham' – whoa there, you nearly missed it. And there's the problem: not a single restaurant you would actually choose to eat in unless your children's lives were at stake.

So, what about hotels? Well there's the Swallow which I can't afford, the Hyatt which is always full and a wide selection where the rooms are too hot and three photocopier salesmen are having a fight in the bar at 2 a.m.

So, what if Hank J. Dieselburger Jnr, main board director of General Motors fancies a little late-night action? Again, a great many pubs which can do you a fist in the a face or a head in the basket, but that's about it.

Here's another odd thing. There are no signposts around Birmingham to the 'City Centre'. There are signs to Kidderminster and Wolverhampton and Stratford. There are signs directing you away from the place, but nothing at all enticing you in.

Which is OK because if you do end up parking where the centre should be, your car, absolutely definitely, will be stolen.

And it will be recovered several weeks later in one of the suburbs which sit like a ring of scum round the empty centre – Birmingham is a rugby team's bath after they've let the water out.

And as a little drop of icing on the cake, the National Exhibition Centre, home to the motor show, isn't even in Birmingham. It's merely near it.

The motor show should be held in London, which fizzes and effervesces with life and action. There are thousands of restaurants and hotels, there are nightclubs to cater for every musical taste, and for Kim Ho Lam, guards that change, big red buses and the Queen.

In the eyes of Johnny Foreigner, Britain is London, so why then, you may be wondering, does the motor show fare no better when it's held at Earls Court?

Easy. Earls Court is a relic from a time when cars were black and people walked quickly.

Sure, they've added an extension but it's still the wrong shape and getting exhibits in and out is harder than getting directory enquiries to give you Salman Rushdie's phone number.

Visitor parking? Er, have you tried Slough?

Plus, you try working in there. Human beings need oxygen to survive but this is the one gas denied to people in Earls Court. If Spock beamed in to the motor show with one of those *Star Trek* atmosphere testers, he'd dismiss the place as uninhabitable.

Last year, my co-presenter Quentin Willson claimed he'd caught consumption, and I saw viruses flying around brandishing knives and forks.

Scientists are said to be baffled as to where ebola lives between strikes. Well hey guys, have a look in Earls Court. After a day in there your skin dries up and you go home all covered in sores and boils.

What the British Motor Show desperately needs if it's to get back on the world stage is a new venue. I hear they're building a million-square-foot monster in the Docklands and that sounds just fine.

Groovy reflective architecture, waterfront bistros, a

nearby airport for Ford's fleet of company jets and a £7 cab ride from Soho's fleshpots. The Koreans will be over in a flash.

Sadly, this project hasn't even started yet, which means that the show, even in such an important year for the British motor industry, is at the NEC.

And as a result, the show-stopper is expected to be the Morris Minor.

Ford has developed a new car which runs on water, does 2000mph, costs £4 and can generate 4 g in a 90 degree bend, but they've chosen to exhibit it instead at the Lubbock show.

And Lubbock, in case you're interested, is a small Texan town about 45 miles from the middle of nowhere.

Last bus to Clarksonville

When I was at the launch of the Escort a few years ago, I never heard anyone on the podium say that it 'handles like a dog'.

History has produced many fools. Alfred may have been great but he couldn't even cook. Dan Quayle couldn't spell 'potato'. Colin Welland thought *Chariots of Fire* heralded the triumphant return of British movie-making. But if you want to see contemporary idiocy on a scale so vast it beggars belief, I urge you to sit in on a Hammersmith and Fulham Highway Committee meeting. Forrest Gump meets Worzel Gummidge isn't in it.

Being Labour-controlled, you expect the town hall to be full of dimwits, but these guys set new standards. Ask them to spell 'potato' and it would come out as 'grpfing'. Let them loose on the roads and all hell breaks loose.

A few months ago they decided it would be a good idea to put a bus lane up the Fulham Palace Road which, as anyone who has ever been to London will tell you, is the busiest road in the world. You will never see an M-reg car down there because nothing has moved since last August.

Anyone with even half a brain could stand on the Hammersmith flyover, gawping at the resultant carnage, and announce that the scheme had been a failure. But not the chaps and chapesses on the council. Oh no. They've gone bus lane crazy. Temporary bus lanes have become full-time bus lanes. Cycle ways are bus lanes. Buses coming out of the station have right of way and their own set of lights. In Fulham the bus is king and the bus driver is Craig Breedlove.

Richard Noble should not worry about American competition for his new land speed record attempt. Nor should he concern himself with technical difficulties. His biggest threat is that, every morning, buses in Fulham are reaching speeds of 900mph.

So why don't I get on board? Well, a) they don't run a service from Battersea to Edgbaston, and b) I don't want to. However, the council is winning. On roads where there are no buses they've built speed bumps. Who cares that they wreck cars, or that they add to pollution as people speed up between them, or that they are a problem for ambulance and fire crews?

Fulham has become so hard for car drivers – and I have to drive through it to get out of town – that I am now seriously considering leaving London for good. Yes, we've started taking *Country Life* and making 'ooh' noises at just what you can get for your money in the shire counties. We even had a practice run last weekend, up in Scotland. We drove from the hotel for five miles before we saw another car. It was heaven. I could have done 100mph. If I'd had a bus, I could have done 900.

For sure, the fields were all green, which is a hateful colour, and there were trees everywhere, rustling and snuffling in the breeze. Then there was the mud, which is what makes the countryside such a foul place. There is no mud in Jermyn Street.

Country pubs are pretty nasty too, full of people in chunky jumpers drinking beer with beetles in it. And I can't think of anything worse than having to get on with my neighbours, or having to talk to the postman in the morning. But the simple fact of the matter is this: you can at least move around, which you cannot do here in London, where my postman could be a green monster from the planet Zarg for all I know.

You can also park. In Fulham residents spend eight hours a day at work, eight hours asleep and four hours looking for somewhere to park. The remaining four hours are spent popping up the Fulham Palace Road for a packet of fags. In the country, people have drives so they can park right outside their front door every night. You can even have a garage without having to sell your children into slavery.

The only way someone can raise enough money to have a garage in London is by becoming a rent boy. Or a stockbroker. Neither of which appeals terribly. Eventually, of course, everyone will see things the same way and the gradual shift to the south-east will be reversed.

Everyone will move back from whence they came and the idiots from Hammersmith and Fulham Council will look down from the top floor of their red-flagged town hall and marvel at what they have done. The buses will have the roads to themselves. But there will be nobody on board.

Land of the Brave, Home of the Dim

My seat was in its upright position, the table tray was folded away and all my electronic games were off.

But despite this, the stewardess was coming down the aisle like an Exocet missile. 'Sir,' she smiled, 'you're going to need to uncross your legs for take off. It's a federal requirement.'

This was a new one on me but comparatively speaking it's a pinprick. Earlier in the day, I heard a security guard in a Las Vegas mall tell a group of weary shoppers to put their shoes back on.

Our cameraman had been dumbfounded in an Albu-querque supermarket when, after asking for a pack of Marlboro, he'd been told, 'This is a family-oriented store sir. We're not allowed to sell you cigarettes.'

Shall I go on? OK, how about this. A sticker affixed to the side of a huge rubbish skip warned passers-by not to clamber on the refuse collecting device. That bit was odd enough but underneath it said, and I quote, that 'It is unlawful to tamper with or remove this notice.'

This means that someone has called a meeting and voted to make it illegal to remove warning notices in the state of Texas. Illegal, you'll note. Not inadvisable. Il-bloody-legal.

But the best I've saved till last. My waitress in Reno said she could not serve me with a second beer until I had finished the first.

Naturally, I asked, despairingly why this should be so and was told, simply, 'It's a rule.'

And that's it. No argument. No truck. You can't line your beers up. You can't cross your legs on a plane. You can't tamper with notices. You can't buy cigarettes in supermarkets.

And the decent Christian folk of middle America just seem to accept it. Now, these people may be fat and their hairstyles need to be seen to be believed but they did invent the space shuttle so they're not stupid.

And yet they're quite happy to put their shoes on when asked to do so by someone in a uniform. Why?

When the French government tried to increase tolls on lorry drivers, they blockaded the autoroutes.

When the Maggon came up with the poll tax, people set fire to Trafalgar Square.

When the Italians are asked to pay VAT they lose all their books, remind the bloke from the tax office that he's 'family' and pop into town for a coffee.

I suspect that inner-city America has become so out of control that the only way you can be marked out as a law-abiding citizen is to obey every rule that comes along no matter how daft or ill conceived.

And if you want the most ill-conceived law of the lot, you need look no further than the new speed limits. This is not some trifling rule about crossing your legs. This is life and death.

An entire generation of Americans has grown up knowing that it's entirely possible to die while driving and coast
to a halt before you hit anything.

Roads that are wider than they are long were subject to
a blanket 55mph speed limit, meaning that you could get
in the back for a snooze or perform complex operations
on your passenger's adenoids without fear of crashing.

Your car could weave from lane to lane but this was
no big deal because the guy behind had all the time in the
world to get out his car's manual, see where the cruise
control off button was and take avoiding action.

But now, most states post a 75mph speed limit,
meaning that it all happens so much faster.

The drivers still allow their hearts to beat once
every fifteen minutes or so, but they don't realize they're
teetering on the edge of a holocaust.

And the cars don't help either. In recent years,
American automotive design has leapt to a standard only
seen before in Italy – the latest Chrysler line-up is staggeringly good-looking – but dynamically, they're still in the
dark ages.

On my recent trip I rented a number of different cars,
which ranged from foul to the Buick Le Sabre.

This compact sedan is 22 feet long and 14 feet wide and
I don't doubt was easily capable of handling 55mph, while
returning four or even five miles to the gallon.

But now, the dowager is expected to heave itself
along at 75 and it just can't cope. The suspension is way,
way softer than marshmallow which means there is no
jarring but even the smallest pebble causes the car to rock
sickeningly for miles afterwards.

Ask it to handle a corner – even a gentle one on an

interstate – and it just won't. I'd have more luck getting my two-year-old daughter to speak Greek. Turn the wheel and it adopts a crazy angle but doesn't really change direction.

After a mile or so, the tyres start to squeal but it's still going in a straight line. No kidding, I've driven hovercrafts which respond more quickly to messages from the helm. This is a hateful car.

The only time I felt even remotely safe in it was outside Arizona's schools, where lollipop ladies erect temporary width restrictions and impose a 15mph speed limit.

As I gratefully slammed on the anchors and wound the car down to this more sedate pace, I even had time to think that here, at last, was a law that made sense.

I also thought that if you're going to America soon, do not allow them to rent you a Buick Le Sabre, and that Montana should be avoided at all costs.

You see, in Montana, they've done away with the speed limits altogether.

Only tyrants build good cars

Last week, Michael Aspel jumped out of a cupboard, holding a big red book, and announced that I was to be the subject of *This Is Your Life*.

My mind was in a whirl. I have no friends who work in television, so they can't wheel any celebrities through the sliding doors. I have no war medals. I do nothing for charity. I'm only 36. I haven't done anything yet.

I was completely at a loss for words which, as it turned

out, was good practice because three days later Suzuki delivered one of their new X-90s to my house.

Ordinarily, the road test starts to form immediately in my mind. With normal cars, I'm starting to think about who might be tempted by such a machine and what sort of things would interest them. Should I major on performance, or style, or economy or roominess?

And that's what bothered me with the Suzuki. Exactly who would be tempted by this weird little car? Me? No. My mother? Absolutely not. The woman in Safeway? Michael Jackson? The Ayatollah? My bank manager?

It's taken a couple of days to work out that in fact no one will be tempted by it for the simple reason that it's the most stupid-looking piece of machinery of all time.

It is almost as though the designer dreamed up the front to a point half-way down the roof and, to save time, did exactly the same with the back. Were it not for the lights, you could drive this car forwards or backwards and no one would be any the wiser.

In essence, it's a two-seat, targa-roofed version of the rather nice Suzuki Vitara. They cost about the same, have the same 1600cc engine and are aimed, I guess, at the same sort of people – hairdressers.

But no hair cuttist I know would dream of buying the X-90. And that brings me back full circle.

Now, Suzuki is a large and clever organization so how on earth did this absurd little car ever slip through the net? Simple. They had a meeting.

If I ran a company, meetings would be banned. Meetings are for people who are under-employed. Have a meeting and you'll end up with the Child Support Agency. Or an ECU. Or Birmingham city centre.

I went to a meeting for the first time in years last week, and was staggered at how little we achieved in five hours.

This was entirely my fault, but then there's always someone like me in a meeting who has an opinion on everything and wishes to share it.

Trouble is, there's always someone with an equally large mouth who disagrees, and that's it. Everyone else is left to do their fingernails while two people call each other names and consider throwing water at one another.

When the bar opened we called a halt, decided we were doing a good job, and no conclusions were needed. Had we been running Suzuki, the X-90 project would have sailed through unscathed.

I remember in my early days as a reporter for the *Rotherham Advertiser*, covering the local parish council meetings. One, in particular, spent 45 minutes wondering whether they should have a glass or a plastic water jug, and ended up deciding to have both.

In the early days of motoring, car firms were run by one man with a vision. Colin Chapman founded Lotus so he could make small, light, agile cars. Sir William Lyons knew exactly what a Jaguar should be. Ferdinand Porsche was a proponent of air-cooled engines slung out at the back. Henry Ford wanted to pile 'em high and sell 'em cheap.

Now look at Japan. With the exception of Honda, all oriental car firms were founded by large corporations whose sole aim is to keep the shareholders happy. Hundreds – if not thousands – of people are involved in every decision, and the management lives in constant fear for its jobs.

Had Suzuki been a dictatorship, the X-90 would never

have happened. Mr Big Cheese would have strolled through his designers' office, seen the drawings, and fired everyone responsible.

But in committee culture, it's a case of the emperor's new clothes. No one dares speak out and the project gains so much momentum it becomes unstoppable.

And if even if someone like me does stick his hand up, the people responsible will fight back. Result: deadlock.

Europe's car firms are in the same boat today. All the founding figures are dead and committees have taken over. That's how we got the Scorpio and the Vectra.

Single ruling figures are out of the equation. No one is prepared to trust gut feeling and instinct any more, because in the meeting the guys from the market research department demonstrated that the Scorpio, or the X-90 or whatever, had clinicked well.

Then the designer stands up and explains how the push-me-pull-you look will be in for 1998. Or that the wide-mouth frog face is a happening thing in 1996.

Rubbish. The motor show finishes today and few would disagree that the most *exciting* car there was TVR's new 7 litre V12 coupé.

It looks and goes like it does because TVR is run by one man who dreams things up on the back of a fag packet and sets fire to anyone who thinks he's wrong.

Democracy. Pah. Never trust anything invented by the Greeks.

The principality of toilets

It seems to me that the closed circuit cameras which are sprouting out of every town centre vantage-point these days are pretty much useless.

Everyone I've ever seen on a still from a security camera looks like Cyrano de Bergerac. And he always appears to be standing at the counter in a bank, brandishing a banana. This, I'm fairly sure, is not a crime. Either that or the supposed criminal has had the foresight to wear a parka with a hood – so we can only see his preposterous nose. And the fruit.

I mean, if you were out robbing, you would be fairly sure that some kind of video recording was being made, so you'd wear a crash helmet or a trilby or anything which would thwart subsequent police enquiries. And if they did come round to your house with some difficult questions, you'd only need say that you were elsewhere at the time but you had seen Gerard Depardieu in town that afternoon, looking a bit shifty. Sure, there are some cameras, way up high, overlooking the most unlikely spots, but the footage from one of these was played the other day on one of the countless new crime programmes and the thieves looked like small mice with enormous conks.

It was all very dramatic, as they ram-raided their way out of a car park with policemen trying to kick in the windows, but the viewer hadn't a hope of identifying the baddies. And that makes the cameras pointless.

Now I'm not one of these weird beard lefties who thinks that Sony is a Luciferian code for some kind of Orwellian police state. If you're just walking along, picking your nose a bit and scratching your backside, who cares if it's all caught on Beta?

The cameras are only there to nail people from the sewers – thieves, murderers and blackguards. But they won't work unless we take a leaf out of Monaco's book.

This tiny principality, just two miles long by as little as 300 yards wide, is watched over by 160 security cameras, not counting the privately run video monitors in car parks and entrance halls.

Coming out of my hotel every morning, there were two which could watch me all the way to the door of the car park and, once inside, there were cameras on every floor and in each of the three lifts. As A.A. Gill wrote in *Tatler* last month, you don't need a holiday camera in Monte Carlo; just stop off at the border on your way out and ask for edited highlights of your visit.

But onanists beware! If they go to the trouble of fitting cameras in car park lifts, you can be sure your nocturnal habits are being monitored too.

Now, there is no crime in Monaco. Half the residents may have made their millions through some sort of rule-bending exercise, but there is no petty theft. One lady regularly walks home alone from the casino after nightfall wearing jewellery worth $3 million. And she's never been touched. People say that this is because of the cameras, but that's nonsense. And neither is it because there's one policeman for every 40 residents. Sure, with hardly any crime to solve, they have nothing to do all day except enforce a dress code. Try walking through Monaco with the hood on your parka pulled up and see how far you get. I'll give you a tenner for every yard you manage before Clouseau interferes. These guys won't even let you shuffle along head down, with your collar turned up. They're like stage managers, making sure you look good for the cameras. And if you refuse to look up, they will escort you politely back to France, where you can convalesce.

Every night, we watched them salute drivers of Porsches and Ferraris and hassle anyone in a dodgy-looking van. Hitchhiking is banned. If you don't look right you don't get in.

And now we're getting nearer the real reason why there is no crime in Monaco – no riff-raff. Before you go and live there, you have to produce a letter from your bank explaining that you have enough money to live on for the rest of your life. And, let's face it, people with £20 million in the bank are not big on mugging. Couple that to the police with their anti-shabby laws and the cameras, and then you get a crime-free state. Lovely. And so simple.

Except for one small thing. Monaco is a lavatory and if I could find the chain . . . I'd pull it.

Clarkson the rentboy finally picks up a Ferrari

Two years ago, I drove a car which made my life hell. The Ferrari 355.

Oh I'd driven all sorts of supercars before, including a great many Ferraris, and they'd been fun. But I hadn't actually considered buying one.

Frankly, even if I could have afforded such a thing, I'd have needed another car to handle the days when it was wet, or when I had to carry more than one person, or when I put my back out. On top of all this, super-cars tend to be as brittle and as vulnerable as baubles on a Christmas tree.

All these things apply, of course, to the 355 but it didn't seem to matter. I wanted one. I needed one. It was like meeting the girl who one day will be your wife. Friends may point out that she has spots and a temper and costs a fortune to run but you don't worry about practicality when you're in love. And I was completely smitten with the 355.

The first step was to leave London. People think I gave up 15 years of fun and games in the capital for the sake of the children but that's not entirely accurate. I did it because I needed a garage.

But the new house meant that my wife had to ricochet between Peter Jones and Osborne & Little, spending what little money we had on curtains and fridge-freezers.

Every day I'd come home and there'd be another cardboard box in the yard, another poignant reminder that the day when I could buy a Ferrari had just been pushed back.

I became desperate. I took my box full of foreign banknotes to the bank and raised £47. I looked down the back of the sofa, and went through old coat pockets. I considered holding up a sub post office. I even started doing advertisements on local radio.

But it wasn't until my wife found me watching a documentary on rent boys in King's Cross that she ordered me to buy the damn car and cheer up.

Fifty-seven minutes later, I was in a Ferrari dealership matching carpets up to bits of leather and wondering how it would all look when teamed with scarlet paint.

It seems just about all first-time Ferrari buyers choose a

red car, even though you can have blue, green, black or yellow. The choice of interior specification is even more limited though.

I'd always wanted cream sports seats, which add £2000 to the price and, despite the salesman's misgivings, that's what we settled on, along with carpets the colour of claret.

And so, after an hour of toing and froing, the order form was brought out and I found myself on my knees, putting the shakiest signature of all time to the document.

And yes, I really was on my knees because there is no chair on the customer's side of a Ferrari salesman's desk.

Rowan Atkinson complained bitterly about being asked to wait until his cheque cleared when he bought a 456 recently but I had no worries as I handed over the £5000 deposit. Now there'd be no going back.

And there wasn't, because in the garage right now is a bright red Ferrari 355 GTS – that's the one with the lift-out roof panel. The GTB is a hardtop, and the Spyder is a full convertible for hairdressers from Altrincham.

The first month with my 355 was, to be honest, disappointing. First, it had come without a radio. Second, it felt strange to be driving my own car after years in press demonstrators.

And third, it needed to be run in, which meant keeping the revs below 4000. That's OK on a motorway, where in sixth gear you can do 90, but on country lanes overtaking was nigh on impossible.

The 355's five-valve-per-cylinder V8 revs so quickly that in a full noise take-off I needed to change gear every half a second.

After I'd covered a thousand miles, the car was taken

back to the dealership for its free first service and for a radio to be fitted. It's an Alpine by the way, and it's £800 worth of junk.

Today, it's heading towards the 2000 mile mark, which is a bit of a worry because I've told the insurers I'll only do 5000 a year. In exchange, they only charge me £850.

That's cheap, but the fuel bills are not – it does 18mpg – and nor will future services be all that Asdaish. In order to change the cam belts, which must be done regularly, they have to take the engine out of the car.

Nevertheless, to date it hasn't put a foot wrong. Oh, the carbon fibre seats squeak against the back of the leather-lined cockpit and the roof panel creaks and groans, but when the revs build up past 6000, you really don't care.

The point is that this supposedly brittle piece of millimetre-perfect engineering appears to have been hewn from a solid piece of granite.

And best of all, I still love it. I've learnt to keep the suspension in its comfort setting, knowing that it automatically switches to sports mode if I start to go quickly.

And I now know how easy it is to graze the undersides on speed bumps, but let me tell you this: when the road is empty and the sun is shining, there is no better car on the planet.

I've always said I'd sell it to pay for the boy child's school fees but I've changed my mind. Sorry Fin but you'll have to go through the state system like everyone else.

Hate mail and wheeler-deelers

A couple of weeks ago, I suggested that Birmingham city centre is a sort of culinary black hole with few decent restaurants and even fewer hotels.

Harmless stuff, you'd have thought, especially as it's true, but the locals went ape. Indeed, I am actually looking forward to the postal strike so the supply of vitriol is halted.

The leader of the city council, who has extraordinary hair, led the charge, suggesting that I shouldn't peddle such insulting rubbish in a London-based newspaper. Ooooh. Touchy.

She went on: 'So, if you ever do venture north of Watford Mr Clarkson, I would be happy to show you round Birmingham and the error of your ways.'

Well, sorry to disappoint you Theresa, but I don't live in London, or even near it. And I spend a damn sight more time in your city than anywhere else. And that's how I know it needs more restaurants.

I will not, as many people have asked, publicly apologize but I do feel the need to get on my knees and grovel at the feet of Eric Ferguson.

Eric reminded me of a piece I wrote back in March, where I made some predictions about the forthcoming Formula One season.

If I'd said Murray Walker would be eaten by aliens, it would have been more accurate. In fact, I said Michael Schumacher had gone to Ferrari because he knew something that we didn't. I insisted he would be the 1996 world champion.

I said that Damon didn't have enough bottle and would be runner-up, that Jacques Villeneuve would make a complete fool of himself, and I dismissed Mika Hakkinen as deranged. All this from a man who, in private, was hurt that ITV never even asked him to get involved with their new assault on Grand Prix racing.

Mr Ferguson suggests I know less about motor racing than I know about motor cars, a point that rankled at first but, having thought about it, he may have a point on that front too.

I, after all, described the Ford Escort, on television, as a terrible, disappointing dog, and it went on to be Britain's best-selling car.

When I reviewed the Toyota Corolla I said it was dull and, to make the point, fell asleep on camera while reading a brochure about it. And the Corolla is now the world's best-selling car.

How about this for a gaffe? I once waxed lyrical about the Renault A610, saying that it was a fabulous car offering hitherto unseen levels of performance for a bargain basement price.

And in the first year, Renault sold six of them.

I am not finished yet. I completely misjudged the Peugeot 306, saying that it lacked sparkle and that it was boring. Ooops. It is, in fact, a wonderful car that I enjoy driving very much.

Then there's the Vauxhall Frontera. On first acquaintance I liked it, but since then I've discovered it's very probably the nastiest new car you can buy.

Yes, Mr Ferguson, sometimes I get it wrong and sometimes I don't. And that automatically makes me a damn sight more reliable than most car dealers.

You see, hate mail comes and goes but there is always a steady stream of letters from people who are being taken to Sketchley's by garages.

Every morning, it's like 'Dear Deirdre' in metal. Today, someone wrote to say they'd spent £16,000 on legal fees, fighting a dealer who refused to mend their car.

Then there's a couple who say Nissan won't honour a warranty on their Micra.

I see in the papers this week that a well-known BMW dealer from Yorkshire, have been fined for knowingly selling fake BMW wheels for real BMW prices.

And my sister, who bought a Mondeo on my recommendation, has vowed never to touch anything with a Ford badge again after the dealer told her a barefaced lie.

The trouble is that, judging by the letters I get, all car dealers are as bad as each other. There's a report in one of the motoring rags this week which tells of a man who bought a £61,000 Daimler only to find it was an out of date model that had been sitting in a field for two years.

When he complained, he was offered a vastly inferior model as a swap.

This whole state of affairs is shambolic. I've spent the bulk of this column apologizing for the error of my ways, and I would like to think that car dealers think hard about doing the same.

I know margins on new cars are tight but a little courtesy and some honesty costs absolutely nothing.

I really do believe that people in the motor industry sometimes forget what a huge purchase a car can be.

My wife bought a new vacuum cleaner this week and was treated like a goddess by the salesman. And yet if she sauntered into a car dealership with £10,000 in her pocket, they'd only just stop short of calling her a bitch for wasting their time.

Here's a tip guys. When a customer comes in, offer them a cup of tea. And if you have premises in Birmingham, offer them biscuits too. It'll have been a while since they ate out and they'll be grateful.

They'll buy a car from you, and be happy, and then they'll stop writing to me. This will free up more time for research and make my pontifications more accurate. As it is, I reckon Damon will be the champ in 1997 and that the Scorpio's a real beaut.

No room for dreamers in the GT40

Back in 1962, Enzo Ferrari was trying to sell his company and Henry Ford was in the frame to buy it. The talks were going well and a deal was only days away when the old man decided that his pride and joy would wither and die under the weight of Ford's global bureaucracy.

Mr Ford was livid and told his Brylcreemed designers to build a car that would make mincemeat out of the Ferraris at Le Mans. He was going to teach that eye-tie dago a lesson he wouldn't forget.

The bunch of fives came in the shape of the GT40 which, in various guises, won the 24-Hours four times.

Now, ever since I was old enough to run round in small circles, clutching at my private parts, I have been a huge fan of Ferrari and especially the 250 LM. But here was a Ford that was beating it. The GT40 became my favourite car and I would plead with my dad to buy a

Cortina, to replace the last one he'd crashed. Ford need the money, I'd argue, to build more GT40s. I had three Dinky toy GT40s and my bedroom wall was plastered with pictures of them. I even sat in one once, when I was eight or so, and decided there and then it would be the car I'd have one day. Like the Lamborghini Miura, which was also built to spite Enzo Ferrari, it came from a time when car design was at its peak. Look at a McLaren or a Diablo today and tell me they have the sheer sexiness of a 1960s supercar.

There have been loads of good-looking cars since but none had quite such dramatic lines as the GT40 – I'm talking about the racers, not the elongated and muted MkIII car.

I was at the Goodwood Festival of Speed earlier in the summer and, though there were many stunning cars squealing up that hill, I maintain that the GT40 was best. Yes indeed, the best-looking car of all time. And fast, too. Nought to 60 took 5.4 seconds and you could get the needle round to the 170mph quadrant on the M1, should you choose. There were no speed limits then, because homosexuality hadn't been invented.

It was also a proper engine. I've always subscribed to the view that there ain't no substitute for cubes and here was a car with 7000 of them in a rumbling V8 package. And there it was, in the grounds of the Elms Hotel in Abberley, fuelled and ready. The keys were in my hand, the sun was shining, the temptation to run round in circles was large. I was going to realize a 30-year-old dream and actually drive a GT40; and I didn't really care that it was a 300bhp, 4.7-litre, Mustang-engined road car with a boot. Ford had only made seven of the things before the

American magazine *Road and Track* said it was a badly made crock of donkey dung and the plug was pulled. And I, the man who loves the GT40 the most, was going to use it to tear up some tarmac.

Actually, I wasn't. For the first time in 10 years of road-testing cars, I had to admit, after desperate struggling, that I am just too tall. And no, it wasn't a Mansell whinge about being uncomfortable. I was simply unable to get my knees under the dash, my head under the roof or my feet anywhere near the pedals.

If you'd put a pint in front of John McCarthy when he stepped off that plane from Beirut and then peed in it when he was about to take a swig, he would have been less disappointed. But now I'm glad. Yes, I'm happy that Ford made the car only suitable for hamsters and other small rodents. I'm happy that my trip to Worcester was a waste of time and that I had to rewrite the item I'd written for the programme. I'm delighted that I shall go to my grave never having driven a GT40. Because the dream will never be tarnished with a dose of reality.

Vanessa Redgrave was my childhood film star idol and now I've learned she is the sort of woman who probably doesn't shave her armpits. Then there was the Ferrari Daytona, another car I'd wanted to drive since I was old enough to use crockery, but which actually feels like it should sport a Seddon Atkinson badge.

So, if you're a child longing for the day when you can get behind the wheel of a McLaren or a Diablo, may I suggest you stand in a bucket of Fison's Make it Grow. Because by the time you're old enough they will have been made to feel old and awful by the hatchback you use every day.

A rolling Moss gathers up Clarkson

You can see him coming from a mile away. He is wearing a blazer and cavalry twill trousers. The tie is undoubtedly regimental as is the stance – either that or someone has sewn a broom handle into the back of his Harvie & Hudson shirt. This guy talks pure home counties with a dash of Queen. He doesn't have a plum in his mouth: it's a banana.

Now, we are not dealing here with a car bore. Car Bore Man has a beard and oily fingernails. Car Bore Man has an MG and drinks beer with twigs in it. Car Bore Man feels a genital stirring whenever you whisper 'track rod end' in his ear.

Whereas Mr Blazer and Slacks would not be able to identify a track rod end if one were to leap out of a hedge and eat his foot. Mr Blazer and Slacks would have trouble telling the difference between a Humber and a humbug.

Mr Blazer and Slacks, however, is even more boring because his specialist subject is . . . motor racing of yester-year. Ask him who set what lap record for what team in the 1956 Cuban Grand Prix and he'll know. In fact, there's no need to ask because he'll tell you anyway sooner or later.

As far as Mr Blazer and Slacks is concerned, real motor racing stopped when tobacco sponsorship and seatbelts moved in. Today, he maintains, F1 is just a business where people with regional accents are paid huge sums of money to do something that's no more spectacular than ironing.

Real motor racers were gentlemen who used their

family's money in the pursuit of the ultimate lap. Real motor racers did the decent thing and died whenever they crashed, which was every weekend.

Unfortunately, I'm a soft touch for these people. They assume that, because I know how much an Audi A3 costs, I must be on first name terms with Archie Scott Brown and Donald Fotherington Sorbet who, don't you know, set the lap record in 1936, etc. etc. etc.

At this point I discover horse-like qualities and manage to fall asleep while standing up.

There is nothing in the world quite so dull as trips down memory lane, especially when the lane in question is Silverstone.

Or so I have always thought. In the last couple of weeks I've been researching a programme I'm making about Aston Martin, and in among the snot-like offal I've encountered some three-quarter inch pearls.

Then I met Stirling Moss who, in less than ten minutes, managed to convince me that Fifties motor racing was more exciting than watching an Apache helicopter gunship trying to get a Hellfire missile up the exhaust of a well-driven Dodge Viper.

This is because you never knew what would happen next. There was a driver in the 1930s who, as night fell, pulled into the pits while racing in the 24 hour race at Le Mans.

He changed out of his sports jacket and suede shoes into a dark suit and formal black lace-ups so that he should be properly dressed. And the following morning he changed back again.

His team, it seems, didn't mind one bit. Indeed, on the very last lap of the race, they hauled him in to the pits again, saying they were nearly out of champagne and did he want the last glass?

And then there's the sportsmanship. Stirling once travelled all the way to Indonesia so that he could engage some long-forgotten adversary in mortal combat on the track.

When Stirling's axle broke half-way through the event, things looked bleak. But the other chap lent him one – a kindly gesture which Stirling repaid by beating him.

This was the old way. In the final round of the 1959 World Sports Car Championship Aston Martin set fire to their pit garage, which would have been curtains. However, the team next door pulled its car out of the race so the hot favourites could carry on.

At around the same time, a driver called Peter Jopp – you simply must know him – suffered a mechanical failure and sought assistance from a fellow competitor who was lounging around on the grass. 'Only too delighted,' said the other chap, summoning his parents' butler. 'Courtney,' he barked at the old retainer, 'after you've poured Mr Jopp a Pimms, perhaps you'd be good enough to mend his clutch.'

The spirit was matched only by the amateur nature of technological developments. When Ferrari developed a flip-up rear spoiler on the back of their racer, they told the other teams it was a device to prevent fuel spilling on the hot exhausts. And everyone believed them.

Cooper found one of its racers wouldn't fit on the trailer so they sawed the rear end off, only to find that it went faster as a result.

Now when you've been brought up on a diet of Schumacher and launch control devices, this is just delightful. Drivers racing for no money. Team bosses helping one another. Pulling into the pits for a glass of fizz. It's all too agreeable.

But what was the motivation? Stirling Moss doesn't even hesitate. 'I did it because I loved driving a good car quickly.'

It's funny. He was standing there in a blazer and slacks. He had a clipped accent and a smart tie. I felt my eyelids getting heavy, but the man takes the era and brings it alive.

Some say he is the greatest driver that ever lived. Well I don't know about that, but I do know this. When he starts to reminisce, I start to feel like I've got a wet fish down the front of my trousers.

Can't sleep? Look at a Camry

By ten o'clock in the evening these days my body is no longer capable of movement.

If you were to use sensitive military equipment you might detect a slight rise and fall of the chest, and perhaps a gradual downward trend in the eyelid department, but that's about it.

If you were to use ordinary medical techniques you'd pronounce me dead, and take away my eyes and liver for transplant purposes.

Tiredness comes in great waves, reaching a point where even speech is no longer possible. Uttering a simple 'uh' is out of the question. I am, quite literally, dead to the world.

It's a condition that lasts right up to the moment when my head hits the pillow, and then BANG: the eyes flash open, the heart begins to beat like a Deep Purple drum solo and my mind could beat a Cray supercomputer at chess.

I write scripts. I think of new story ideas, and already this year I have six plots for new books. As the digital clock continues its remorseless march past 4 a.m., I'm sitting up bathed in sweat, wondering why the vicar had popped out of the wardrobe at that precise moment.

And what were Genesis thinking about when they decided that they were lawnmowers and it was time for lunch . . . wait a minute. I wonder if anyone knows what the car was on the cover of Peter Gabriel's first solo album? I could do a story about that.

The story is then written and mentally logged by which time it's 5 a.m. and I'm starting to get angry. In 90 minutes, I shall have to get up and go to work. I can't do a day's work on 90 minutes' sleep. Not when I only had 34 minutes last night.

I've tried everything. I've done the unspeakable and taken up decaffeinated coffee, which is like liquid lettuce. I've tried drinking huge quantities of Scotch. I've counted sheep, but that all went terribly wrong when I started to wonder whether other farmyard animals could bound over fences. Can pigs jump? That's a big, big question.

I've tried herbal remedies, though they also keep me awake, worrying that someone will find out. Clarkson's on herbal medicine. Must be a poof.

The problem is that I will not use prescribed drugs. Once, on a long, no smoking flight from Beijing to Paris – don't ask – I took a Mogadon and was still wondering how such a tiny, tiny tablet could possibly work on a 15 stone adult . . . when I went unconscious.

I was in a coma all through the stopover in Sharjah, and at French customs I thought I was the captain of a federation starship. Do NOT take a sleeping tablet, unless you have nothing on for about two weeks.

The worst thing about insomnia is that no one sympathizes. Tell someone you can't sleep and they'll give you chapter and verse on how easily they nod off. Why do they do this?

When I meet a blind person, I don't tell him that I can see just fine.

But now it doesn't matter any more, because for the past week I've been getting the full eight hours a night. I've been waking up each morning well able to handle all manner of heavy machinery.

The cure is not, I'm happy to say, a dangerous and addictive drug. It is not some dubious root from Mongolia. It is not alcohol either, unfortunately. No, the cure comes from a most unlikely source – Toyota.

You only have to mention the word 'Camry' and I'm long gone. Indeed, I had to get a colleague to type 'the C word' just then because if I'd done it, I'd have been unable to finish the story.

I want to make it plain that this is by no means a bad car. For the money, you'd be hard pressed to find a better built machine on the road. It's quiet. It's comfortable, and it's incredibly easy to drive.

But all this engineering whiz kiddery is shrouded in by far the dullest shaped body I have ever seen in my whole life. There is no single feature that is in any way even slightly outstanding. There is a bonnet because you need one to hide the engine. There is a passenger cell where people sit, and there is a boot for luggage.

All I need do now is think about the shape and I come over all drowsy. If, while cleaning my teeth at night, I glance out of the bathroom window and see it in the yard, that's it – I'm a goner.

Now obviously, we can't all buy a C★★★★ just to help us sleep – I mean the 2.2 litre base model is £19,000 and that's a hill of money. But I suspect a photograph of such a car pinned to your bedroom ceiling would work.

Or cut out this next bit of the story, and read it before you go to sleep every night. The C★★★★, you see, has HSEA glass to reduce glare and eye fatigue and cut down on heat build-up. HSEA cuts ultraviolet by 86 per cent and solar energy by 74 per cent. The stereo has autoscan . . . feeling drowsy yet?

What if I tell you the engine and transmission is mounted in its own cradle-type sub frame and that the suspension geometry has been fine tuned to raise the rear roll centre? Gone yet?

OK then, and this is all you'll need. The rear wheels have been set with initial negative camber.

It's marvellous; the first ever car with medicinal properties. But it's not alone out there. Next week, I'll tell you all about the Nissan QX and how it put me into a deep, hypnotic trance.

Big foot down for a ten gallon blat

After a million or so years doing nothing, man really seemed to be coming along in the last hundred or so. He motorized his wheels, sprouted wings, went to the moon and, best of all, he invented the fax. But in the last 20 years it all seems to have stopped. Where's the follow-up to Concorde? When are we off to Mars? What comes after rock 'n' roll?

I blame miniaturization. Clever people have stopped inventing things and started making what we've already got, smaller. When I had a hi-fi system in the 1970s, it was a massive, teak thing with an arm like something from the Tyneside docks. But today, you need tweezers to hit the buttons and Jodrell Bank to see the read-outs.

Then there are cameras. I saw a guy in the States last month with a device that was actually lighter than air. Had he dropped it, it would have floated, which is perhaps a good thing, but honestly, you can't beat my Nikon which needs its own team of baggage handlers at airports.

And then there's Kate Moss. Well look. I like big breasts, a big amount of food on my plate and I'd much rather watch *Terminator 2* at the cinema than on video. I also like big cars, a point rammed home this month when I drove Big Foot. First of all, its nine-litre V8 gets through alcohol at the rate of 5 gallons for every 300 yards. This is good stuff. This is 29 gallons to the mile and that makes it by far the least economical vehicle in the world. It's fast too. No one has ever done any performance tests, but having done a full-bore, full-power standing start I can report that we are talking 0 to 60 in about four seconds. This is impressive in any car but it's especially noteworthy in something that has tyres that are over 6 feet tall. To get in, you climb up through the chassis, emerging into the cockpit through a trap door in the Perspex floor. Everything about the pick-up truck body perched up there on the top is fake. It's just a plastic facsimile of a real Ford F150 – not even the doors open. There's just one centrally mounted seat with a full, five-point racing harness and about 2500 dials in front. There are warning lights too, each of which was carefully identified by my tutor before I was allowed to set off. But I didn't listen to a word he said. Nor did I pay attention when he talked me though the gearbox. It's an auto but, though there's no clutch, you do have to pull the lever back each time you want to shift up. And that was the problem: pull is the wrong word. You are supposed to wrench it back, as I'd soon discover.

With the lecture over and my neck brace in place, the instructor was disappearing through the trap door when he turned and said: 'Have you ever driven a fast car before?' I told him I'd driven a Diablo and he left, wearing a peculiar smile on his face.

To fire that mid-mounted tower block of an engine, you just hit a big rubber knob and then thank God you're wearing a helmet. It's loud like a hovering Harrier and when you hit the throttle it sends your vision all wobbly.

About one second into what felt like an interstellar voyage, various dials and the noise suggested a gear change might not be such a bad idea, so I eased the lever back. Nothing happened. The revs kept on building, so I tried again. Nothing, except this time a selection of warning lights came on. By now I was in a temper so I yanked the lever back and the truck just seemed to explode forwards. This could catch a Diablo and run over it.

And even though it was on wet grass, it seemed to 'dig and grip' pretty well. I never did find out how well, though, because by then I was struggling for third and may have hit first instead. I was in Vermont but people in Gibraltar heard the bang. They gave me another five minutes before the people from Ford hit their remote shutdown button and the engine died. I was going to give them hell but decided to run away instead when I noticed the rev-counter telltale said I'd taken the £100,000, nine-litre motor to 10,000rpm.

I didn't stop running until I was in Chicago, where I decided that Big Feet (is that the plural?) are wasted at exhibitions, jumping over saloon cars. We should use them for trips into town. I'm about to move to Chipping Norton where, I'm sure, it'll go down a storm.

Car chase in cuckoo-land

Over the years James Bond has been Scottish, Welsh, Australian, and English. I tell you this because for no particular reason I was lying in the bath this morning thinking about Pierce Brosnan who, of course, is Irish.

I guess he was on my mind because like every other small boy I was given *Goldeneye* for Christmas, a 007 video in which Sean Bean attempts to break some computers.

The film is not bad, actually, but there are two problems. First, Brosnan delivers all his lines in a curious high-pitched squeak, making him about as frightening as Willie Carson.

And second, *Goldeneye* plays host to the most preposterous car chase of all time.

Bond, in an old Aston Martin DB5, duels with a Russian fighter pilot in a Ferrari 355 and the two screech, neck and neck, through Alpine passes in a flurry of tortured rubber and wailing engine notes.

Well now, look here chaps. If you put Tiff Needell, who is the best driver I know, in a DB5 and Stevie Wonder in a 355, Stevie would be tucked up in bed at home, after a good supper, long before Tiff got into second gear.

To make things even worse, the music was all wrong and the whole thing was intercut with a series of glib one-liners from Brosnan, which were only audible to dogs.

There are so many things to love about Bond films, but the car chase sequences are always wrong. Who can forget that speeded up nonsense in *Goldfinger*, or the way his trick Aston V8 skied its way out of trouble in *The Living Daylights*?

The first thing a decent car chase needs is plausibility. I mean, look at *The Rock*, a Hollywood blockbuster in which Nick Cage, a timid little man with pipe cleaners where his arms should be, leaps into a Ferrari 355 and sets off in pursuit of Sean Connery in a Hummer.

Now even though Sean had been in jail for 30 years and would never have driven anything remotely similar to a Hummer before, and even though Cage had a vastly superior car, the 355 ended up crashing out of contention.

It's the same deal in Ryan O'Neal's film – *The Driver*. I know the Pontiac Firebird is slow and truculent but there is no way, no matter how good you are, that you could keep up with one if you happened to be in a pick-up truck at the time.

Why, I whisper to myself, do they not put the combatants in similar cars? That's exactly what they did in *Bullitt* and that's one of the many reasons why this is still regarded as THE best car chase of all time.

The baddies have a Dodge Charger and Steve McQueen has a Ford Mustang – two cars which are evenly matched. Both have V8s, which provide all the aural backdrop you could possibly want, and as a result the director, Peter Yates, decided no musical accompaniment was necessary.

When the baddies finally went off, into a garage, which explodes, it was for a very good reason and not because the driver had suddenly decided to apply the handbrake.

Why do they always do that? Why, when there's a corner to be negotiated, does the baddie always travel a little way beyond the apex and then attempt to turn?

This is forgivable if you're pottering along at 30, listening to *The Archers*, but when your life is on the line and you're doing 90 down Regent Street, I suspect you'd be concentrating pretty damn hard on where the road goes next.

The point is, of course, that a car skidding into an ammunition dump is good cinema. A car skidding into a pile of boxes is good television. But a car just stopping is what happens when directors try to stage a car chase for £4.50.

When the two drivers career into a side street and every parked car is a 1972 Hillman Avenger, you know someone has been skimping.

This is at its best in *The Bill*, when from time to time Metro panda cars are to be seen in hot pursuit of a suspect in a stolen Montego.

Now we've seen enough Michael Buerk *999* crash emergency paramedic programmes to know that in real life thieves don't care two hoots about the car they've nicked. They'll ram anything that gets in their way.

But in *The Bill* they indicate when turning left, stop for old ladies and, when there are width restrictions ahead through which they can't quite fit, they'll stop and run off on foot. That makes the subsequent arrest dull but much, much cheaper to film.

Car chases should only be attempted when the producers have found a spare million down the back of the sofa.

But, that said, money is no guarantee of success. You see, *Days of Thunder* with Tom Cruise and *Vanishing Point* with Barry Newman both featured *the* classic car-chase mistake.

Here's the scene: the road is straight as far as the eye can see and both combatants are flat out, alongside one another.

It's stalemate, or so you'd think, but wait, what's this – the goodie has just changed gear and whehey, his car has roared ahead.

What?! . . . did he just forget there was one more gear still to go?

Small wonder joy-riders feel a need to go out at night provoking the police to chase them. If only they'd start putting decent car scenes in films again, there'd be no need to do it for real.

Frost-bite and cocktail sausages up the nose

In 1996 I remember hearing that the commandos are having to lower their standards in a bid to attract new talent.

Spotty sixth formers were asked why they didn't want to pursue a career in this most élite division of the army. They were told that it would involve getting up at 4 a.m. and running to Barnsley with a 200 lb pack on their backs. They'd be expected to eat leaves, wipe their bottoms with smooth stones and shoot down helicopters a lot.

Having listened with a gormless expression that only teenagers can muster properly, one said, 'It sounds very tiring'.

And I reckoned he had a point. Why push your body to the limits when you can get a nice job in a bank, and flirt with the cashiers all day?

It gets worse. Did you read all that stuff by Ranulph Fiennes in the News Review section a couple of weeks ago? The man was trying to walk across the Antarctic, towing a sledge that was even heavier than a night storage heater – but less useful. He had pus streaming down his chin and every night he'd pull off his socks to find another toe had come off.

Then there was whatsisname in that upside-down boat. He's my new hero for eschewing counselling, saying he'd rather go to the pub for a couple of pints, but that still doesn't explain what he was doing down there in the first place, attempting to get a sailing boat through seas that could smash Manhattan.

And what about Branson? Richard, my dear chap, you

have it all. Why risk life and limb trying to get a helium balloon round the world when you have a fleet of 747s which are so much more appropriate?

And the same goes for his American competitor, Steve Fossett. Now I met this guy last year, and he's done everything; swum the Channel, raced twice at Le Mans, climbed six of the world's seven highest mountains and sailed across the Pacific faster than anyone else.

But he has $600 million in the bank. Can't he just learn to play the piano?

While driving to work the other day I was thinking about all this, wondering what drives people to go further and faster, to boldly go where no one has wanted to go before. And then I turned on to the M40 at junction 15 and cursed slightly. It had been 31 minutes since I left home and that's a very average time indeed. On the way home, I'd try and do better.

Aaaaaaargh. It hit me like a juggernaut. In my own small way, I'm no better.

When you do the same journey, day in and day out, you start to set yourself little targets – can I be in Shipston by twenty past? Damn, I'm late and now I've got a minute to make up before I get to Halford.

The only rules of engagement are that I don't exceed the speed limits in villages but that's it. In between the built-up areas, I drive like I've accidentally set fire to my hair and tried to put it out with scalding hot water.

I stir the gear lever like I'm trying to beat an egg and stamp on the pedals as though they're funnel web spiders. I know this is dangerous, but there is a feeling of elation when you arrive at a predetermined point in the fastest ever time.

I've tried pointing out to myself that it doesn't matter; that Chris Akabussi and Sarah Green won't be waiting with *The Record Breakers* team. I've tried considering the cost, telling myself that when you really give an XJR some stick it uses a pint of petrol every 3000 yards.

At full moo, the fuel injectors begin to look like a collection of firemen's hoses, but it doesn't matter. And nor does the pointlessness of it all.

In my Ferrari, in the middle of the night, I've done the trip in 29 minutes. But yesterday I was in a diesel-powered, left-hand drive Renault Espace. I didn't have the power to get past lorries, but this didn't matter because I couldn't see past them in the first place. And the time: 30 minutes, 22 seconds.

The most important thing is that every day we need challenges. For some, that means chatting up colleagues at work or making the ultimate ratatouille. Others though need to balloon round the world or walk to Mars, or drive faster than anyone else.

Richard Noble is working right now on a new car which, later this year, should break the sound barrier on land, achieving a speed in excess of 700mph.

It is powered by two jet engines, which are mounted right up at the front to make the car-nose heavy like an arrow. Driver Andy Green is going to light those after-burners and sit there in what is basically a controlled explosion on wheels, hoping that he'll get a three-line entry in *The Guinness Book of Records*.

He knows, too, that just a few weeks after his attempt it may well be beaten by an American outfit headed by Craig Breedlove, and that his fifteen minutes of fame will be just that.

But he's going to try for it anyway, and in so doing he's going to make our lives seem just that little bit more puny.

Certainly, he will make my efforts to get to the M40 quickly look ever so inconsequential, which is why I'm going to pack it in and do something constructive. Tonight, I'm going to see how many cocktail sausages I can get up my nose.

Bursting bladders on Boxing Day

Christmas is a religious festival where Christians celebrate the birth of their spiritual leader by getting together with their families, giving one another socks and arguing. There is usually finger pointing over the turkey, and after lunch warring factions gather in different parts of the house, whispering about how they never want to see one another again. Mostly though, families can keep smiling through gritted teeth, never actually saying 'Auntie, I hope this cracker blows your hand off'. On Boxing Day, everyone climbs into their cars and heads for home.

This is when it starts to get difficult. I have driven up the side of cliff faces in Iceland, and I have survived the Bombay to Pune highway in India, but for sheer lunatic driving you can't beat the M1 on Boxing Day. Husband is sitting there in his brand-new woolly pully, telling his wife that he never wants to go to her parents again. After 15 years, he has just admitted that her mother is a fat, interfering cow who, he hopes, contracts BSE very soon. She is crying and accusing him of not making an effort:

'You know Daddy hates it when you call the Queen a lesbian.' The upshot is that he is not paying the slightest bit of attention to the road and hasn't noticed that visibility is down to two inches. He is still doing 90, relying on the glow from his new jumper for guidance.

No kidding. Last year I was crawling down the inside lane doing 30mph, and there was a constant stream of over-burdened Volvos screaming past doing 90. And then, south of Northampton when the fog lifted, I was making up lost time, going past the Volvos; and they had the audacity to indulge in some major-league finger wagging. Well, they would have, except the row by this time had gone nuclear. She was on the mobile to the solicitor and he was admitting to eight affairs.

Even if there was a lull in the fighting up front, it would be filled with squawking from the back. Daughter had just broken the son's train who, in return, vomited on her new doll. Further back still, the boot was loaded to the point where the car weighed more than an Intercity 125.

I find it little short of amazing that we can't drive while drunk, because alcohol impairs our judgement, and yet we are at liberty to drive around while getting divorced. You are also allowed to drive while wanting a pee. When I go past a sign saying 'services 30 miles' and I need to go, I will admit here and now that I will let my speed creep up to 130 and I will overtake on whichever side of the road I see fit. It becomes all-consuming to the exclusion even of life preservation. And when I finally make it, I will screech to a stop in a disabled parking bay. I do not think I am alone in this.

And I am certainly not the only person ever to have

driven while suffering from hay fever. Last summer, I drove an 850bhp Nissan Skyline GTR even though I was virtually blind, a condition that became complete when I sneezed every four seconds. Here's a fact: if you have a three-second sneeze at 60mph, you are blinded for a staggering 264 feet. I will also admit to driving around while enraged by something on the radio and yes, I've turned round to check my daughter is OK on the back seat. I have also driven with a splitting headache. Only the other day I had to pull on to the hard shoulder of the M40, where I fainted.

In fact, the days when I climb into a car feeling refreshed and ready to cope with diesel spills and people in hats are pretty few and far between. And what about old people? If good driving is all about awareness and speed of reaction, then they should surely be taken off the roads. A 17-year-old youth just over the legal drink limit is going to be better able to deal with an emergency than the average, sober 70-year-old. But there are no laws about driving while under the influence of Anthony Eden, or having a hay fever attack, or with a bursting bladder.

Which is why, when I hear the police are cracking down on people who drive around while talking into mobile phones, I laugh. In the big scheme of things, this is not really so bad.

Lies, damn lies and statistics

I've just spent two delightful weeks in Barbados where the sun shone, the diving was fine and the jet skis were fast.

But I had time for none of this nonsense because I had taken by far and away the best read of the year so far.

While everyone else on the beach was tucking into the new Patricia Cornwell, hoping that Lucy will get into a juicy lesbian affair some day soon, I was ensconced in the annual *Lex Report on Motoring*.

Written after exhaustive research with 1209 drivers including 160 truckers, it is intended to show how our attitudes change from year to year.

I'll start you off with a simple, and obvious, one. Eighty-one per cent of all those questioned said the newly introduced written part of the driving test is a good idea. Of course it is: 'those questioned' don't have to take it.

Then there's the old chestnut. Thirty-three per cent of motorists, says the report, think that driving standards in Britain are bad or very bad and 36 per cent think they're only average.

However, a whopping 74 per cent said they were good or very good. Only one per cent admitted to being hopeless behind the wheel.

That doesn't add up, until you consider another statistic that floated across my desk a couple of years ago. It said that something like 80 per cent of the British population had never been on an aeroplane, which means that the vast majority of those questioned are comparing our driving standards to . . . what?

Had they been to India, or Greece, or New York, or Italy, or pretty well any damn place, they'd know that the standard of driving in Britain is outstanding and that about 90 per cent of our drivers rank as either superb or unbelievably gifted.

Speed rears its ugly head in the report too, with some

people saying they never break the limits. Presumably, they form part of the 11 per cent who were not aware of the existence of speed cameras.

My favourite bit on the question of speed, though, is that 66 per cent of motorists did not feel speed bumps would slow them down. Well now, look here chaps; would you please write and let me know what sort of cars you have, because if they can get over those ludicrous humps without shaking the dashboard from its mountings, I want one too.

Here's another one. Sixty-four per cent of motorists wouldn't slow down if heavier penalties were imposed, even if there were a three-month ban for exceeding the urban limit by 10mph or more.

I see. So if the penalty for speeding were the loss of your eyes, the burning of your house and the rape and pillage of your children, you'd still sail through villages at 50 would you?

People are telling porkies here, a point that becomes obvious when you get to the section on rubbernecking. Forty-three per cent of drivers admitted to slowing down to look at an accident with 37 per cent saying they look without slowing down. Two per cent say they'll go so far as to change lanes for a better view. Which leaves 16 per cent who can sail past what looks like the conclusion of a Sam Peckinpah film without gawping a bit. Sorry guys. Not possible.

If you do believe in statistics the section on drugs and drinking makes for alarming reading, because it says two million people have been in a car when the driver was over the alcohol limit, half a million have been driven by someone on cannabis, 250,000 when the driver had taken speed and 100,000 where ecstasy, cocaine or heroin was involved.

Well now, that is extraordinary. All those people charging around either asleep, very wide awake indeed, or being chased by giraffes on surfboards, and so few accidents. We're even better drivers than I thought.

I do wonder, however, whether if I did a survey into market research I'd find that a majority of respondents were drunk or stoned out of their minds at the time.

Or reading a map while on the phone. Here's a good one – only 13 per cent of people admit to driving while talking on a hand-held phone. They're lying because, as we all know, mobile phones don't work.

Back to the story. Only 9 per cent of drivers said they'd been in an accident in the last year, but there are some gender and sex differences here. The figure shoots up to 13 per cent when you're dealing with 17 to 34-year-olds and 14 per cent when company car drivers were quizzed.

Here's the best bit though. Only 35 per cent of women drivers have ever had a crash compared to 53 per cent of men.

And yet, when asked to sit the new written driving test, only one in five women passed compared to one in four men. It seems then that women don't know what they're doing, but they're doing it very well.

It's easy to get depressed by some of the findings but we should all remember that market research said, categorically, that Labour would win the last election.

This is why I always treat any form of survey as a work of fiction. On that basis, the Lex effort easily beats Tom Clancy's new book, *Executive Orders*.

To save you the bother of reading it, the two hillbillies

don't get their bomb to Washington and the ebola virus goes away all on its own. Oh, and the Arab baddie winds up dead.

Radio Ga Ga

So, you're settled in front of Des O'Connor and whoops, half-way through, along comes a car commercial which costs about twenty times more than the programme.

But when you've finished watching that Audi A3 charging around in Iceland, or the Ka in la-la land, or the Volvo in Palm Springs, what have you learnt?

Well you might have an idea what it looks like, though this is by no means certain, but that is it. You don't know how much it costs, whether it has a warranty, whether it is safe or whether it is fast.

In the past this hasn't mattered. The agency bought space on television to let us know the car exists, hoping that we'd all be captivated enough to read more detail in newspaper and magazine advertisements.

But they don't do this any more. My favourite press advertisement at the moment is for Peugeot's wonderful 106GTi which is seen about 500 feet in the sky, and still heading upwards, having driven over a humpbacked bridge.

There is a hint here that the car is exciting and fast, but you need Jodrell Bank to read the small print where a price is quoted.

Same goes for the splendid press advertisement for the Mercedes SLK. The car is parked at the side of the road, alongside a series of skid marks, the idea being that everyone is slowing down for a better look. Great, but are there any seats in the back? And how long's the waiting list?

We are not told; well not in print anyway. No, that is now the job of local radio, in which the entire technical specification of the car, the price and the discount offers are crammed into 26 seconds of garbled nonsense. That leaves four seconds for the jokey pay-off.

At the moment they're running one round these parts for the Fiat Ducato van, in which a Scouser is trying to buy the said vehicle from a homosexual gentlemen's out-fitter. The customer asks whether he can fit his drainpipes inside, and we think he's talking about trousers. But he isn't. It's hysterical; nearly as funny, in fact, as having a vasectomy.

Then there's an advertisement for a Saab, in which a salesman is telling his colleague all about his new 900 and how it's just set a world endurance record. But all the way through, the colleague thinks he's talking about 'Melanie', the office crumpet.

Brian Rix couldn't stretch a joke further, but you haven't heard the punch line yet.

Ready? Saab-man says he's going to show his new car to the aforementioned Melanie, and his colleague is heard to remark, in an amusing, sceptical voice, 'Must you now?' I nearly crashed I was laughing so much.

These days, television and press advertisements treat the viewer and reader with respect. We are shown subtle images and are expected to work them out for ourselves.

But the banal is acceptable and even welcome in a medium that plays Phil Collins all day long, a medium that whooped for joy when it heard about the balding one's latest marriage rift. It means more soppy songs, and more excuses to never play anything else.

And then there are the otherworldly disc jockeys. Why is the breakfast presenter always having a suggested affair with the bimbo in the traffic helicopter, that's what I want to know?

And why can't they ever say something a teeny bit controversial? Actually, to be fair, they do from time to time. But after Phil has sung about leaving a wife, there's always a little apology.

Just yesterday, a lad phoned in from his bedroom to request 'You Can't Hurry Love' and the presenter berated him for running up his parents' phone bill. But, after the tune, she was back to say that calls to her radio station were of course free, so everyone was happy and no one's armpits smell.

Local radio presentation is so bland I bet they never get any letters at all. I bet that if you asked a postman where the town's FM station was, he'd have no idea.

He certainly won't be listening any more because he'll have been annoyed by the car advertisements and completely bamboozled by those from local garages.

It's bad enough when a major car firm, using the very best Soho agency brains, turns to local radio, but when the advertisement is paid for by the managing director of Rotters Autos, who thinks he can do the job just as well, the results are catastrophic. And they're even worse if he tries to write a jingle where 'autos' and 'customer' are somehow made to rhyme.

He knows that in the real world people are interested in price, and he feels it's entirely appropriate to tell everyone about all the deals he can offer. There are quite a few, but if he speaks quickly he can get them all in.

Then he learns that by law there have to be disclaimers after any advertisement where money is mentioned, and that another voice must explain that offers are subject to status, that written details are available on request and that the typical APR is 14.3 per cent.

He therefore has to speak so fast that he'd even make a Noddy story sound like mumbo jumbo.

I've been listening to local radio for a couple of years now, but the advertisements in general, and the car advertisements in particular, have forced me back to the BBC. I even thought about giving Chris Evans a try, but he seems to be on holiday.

Anyone know when he'll be back?

Spooked by a Polish spectre

I don't doubt that you go a bit red round the gills every morning when you find that Postman Pat has filled your hall with junk mail. You don't want to win a tumble dryer, you don't need an Amex card and you'd rather buy *Razzle* than *Reader's Digest*. But consider for a moment what life would be like if you actually had to read everything that came through your door. Imagine if you were forced to open bank statements and bills, rather than simply feed them to the waste disposal unit.

Well, that's what happens at Telly Towers every morning. I have to scoop up the debris that Pat has fed to my doormat, and read it. I'm talking about press releases from the world's car companies – tomes that redefine the concept of dull. They are more boring than a Jane Austen novel, more shiversomely tedious than a parish council meeting. Just last week, Nissan changed the radiator grille or something on the Micra and poor Pat gave himself a hernia lugging the press pack up my drive. Seventeen pages in, I'd already worked out that the whole thing could have been done in one sentence: 'We've changed the Micra a bit.' But today, in amongst the encyclopaedic volume on the Corsa's new engine and a gushing diatribe about the new Hyundai Lantra Estate, was something that stopped me dead. FSO is not dead. The Polish car company has managed to survive the transition from Communism to Lech and back to Communism again. And more than that, the cars are still being imported to Britain. Oh no. I still maintain that the Nissan Sunny was the worst car of all time. It had no redeeming features; nothing that you couldn't find better and cheaper elsewhere. But the worst car in the world to drive was the FSO Polonez.

It did have a redeeming feature – it was cheap. But it had to be, because it was a car that wasn't really a car at all. It was a box under which the careless car-buyer would discover a 1940s tractor. The styling was enough to put most people off, but it only had to compete with the Wartburg and the Trabant, neither of which will ever feature in a book called 'Beautiful Cars' by Jeremy Clarkson.

You cannot begin to imagine how bad the ride was on this truly awful car, and just as you were marvelling at its ability to bounce so high off the ground, you'd find its steering didn't really work because the front wheels had been concreted on. If Karl Benz had invented its engine, he'd have given up on the whole concept of internal combustion. The noise frightened birds and the fuel consumption read like the spec sheet from an Intercity 125.

The last time I actually went in a Polonez was last year. It was a minicab and it broke down in Heathrow's tunnel. Then I had an argument with the fat driver when I point-blank refused to pay. But since that long, fume-filled walk to the terminal, I've not heard anything about this wart on the bottom of motoring. Until now. It seems FSO has a new car called the Caro which has met with some success in Britain. Last year 480 were sold here, but I can only assume that the owners limit forays onto the road to the hours of darkness. I've certainly never seen one.

I'm sure that it's a pretty hateful machine. But there's no denying one thing. At £4527, it is cheap. Also, it can be ordered with a 1.9-litre Citroen diesel engine and it will eventually get ZF power steering and Lucas brake systems. It may then become a half-decent car, but I'm also sure its price will rise. They'll end up with a half-decent car at an indecent price. Except they won't, because this press release says that Daewoo has taken a 10 per cent stake in FSO, and that in the next five or six years the Korean company's share will rise to 70 per cent.

The idea is simple. Daewoo will ship bits of old Astras and Cavaliers from Korea to Poland where they will be nailed together to form a vague, but inexpensive, interpretation of what motoring should be all about in the next millennium.

We know all about that already, of course, because Vauxhall has shown us. No more fast cars. Birds in the trees and the good people of the world transported to and from work in Vectras. God Help Us.

Boxster on the ropes

I should make it plain, right at the outset, that I was born in Yorkshire, but don't worry. Because I can't play cricket, I don't suffer from that most hideous of diseases – Professional Yorkshireman Syndrome.

Even so, I do love the place. I think it's hard to find scenery anywhere in the world quite so inviting as the Dales, the people are chummy and Leeds is just plain outstanding.

The best bit of the county is to be found by going to Hull and turning left; up through Driffield and on towards Scarborough and Filey.

If you removed all the people and their yellow houses from the Cotswolds, you'd be getting near the mark. It is chocolate-box pretty and you don't even have to fight with a coachload of American tourists for the hazel nut crunch.

Now this, of course, means you can unleash the beast under your bonnet on some of the smoothest, best maintained and almost completely empty roads. East Yorkshire is petrolhead heaven.

And I was in a car to suit the mood – Porsche's new two-seater, mid-engined convertible – the Boxster.

It was fabulous. There is a delicacy to the steering that you simply don't find in lesser machinery, which meant that on the moorland roads north of Pickering the car was a dream.

And the roadholding is absolutely sublime. Even if I stabbed the throttle midway through a corner, the inside wheel just spun the unnecessary power away with the careless disdain of an Australian bushman whooshing away a fly.

As far as ride comfort is concerned, it's just unbeliev-able. The Boxster's suspension is like the perfect secretary, dealing with the dross and making sure that only the really important information reaches its boss.

You can have big fun in this car, revelling in the crisp bark of that 2.5 litre, six-cylinder engine as it rasps past 5000rpm, snicking the gear lever from fourth to fifth, and then down to third for the next bend.

The brakes wash away the excess speed, the big tyres grip, the sports seat holds you firmly in place and then, as you marvel at the complete absence of scuttle shake, it's time to nail the throttle again. And don't worry: even if you drive like the illegitimate son of a madman, you'll still get 25mpg.

You'll even get home with enough energy left to press a little button and put the roof up.

A clean bill of health then for Porsche's mainstay into the next century? All you people on the year-long waiting list can sit back, safe in the knowledge that the car you've chosen is spot-on? Er no, not really.

You see, it's all very well thumping round East Yorkshire in a Boxster but don't forget, you have to get there. And that's not so much fun.

First of all, you've got to buy it, which will set you back £37,000 if you specify big wheels and leather seats. This is a lot of money, especially when you remember that an MGF VVC is less than £20,000.

And no, I'm not being silly. The MG is also a mid-engined, two-seater convertible which, in the real world . . . wait for it . . . is actually faster than the Porsche.

The Boxster's biggest problem is that unless you drive like your hair's on fire, it's JCB slow. In-gear acceleration times show an ordinary Rover 620i is faster.

And to make matters worse, the Boxster has an irritating habit on the motorway of gradually slowing down. Drive any expensive car down the M1 and you'll find it gets faster and faster. At turn off 32, you're doing 70. By turn off 28, you're up to 75. When you go past Leicester, you're up to 80. And by Watford, your sonic boom is breaking people's windows. Next stop, Hendon magistrates court. I know. I've been.

But in the Porsche the reverse happens. I started out at 70 and would find, ten miles later, that I was down to 50. Half-way up the M18, you would have needed a theodolite to ascertain that I was actually moving at all.

And on the M62, I think I started going backwards. It was hard to be sure, though, because the front and the back of a Boxster look like Siamese twins.

I think this licence-saving trend is the result of a wonky driving position which means your right foot is never quite comfortable on the accelerator – it tends to lift, little by little. You have to concentrate, all the time, on maintaining a single speed, and this gets wearing; so wearing, in fact, that on the way home I ended up with a headache in my back.

Overall then, the Porsche was a disappointment. On its day, it's as much fun as an evening with Steve Coogan, but most of the time it's as dreary and as plain as his alter ego, Alan Partridge.

The good news for Porsche fans is that it will stay in production for 20 or more years, and that in the fullness of time they'll up the power, fix the driving position and lose the push-me-pull-you looks.

This is great if you're planning on buying one in the next millennium, but if you want a sports car now there's the Mercedes SLK for pansies and the TVR Chimaera for people who like their meat still mooing.

The trouble with the TVR though, is that it's made in Lancashire, which is on the wrong side of the Pennines, the side where they can't play cricket and ... cont. in Michael Parkinson's column every week.

Concept or reality?

I'm thinking of having Rover's board over for dinner and, when they're all seated, I shall produce braised pork with apples and cider.

I shall regale them with tales of exactly how this mouthwatering concoction had been made; which tiny little specialist shops in Soho had provided the juniper berries and how I'd marinated the meat for a week. I'll even give them the actual name of the pig that had found the truffles.

But then, just as they pick up their eating irons, I'll whisk their plates away and produce, instead, a baked potato that hasn't been in the microwave quite long enough.

Perhaps then they will understand the folly of producing concept cars that they have no intention of making.

Everyone knows that a new Mini is on the drawing

board, and we are sure that it will be bigger and more expensive than the current incarnation.

But we have no idea what it will look like, what sort of engine it will have or even where that engine will be. Somewhere in the car is a safe guess, though.

So Rover is teasing us. A few months ago they wheeled out a sporty-looking little thing in Monte Carlo rallying colours, and people in the specialist motoring press had to spend the entire day in the lavatory, whimpering gently.

When these guys came out some were blind, but Rover admitted the concept was just that. It will never be made. They were just fooling around.

And now they've done it again. At the Geneva motor show they whipped the covers off another new Mini, which they explained was a mid-engined, rear-wheel drive sportster that, in five-door form, had more space on the inside than an S Class Mercedes Benz.

'And,' they added, 'it could be a production reality tomorrow.' Everyone was impressed, right up to the point when they announced that it won't ever happen. So what's going on here?

It seems that Rover's new masters at BMW were worried that the intriguing Mercedes A Class would steal all the headlines at Geneva. Here was Merc's first-ever front-wheel drive machine, which is even smaller than a Ford Ka. It would be a big story.

So BMW decided to relieve themselves all over Merc's bonfire by instructing the bright young designers at Rover to come up with something even cleverer. And it worked. Merc's A Class, which is going on sale in the next few weeks, was eclipsed by a car that's nothing more than a hallucinogenic vision.

However, I suspect Mercedes will enjoy the last laugh because in three years' time Rover will have to stop giving us concept cars and unveil the real thing.

They've been showing us they can make the most amazing hollandaise sauce, but in reality they're going to serve up a slightly underdone potato. Marvellous.

Now this is bad, but it's worse when a manufacturer shows us a concept car that can't be made.

My favourite was a Peugeot that did the rounds ten years ago. It was startling. Small boys were captivated by its swooping lines. But the reason they were so swoopy is because there was no room within the framework of the car for an engine.

Now I'm sorry, but if I were running a car firm and someone brought me a design which precluded the fitment of a motor I would sack him immediately. I would not instruct him to go away with several hundred thousand of my shareholder's pounds, demanding that he builds a mock-up of the damn thing.

And even if I did, I certainly wouldn't let the public see it because then the shareholders would sack me. A car with no engine Jezza? Great work, idiot. Now get out.

Then there's the fascination with the rear-view mirror, a device that I think works rather well.

But no. It is almost always ditched on a concept car and replaced with an expensive television camera that feeds its signal to a screen mounted on the dash. Mmmm? I see, but what happens when the lens gets dirty? People who keep plugging away at this ludicrous idea are obviously mad and should be quietly murdered.

So what about doors? The hinge is a good idea. It was invented several years ago and works in all sorts of different applications – your fridge, your rabbit hutch and your canal.

But car designers are obsessed with alternatives. There was a Cadillac I saw at the Detroit motor show in 1989 where the door popped away from the body, electrically, and then slid forwards to let you in. It worked . . . but only in the same way that you can use a urinal while doing a handstand.

There was a car at this year's Geneva show whose name has gone from my mind. It had chunky four-wheel drive wheels and, I suspect, the running gear from a Mercedes G Wagen. But instead of a boxy, practical body it had a sleek, two-seat layout – sort of Robbie Coltraine in lace panties.

I know it's vital that car designers keep an eye on the furthest horizon, and that it's a good idea to constantly challenge accepted wisdom.

But can't this be done quietly, behind the scenes? That way, when something turns out to be impossible they can forget it, and when something, like the latest Mini concept, turns out to be a goer, they just go right ahead.

I mean the old Mini has now been around for very nearly 40 years. I really don't think anyone would accuse Rover of rushing into things.

Top Landing Gear – Clarkson in full flight

Go on, ask me whether to buy one of the new Peugeot 406s or a Rover 600 diesel and I'll surprise you with my answer. I haven't driven either of them. And nor, I'm ashamed to say, have I tried a Daewoo Espero, a Jeep Grand Cherokee, a Nissan Almera or a Hyundai Lantra. Wanna know about the Seat Cordoba? I'm not your man. So it's a toss-up between the Toyota RAV4 and a Vauxhall Tigra. Well I've seen lots but, to date, neither has been to Telly Towers for evaluation. But ask me whether to fly to the USA on Virgin or Delta, and I'll be in like a shot. Ask me how a smoker can get to Australia without eating their own seat and I'll have a starter for ten. In the last two years I've been so busy making 12 *Motorworld* programmes that I've rather lost touch with what's what in cardom. But at 30,000 feet I'm on *Mastermind* with no passes. Did you know, for instance, that if a fresh-air fanatic sits in a smoking row on an American airline that row, under federal law, becomes no smoking?

And I have worked out why Australia failed to beat us in the rugby world cup. They're all wimps. I know this because under state law baggage handlers are not permitted to accept any suitcase which weighs more than 32 kilos. And though you may know how to change the plugs on a 1983 Citroen CX, I know how to smoke on a Cathay Pacific 747. It's not terribly dignified, but what you do is bury your head in the lavatory, keeping your knee on the vacuum flush button. That way, the smoke is sucked into the bowl and away from the infernal detectors. This is important stuff. Well, as vital as knowing how to turn the wiper off in an Audi Coupé, and even the road testers on this magazine can't do that.

As far as quality is concerned, British Airways is simply head, shoulders, torso and thighs above the competition. No matter where you are, when you step on a BA jet it feels like you're home already.

If I can liken airlines to car companies, BA has the efficiency and reliability of Mercedes Benz with the quiet dignity of Bentley. The Far Eastern carriers used to have things sewn up with their devastating stewardesses and tasty titbits. But today MAS – the Malaysian outfit – is the only one worth writing home about. If you need to get to the Far East, and British Airways is full, go via Dubai and use Emirates. I can't say that I care very much for the tan and red uniforms, or the decor, but they have television screens for everyone; even in the back, with the cattle. If you're going the other way, to America, the first thing you must do is ignore any US carrier. Without exception, they are brash and their stewardesses need Zimmer frames.

South Western, from Texas, has a remarkable ticketing system which makes most airports look like bus stations, but when you get on board and are served a cup of warm brown water by a woman in specs the size of a Triumph Herald, you know it's doing it all wrong. However, even Americans are better than the Third World. In Vietnam, the pilot made a number of attempts to hit the runway in Hué, finally opting to land his jet near it instead.

In Cuba they fly planes that would be rejected by Fred Flintstone. One had no windows, and filled with smoke 15 minutes after take-off. Another had windows but was flown by a fully paid-up member of EXIT. He knew his engines were on their last legs but, even so, he flew right into the biggest thunderstorm I've ever seen. We went in at 2000 feet and came out through some bushes. However, while Cuba may be the FSO Polonez of airlines, it is not the worst. The Nissan Sunny award for hopelessness goes to . . . Qantas. They are incapable of getting a plane off the ground on time. The staff are ruder than French waiters and the food is inedible. Even the appropriately named CAAC – China's airline – which shows 12 hour animated kung fu films through loudspeakers, has them licked.

I'm sorry if you think you've been reading *Top Landing Gear* this month. However, fear not. Judging by my drive, which is now full of cars, and by my diary, which shows no trips abroad ever again, normal service will be resumed shortly.

A fast car is the only life assurance

Between 1982 and 1985 I used to play a great deal of blackjack, and it almost always made me miserable when I lost. Which I did. All the time.

Nowadays, however, I have learnt to approach the table fully expecting an hour's cards to cost me a hundred quid. Which it does. All the time. By abandoning hope, I've removed the despair.

This is a very important philosophy when you're confronted with someone who's trying to sell you a pension. Do not listen. As soon as they open their briefcase, put your fingers in your ears and hum Bruce Springsteen ballads.

A pension is by far and away the most stupid thing ever to hit the civilized world. You scrimp and you save for 30 miserable years, hoping that you'll live to reap the rewards.

And what, pray, are you going to do with those rewards at the age of 92? Buy a gold-plated Nebulizer? Luxuriate in some silk-lined incontinence pants?

They say you're investing for the life you don't yet

know, but that can work both ways. What if, after a life of deprivation, you are eaten by a tiger? Or what if, just a week before the big pay-out, you get an even bigger one from Camelot.

Pensions are all about planning, and planning is all about hope. And hope, invariably, leads to despair.

Tonight, I'm off to New Zealand to race a V8 jet boat up some rapids, and that's an experience to beat any 4 per cent growth on capital, believe me.

They say pensions are tax-efficient but, quite frankly, they're so dull I'd rather give my money to the government. So long as they promise to spend it on F-15s and nuclear submarines. Things I can be proud of.

The whole point of having money is to have fun. That's it. There is no other reason, which is why you must also slam the door on anyone trying to tell you that a PEP or a TESSA is a good idea. It is not.

You give your cash to someone who, in return, sends you statements once in a while saying that you now have more money than you had when you started. But you haven't, because it's locked away.

Some would argue that the stock market provides a better alternative because the money is always available. But I'd be grateful if all talk of EC1 were kept out of the equation just at the moment. Even the most idiotic gambler can see the Footsie is at its highest level ever, and that Mr Blair is at the gates. The only way is down.

I have thought about this quite carefully and it seems to me that all investments do nothing to enrich your soul. And I don't care what the grinning salesman says, they're all risky. Remember, pensions have a habit of falling off boats.

You will spend your life hoping this doesn't happen. You'll wind up frightened and alone, shivering in the corner of a one-room bedsit, unable to afford a single-bar fire because some City institution is playing the silicone gee-gees with your cash.

My advice is simple. Remove the risk. Invest your money . . . no wait, *spend* your money, on something you *know* will lose. A car.

Now at this point, a few people will raise their hands and draw my attention to the Mercedes SLK, which, when it was launched at the beginning of the year, came with a two-year waiting list. Secondhand values went through the roof, up the chimney and in some cases right to the top of the television aerial.

SLKs were being advertised at £45,000, which is £10,000 more than they'd cost a patient man. But pretty soon everyone was trying to sell their Merc, and in a matter of days prices settled down again. The SLK was a freak.

And I see no new cars on the horizon which will perform a similar trick; certainly not the BMW Z3. Last weekend, I counted 26 in the secondhand columns, giving it the exclusivity of a packet of biscuits.

Cars were an investment once, but too many people walked away from those late-1980s boom years nursing fingers that were burned through to the bone. However, and this is the key, they may have paid £100,000 for a Ferrari 308 GT4 which is now worth £20,000, but at least they still have a Ferrari.

I know of two chaps who clubbed together in the height of the madness and bought an Alfa Romeo SZ thinking it was their passport to a life of rum punches in

Barbados. However, as they took delivery the bubble burst, and they've been wearing margarine trousers on a slide into oblivion ever since.

But who cares? If it had been a stock exchange deal that had gone wrong, they'd have a worthless piece of paper. But they've still got a Group A racing chassis, those wonderful looks and that magical 3.0 litre V6 engine. They did all right.

I was flicking through the small ads in this paper last week and found that for, say, thirty grand you could have a Jaguar XJR or a low-mileage BMW M5 or, staggeringly, a Bentley Turbo R.

Sure, when you're 90, a pension would keep you in panty pads and a Bentley won't, but at least you'll have had a life. You'll be seen as wicked and interesting, and as a result no one will care when you simply wet the chair.

Rav4 lacks Kiwi polish

Last night, I found myself at the Auckland Travelodge, tucking into a bed of wilted leaves and chicken served with 'jus'.

To be honest, there's nothing much wrong with Travelodge – except they always put me in a room that's two time zones away from reception – but I would like to know how on earth the word 'jus' wormed its way on to one of their menus.

In the very recent past, 'jus' was only to be found at the very finest restaurants in France, but in just a few years it's filtered down the food chain, through bistroland and onwards until it ended up in New Zealand, in a Travelodge, under my chicken.

Where it tasted pretty ropy actually. I knew it would, because it just doesn't belong. I go to a Travelodge when I want a pasta salad, and I go to Château du Domaine St-Martin when I want 'jus'.

In carspeak, this is even easier. I go to Land Rover if I want an off-roader, and I go to VW if I want a hatchback. So would someone please explain to me what the Toyota Rav4 is all about?

When it was launched in Britain a couple of years ago, I think I was going through a rough patch with Toyota. I wasn't speaking to them, or they weren't speaking to me, but either way I never actually drove one.

I heard it was an attempt to marry the Camel Trophy to Marks & Spencer, which seemed a bit unlikely, but press reviews were favourable and it sold pretty well.

Thus, when I came to New Zealand last week, and needed a car that would travel great distances and do some off-roading, all on a BBC budget, it seemed like the ideal solution. In fact, I'd have been better off renting a space hopper.

Most of the roads here are single-carriageway, so you need plenty of overtaking punch which the 2000cc engine simply can't deliver. When the road ahead is empty, you drop a cog, mash the throttle and yes, yes, yes, you are gaining and you're pulling out, and a few minutes later you're alongside, but you're never going to make it before the next bend, so you drop back again, a beaten man.

I know of no road, anywhere in the world, where there is a long enough straight for a fully laden Rav4 to strut its stuff in the overtaking lane. And the roads in New Zealand are bent like the wires in a Brillo pad.

This, however, does mean I had a chance to check out the handling. And yes, I'll admit that for an off-roader it steers and rides and corners with a surprising degree of comfort and agility. In saloon car terms, it's right up there with, say, a Lada Samara.

Now, of course, all these drawbacks are to be expected in a car that's been designed to drive through swollen rivers and up sheer cliff faces. But the Toyota, sadly, can't do either of those things. Indeed, with its road tyres, even a gently undulating grassy field proved too much.

I stabbed away at the centre differential locking button like a dying man trying to restart the engine on his crippled submarine, but it was to no avail. With insufficient grip, and even less torque, I was merely digging a hole that eventually would have taken me back to England. I'm afraid that despite a high ground clearance, the Rav4 is about as much use in backcountry New Zealand as an aqualung made of cheese.

Toyota has simply tried too hard. By trying to make an off-road 'car', they've ended up with something that's no good at anything.

Now this applies equally well to Suzuki's Vitara, but at least this makes up for its numerous shortfalls by being handsome in a hairdressery sort of way.

I'm forced to say the same applies with the three-door Rav4 too, but I had a *five*-door version, which is a terrible, terrible mutation that looks like it was styled by a World War Two plastic surgeon.

And the interior was done by someone who obviously works in a poorly lit room. The dash is so bland that my colleagues resorted to drawing extra dials and switches on it with chalk. And because we couldn't overtake anything, and therefore each journey took an age, we ended up with six boost gauges, a rev counter, an eight track, a CD autochanger and, if memory serves, a fart counter too.

Worse than the tedious innards, though, is that, unlike any other off-roader, it doesn't have a high driving position, so you can't sneer at other drivers. Not that there's much to sneer about in a car with almost no redeeming features.

Certainly, you can't sneer about the price. The five-door Rav4 is an almost unbelievable £17,000, making it the most preposterously overpriced piece of under-powered, nausea-inducing nonsense ever to hit Britain's roads.

Hard words, but just to make sure Toyota and I don't have to do pugilism again, I've now swapped it for one of their Land Cruisers – a huge diesel automatic, and I love it. Sure, it won't go round corners, but each time it ploughs off the road it just ploughs *through* whatever it hits.

It knows its place in the world. It doesn't try to be something it's not, and concentrates instead on simply being big. If the Rav4 is 'jus' in a Travelodge, the Land Cruiser is gravy in a transport café.

Cuddle the cat and battle the Boche

Some time between the seventh and eighth grappa, Tiff climbed back into his chair and announced that he wanted to buy a BMW M5. At first, we thought he was a little more tired and emotional than usual, but his arguments seemed rational. 'Its engine is so good and I love its looks and it feels so right and you can pick one up for £15,000 or so,' he said, before falling off his chair again. Mr Editor Blick and I didn't notice though, because we were deep in conspiratorial mutterings. We've got to stop him. We've got to demonstrate that the supercharged Jaguar is better.

The next day, Count Quentula was out parking cars for his village fête when I called with the news. 'Tiff wants a Bee Em,' I said. 'Oh Christ,' said Quentin. 'The poor deluded fool. I'd better let him have a go in my S Class.' And therein lies the problem. At this level in the market, people have nailed their colours to the mast and almost nothing will shake them loose. Tiff likes BMWs. Quentin likes Mercs. And I like Jags. When I start banging on about my XJR, Tiff will look up from his 24th grappa and ask if I'd like another gin and tonic. 'And how are the Masons these days?' When Tiff is in mid-soliloquy about the smoothness of a BMW 6, Quentin will interrupt to ask if he's run over any old ladies yet.

And when the Count tells us about the unburstability of a 500, Tiff and I wonder how we managed to miss his 50th birthday. With cars like this it doesn't matter what they look like, or how fast they go, or whether they do 12 or 200 miles to the gallon. It's an image thing, pure and simple.

The data is confused, but some figures suggest that up to 90 per cent of Britain's executives never change marques. If they start out in business with a C Class, they are in Merc's web and there is no escape. During the late

1970s and early 1980s, a great many bosses did the unthinkable and deserted Jaguar's leaking ship. But the big cat was in their soul, and now the cars are made properly again many are coming back to the fold. This, of course, means that if BMW wants to maintain its healthy market share, the new 5 Series only needs to be as good as the old one. Tiff will want one no matter what. Thus, as one magazine has called the new boy 'close to perfect', you could accuse the Hun of overkill.

Certainly, I can't remember driving any other car which does quite so many things quite so well. The £30,000 528 I tested was truly fast and yet eerily efficient. It has room in the back for a small tennis tournament and yet it handles with an aplomb that leaves you breathless. Then there are the details, the best of which is the interior lighting. You get the usual red instruments, which BMW says provides a restful get-you-home environment, and I'd have to agree. But in the new 5 Series they've gone further, because next to the mirror are two tiny red spotlights, providing a stylish red glow around the centre console. It gives the whole dash an exquisite 3D effect and, in addition, you can find your phone and fags.

For the Tiffs of this world, for all BMW drivers, this car is better than close to perfect. It's a solid 10. If it had been crap you'd have loved it, but it's brilliant, so I dare say you'll want to spend your evenings in the garage with it and a bucket of KY jelly. Me? I couldn't wait to see the back of it. And Quentin is hardly jumping up and down, clutching his privates, as he waits for a go. Dynamically, it is superior to anything for the same sort of money made by Jaguar or Mercedes. But we don't care. When I overtake someone in the Jag, you can feel the warmth of approval.

People point and coo; they're talking about how good it looks and how quality is better these days. Middle England wants a Jag. Now try the same overtaking manoeuvre in a 528 and feel the hate. There goes another pushy yuppie, hoping to hit a tree before his ticker gives out. Gaps that open for Jag Man are closed when you're in a BMW. People don't like them.

I tried this argument on Tiff but got nowhere. 'Look,' he said, pouring another grappa. 'You can go faster in a BMW than you can in a Jag or a Merc.' And then he fell off his chair again.

Secret crash testing revealed

When you read a road test report in any newspaper or magazine, you will learn how a car handles at its very limits of adhesion.

The reporter will tell you that on a twisting mountain road in the South of France he hurled the new model into a series of fast sweeping bends, and felt the front tyres fighting for grip under acceleration, and the back swaying this way and that under braking.

Amazing. The guy has flown out there, climbed into a car that he's never even seen before, and within hours he's taking it right to the outer reaches of its performance envelope . . . without crashing.

Formula One drivers test their cars week in and week out. They're on first-name terms with every nut and bolt. They could drive each corner blindfolded. And yet even the great Michael Schumacher is capable of flying off the track backwards from time to time. So what's going on here?

Well a motoring journalist must try to convince his readers that he is, in fact, a great deal more talented than Michael Schumacher, and that the only reason he isn't out there in an F1 Ferrari is that he's too fat – or in my case, tall *and* fat.

So, if we crash, – and we do, a lot – then it is important to keep the fact hidden from our readers.

Did you, for instance, ever hear about the chap who missed a signpost while driving a £30,000 Mercedes G Wagen alongside a river in Scotland? I was following him at the time and remember well the moment when it stopped bouncing along the bottom and began to move in a serene and graceful way ... like it was floating. Which it was.

It bobbed along for some time while the public relations man hopped about on the bank wondering what on earth to do. Either he could get the ghillie to pull it out with his Land Rover, in which case the pictures would appear in every newspaper the next day. Or he could let it sink so no one would have anything to point their cameras at.

He let it sink.

Then there was the guy who stuffed a Ford RS200 into one of Scotland's more pointy parts. He claims he went off the road in this £50,000, mid-engined supercar to spare the life of a £40 sheep which had wandered into his path.

So what about Quentin Willson, my colleague on *Top Gear*, who, while going the wrong way round the first corner at Silverstone in a £60,000 De Tomaso Pantera, got two wheels on the grass? He hit the barrier, bounced into the pit wall and would have hit the barrier *again* but there was nothing left by then.

And surely, no one can have forgotten about the *Guardian*'s man who changed into first while doing 90 or so in the then new Jaguar XJ220. They had to take the engine back to Coventry in a Hoover bag.

But the only reason we heard about this is because it was reported by the man from the *Mail* who, just weeks later, quietly crashed a £200,000 Bentley Azure.

I'm in the hall of shame too. A few years ago I rammed a Porsche 928 under an Armco barrier just outside Cwmbran, and then peeled the bonnet off like it was the lid of a sardine tin while reversing it out again.

Now I am a man who, at school, could worm his way out of all kinds of trouble by coming up with preposterous excuses, usually involving tigers, but after crashing the Porsche I had to stand up like a man, and admit to its owners that I'd been a fool. Not in print though. And definitely not on television.

Only this week, I had a minor 'off' in a new type of ultra-racy Vauxhall Vectra. I think I may have bent a steering arm, so that it now drives like a crab, but will you see how I did it on *Top Gear* this Thursday? No chance.

Now here's my point. Why don't we report these accidents? They're big news. I mean, if you have a prang your car is off the road for weeks while the insurance company squirms and wriggles. The subsequent repairs will send your premiums into the stratosphere and badly affect the secondhand value of your vehicle.

And then there's motor racing. You don't care about deft overtaking manoeuvres or whiz–bang pit stops. No, you like the crashes and the fireballs. That's why you all slow down to gawp at mangled metal on the motorway.

So perhaps then, it's time for us motoring journalists to swallow our pride and understand that the size of a car's ashtray is maybe not that important. People are more interested in how we managed to leave the road at 100mph, backwards.

The trouble is that when we crash it's like Barry Norman spilling his popcorn. Or A.A. Gill dropping some butter on the carpet. We just ring the manufacturer and a tow truck comes. We fill in an accident report form and nothing more is said. We don't think of it as a big deal.

I once tore the front end from a Daihatsu Charade GTti after plonking it in a ditch at 80mph. And the press officer merely shrugged it off saying, 'Don't worry. We make one every 23 seconds.'

Well, good for you matey, but when I'm sitting here struggling to think of anything to say about the latest dull car that's parked outside, I've just realized that a good crash can fill several column inches.

That's is why I'm going out right now to ram a Toyota Corolla into a tree.

Diesel man on the couch

A policeman once told me that if there is room to overtake someone on the inside, then there was room for that person to have pulled over. Wise words, but don't bother using them in court. Undertake someone, and in the eyes of the law you're a mugger with a crack habit.

Now in the normal course of events this doesn't really matter, because all three lanes of every motorway are full and you just drive along at whatever speed the traffic happens to be doing.

The trouble is that this lulls people into a sort of never-never land where your heart is beating and your eyes are open but you are not really awake. A leprechaun could jump on to your bonnet and make a wigwam out of your windscreen wipers but you wouldn't even blink.

Consequently, you don't really notice that it's getting late and that the traffic has thinned out. You are in a deep, deep coma.

But then, suddenly, your rear-view mirror melts as it is assaulted by a 400 gigawatt burst of light. You come to realize that someone is behind and you pull over feeling a bit sheepish . . . unless you are driving a diesel.

This is the first trend I've ever spotted. We've had Essex Man and New Man, and only a couple of weeks ago the Freight Transport Association came up with Van Man, a 19-year-old plumber who genuinely believes his Astramax can break the sound barrier.

Well now, I'd like to introduce you to Diesel Man. Diesel Man is less well defined than the others in that he could be 17 or 70, blue-collar or middle management. Strangely, Diesel Man might even be a woman.

He's not easy to spot in ordinary life because he behaves just like you do. He's ordinary. He blends . . . right up to the point when he climbs into his diesel-powered car. And then he is more bitter and twisted than the lemon you put in your gin and tonic last night.

In the past, it was hard for Diesel Man to fall into a catatonic state while driving up the motorway because of the engine noise, but these days diesels are pretty silent at speed, so he nods off as surely as you and I.

However, when he becomes aware that another car is keen to come by, he reacts in an unusual fashion. He drops a cog to get that hideously inefficient engine into the upper echelons of its miserable power band, and floors the throttle.

From behind, it's hard to tell he's done this because, obviously, there's no discernible change in pace. Put your foot down in a diesel at 70mph and it can take ten or twelve minutes for you to be doing 71.

However, there will be a puff of carcinogenic smoke from the exhaust, and that's the sign. Diesel Man is going to prove that his car is just as fast as yours.

Psychologically, it's easy to see what's happening here. His boss has heard that diesel engines are more economical than their petrol-powered counterparts, and that because they tend to be less powerful, accidents happen infrequently. So he decides that his staff, from now on, will have diesels.

Now we all know that you can call a man's baby ugly and he won't mind. We know that you can take a man's wife to bed and it'll all be forgotten in a week or so. But laugh at a man's wheels and you're in serious trouble.

Diesel Man is well aware of his car's shortfalls. He knows it's pitifully slow and that it makes the most Godawful din when he starts it up in the morning. He also knows that he doesn't benefit one jot from the lower running costs. Basically, he knows the car is a worthless pile of junk, but is he going to admit this in public? Hell no.

To admit that his diesel is a step down is tantamount to

admitting that he has taken some kind of demotion. So he's going to prove, no matter what the cost, that his diesel is superior in every way to a petrol-powered car.

And it's the same story with private buyers who've been enticed by the promise of 45mpg only to discover that the downsides easily outweigh the few pence that are saved each week. But are they going to say so? Only after they've owned up to being hung like a maggot.

So what's to be done? How do we get past? Well you might argue that the speed limit is 70mph on the motor-way, and it is. You may say that all I'm doing here is encouraging people to break the law, but we all know the score. The speed limit is 70, so we can all do 85.

Except we can't, because Diesel Man is having an ego crisis right in front of us.

There is, I suspect, only one solution. Car manu-facturers must refrain from putting any form of diesel logo on the back of a car. The BMW tds, Citroen 1.9D, the Rover SDi. Diesel man knows we can see this little 'D' and suspects we may be laughing at him. That's why he puts his foot down.

But if the 'D' were replaced by an innocuous 'p' or 'z' or whatever, he could simply get out of our way, happy that we'll sail by unaware of the aberration under his bonnet.

Or he could, of course, go out there and remove the 'D' himself, but I've just thought of a much better idea. Grow up.

Stuck on the charisma bypass

The new Maserati Quattroporte is, in many ways, a breath of fresh air. Here, at last, is a car that's truly, madly bad. Armed with a ridiculous price tag, it wades into battle with a slightly bent peashooter and adaptive suspension that doesn't work. It is ugly. It has an engine that sounds like it's trying to mix cement. The leatherwork is shoddy. It is badly equipped and it has a clock shaped like women's bits. You wouldn't want to buy it, but at least you can discuss it, with much finger-pointing and shouting, over a beer. That automatically makes it better than some of the dross I drove last week. My God, there are some boring cars out there.

Bring the Hyundai Lantra Estate up in a pub and it would have the same effect as putting a Mogadon in everyone's drinks. We all know someone like this car – someone who tries to disguise his innate and inbred ability to redefine tedium by wearing a stripy orange and brown tank top. The car is quiet, it will rarely break down and I'm sure it would buy its girlfriend – a librarian – chocolates on her birthday. At work, it would have a sign on its desk saying: 'You don't have to be mad to work here – but it helps.' What a wag. What a git.

Then there's the Rover 400 Saloon, a Honda Civic with delusions of grandeur. It's someone who's made a few bob and thinks that by shopping at Hackett and wearing brogues he'll be accepted by the county set. Volkswagen has cocked up too, with its new Polo saloon. What a heap of steaming manure this is. The hatchback is a charming and funky little device with cool graphics, a wild range of colours and lots of street cred. But by putting a boot on, the designers have put the boot in.

Could this car really be worse than the old Derby? I think so. Could it be worse than the old Vauxhall Nova saloon, with the elephantine proportions and the unicycle wheels? No, that's ridiculous. I'd like to tell you about the Daihatsu Charade at this point, but nothing springs to mind. It's a glass of water on wheels. Hey, what's this? It's the new Audi A4 rattling into view. Now this is some car, beautiful to behold and made with the sort of care normally reserved for space shuttles. But wait. What's that under the bonnet? Oh no. It's a diesel. Start it up and there's the familiar clatter which can give old people arthritis. But this one has a turbo, so when you put your foot down, especially at low revs, there's some serious grunt. The trouble is that the power band is so narrow you only need blink and it's all over. 'Dear Deirdre, My car suffers from premature ejaculation. What should I do?'

Deirdre replies: 'This is a common complaint which is getting worse as more and more people fall for the turbo-diesel sales patter. Leave your car now and go for a real man: one with a petrol engine.' This is not to say that unleaded is the cure for all our ills. Witness the VW Passat and the Seat Toledo, cars which, if they were ovens, would cook food.

Then there's the king and queen of horror – the Toyota Corolla and Nissan Almera. Styled by adding machines with interior trim by BHS, this duo leave me so cold hypothermia starts to set in. After a spin in either, even the Vectra starts to look like a Ferrari F512. But it takes more than casual comparisons to enliven the Astra. Like the Escort, this car is barely fit to be a pox doctor's clerk. It isn't especially good value for money. It isn't handsome. It isn't noteworthy in terms of performance and it doesn't have microwave reliability either. I could fill up the rest of this magazine with cars that just don't make the grade. I'd need 44 pages alone for the Nissan Serena Diesel, which takes an almost unbelievable 26 seconds to heave itself from 0 to 60.

The best thing is to list the worthwhile mainstream cars. It won't take long, so here goes. At the bottom we have the Ford Fiesta and Nissan Micra. In the middle, the Fiat Bravo and Renault Migraine. Up a bit and the Mondeo and Pug 406 dominate. Further up, say hello to the Audi A4 and the Honda Accord. And at the top, the BMW 5 Series makes big sense. Though Jaguar and Mercedes also do something pretty special for 30 grand.

If it isn't in this list, frankly, it isn't worth the metal it's made from.

Travel tips with Jezza Chalmers

If you were to be wrongly charged with murder while in Thailand, I think it fair to assume that you wouldn't conduct your own defence.

You don't speak Thai and you don't understand the nuances of the legal system.

And yet when we go abroad on holiday, we're all quite happy to pile into a rental car even though we can't read the signposts, we don't understand the customs and, more often than not, everyone is on the wrong side of the road.

In Britain, if someone flashes their lights it means

they're waving you through, but elsewhere it means 'Look out, I'm not going to stop.' However, you won't realize this until you're half-way through your windscreen.

So what I've attempted to do in the limited space here is try to offer some advice for those who will be driving abroad this summer.

Let's start in America, just outside Miami airport, where the slip road joins the Florida turnpike. You've been rammed, gently, from behind and you've climbed out to inspect the damage.

This was a mistake because the man who ran into your rear end is a Colombian drug dealer who will now shoot you, your wife and your children. Then he'll help himself to all your belongings.

Locals say that if you're rammed, you should drive around until you see a policeman . . . but this is not sensible either. You see, if Plod is confronted with a hysterical Limey babbling away in an accent he doesn't really understand, he will shoot you so he can get back to his seventh doughnut of the morning.

It is possible, just, to get out of the Miami district without being murdered, but then you face an altogether new problem. The road may be wide and straight, but do not, however tempting it might be, exceed 70mph.

Your American rental car is simply not capable of high-speed travel and will, if you push it, bounce off the road into a swamp. Whereupon you will be eaten by an alligator.

Other tips: you must pay for fuel before filling your tank, which is stupid because you have no idea how much your tank will take. Nevertheless, don't argue or try to buck the system because most petrol pump attendants in Florida are daft and armed – a lethal combination which will result in you springing a leak.

It's also worth remembering that in America most establishments have valets with massive teeth and idiotic red waistcoats who will volunteer to park your car. When they return it, they will expect a tip, but you can get round this by pretending to be Icelandic.

Whatever you do, do not claim to be Romanian or Czechoslovakian because the valet will think you're a commie and may try to shoot you.

The best piece of advice I can give to anyone thinking of driving in America this year is … have you thought about Europe?

Italy is perfect, but be aware that your rental car will be a wreck with a shagged engine. This will make any foray into the autostrada's overtaking lane tantamount to dancing with the devil.

In Italy, they don't wait patiently for slower cars to move over, and nor do they flash their lights or attempt to get past on the inside. They just ram you.

Life is a lot more disciplined in Germany, of course, but I can't think of a single reason why you would want to go on holiday there.

France is much nicer, but whatever happens do not take your own car across the Channel. There is something called ferry psychology which means that as you approach Calais to come home you will definitely be driving too fast.

This is either because you are late for the sailing on which you're booked or you're miles too early, in which case you're hurrying to catch an earlier boat. Either way, you'll be caught speeding and made to pay a fine so massive that when you get home, your house will have been repossessed.

And don't try claiming you can't pay, because then they will take your car, and your wife . . . out to dinner where she will fall for their Gallic charm and leave you to a life of meths and shop doorways.

The most popular foreign destination for British tourists is Spain, which is one of those strange facts that I can never understand. Like why is the motorway central reservation always littered with shoes?

I really don't like Spain, but I will admit that their roads, these days, are simply superb – smooth, wide, fast, and free in large chunks from too much in the way of traffic. If you've ever wondered where the European Union spends all its money, have a look at Barcelona's motorway network. And then get out there and enjoy it. You paid for it.

Briefly, because I'm running out of space here, I should warn you that wildlife is a massive problem elsewhere in the world. In Britain, we are unused to rounding a bend to find the road blocked by half a ton of snorting muscle, but in Australia, camels and kangaroos regularly play chicken. In India it's cows, and in Sweden even more people are killed by errant elks than by razor blades and Mogadon.

In the Caribbean, mercifully, large animals are scarce but then the cars they rent you would lose if they went head-to-head against a breeze. In Barbados be very careful indeed to avoid what is basically a Suzuki Alto with no bodywork at all.

In fact, it's probably best to avoid going abroad in the

first place. Me? I'm off to the Isle of Man for some fresh air, some invigorating scenery and the joy of being able to drive through it without the burden of speed limits.

Capsized in Capri

I've spent the last year working on a new television series all about big boys' toys. This means I've shot the rapids in New Zealand in a 100mph jet boat. I've flown an F-15 fighter jet. I've done the Reno air races in a 1942 Mustang P-51 and in Sweden I lost my liver to a drag snowmobile that could do the standing quarter in 6 seconds . . . whilst wearing women's clothes.

But the high spot was to have been my time in the world of Class One offshore power boat racing.

This, as far as I'm concerned, is about as good as sport gets. The 4 ton boats are a subtle blend of hydrodynamics and aerodynamics so they skim along the surface of the sea, with just the bottom half of their propellers in the water.

Each uses a *brace* of 1000 horsepower, 8.0 litre Lamborghini race engines which make a noise that can curdle blood at 500 paces. Only once have I heard a sound to beat it: Bob Seger at the Hammersmith Odeon in 1976.

In rough water the drivers simply go as fast as they dare, which means that all the expense, all the technical development and all the power is wasted if the guy has nothing in his pants. It's not like car racing, where you can do a crossword while going down the straights. You're on water which moves and wobbles, and you are being beaten to death inside that Kevlar cockpit. For days after a rough race the two-man crews pee blood.

To get to the bottom of power boat racing we filmed one of the boats being made, and I went out for its first ever shakedown run on the Solent late one Saturday evening in May. It was good. We hit 148mph which, technically speaking, is pretty fast.

And then I faced the agonizing choice of deciding which race I'd watch – St Petersburg, Beirut, Tunisia, Norway, Dubai or the first round, which was to be held somewhere I've wanted to go for 25 years.

Capri: an island in the Mediterranean which was home to the emperor Tiberius. And then nothing much happened there until the 1930s, when Gracie Fields arrived.

Some say it was named after *capreae,* which, as everyone knows, is the Latin for goat. Others, having found strange skeletons embedded in the limestone, say it's more likely to be derived from *kaprus*, the Greek for wild boar. Either way, it seems to have nothing to do with a crummy Ford.

In recent times, Claudia Schiffer and Naomi Campbell have been holidaying there so I figured there'd be a jet-set backdrop to just about the most glamorous and dangerous sporting spectacle the world has ever seen.

But getting there is not easy. You have to fly from Gatwick, which is always depressing – all those lard arses in towelling tracksuits and tight perms off to Torremolinos. Then you land in Naples – the only city on the planet where a red traffic light means go.

And then you have to reverse onto a car ferry which only has three tins of beer on board. And when you get to Capri they won't let you off unless you can prove you live there. This was hard, because I don't. But if you argue with an Italian in a uniform for long enough, he will eventually tire of the exchange and allow you through.

So we were in Capri at last. Er, not exactly. We were on the harbour wall – which was just wide enough for one car – facing a stream of traffic coming the other way, trying to board the boat we'd just left.

Either I backed into the boat and went back to the mainland again or 30 residents reversed out of my way.

So it was back on the boat, and after a round trip via Naples again I found myself on the most spectacularly beautiful island in the entire world. I've been to the Maldives and the Caribbean. I've explored Mauritius and Orpheus on the Great Barrier Reef. I've seen the north coast of Majorca and once spent a holiday in Sicily, but you can forget all of these, and Sardinia, and Corsica and all those hideous lumps round Greece. Capri is heaven.

Every villa and every bit of cliff is smothered in that purple stuff which isn't buddleia but looks a bit like it if you squint. Gorgonzola? No, Borden villier. Something like that. Damn pretty anyway.

And I should know because I got very, very close to it while trying to squeeze past traffic on the three-mile drive to my hotel. It was hopeless and eventually I would have to back up to the nearest passing point, which was usually the harbour.

Eventually, I learned to reverse everywhere so that when I encountered traffic coming the other way they'd assume I was backing away from a bus and would begin to reverse as well.

Still, it would all be worthwhile because I was going to see nine face-distortingly fast boats in aquatic combat. I was so excited that I clean forgot Capri is a part of Italy and that therefore nothing should ever be taken for granted.

This may have been the first round in the championship. The drivers may have come from all over the world, and the boats had been brought on trailers on THAT beerless ferry. But, apparently, there was a bit of a mix-up with the coastguard over timing, so the event was cancelled.

It took 14 hours to get home.

Noel's Le Mans party blows a fuse

Last week I went to Le Mans, where, for the first time, I was introduced formally to the world of big-time motor racing. And I've worked something out.

We tend to think of motor racing as being a driver thing, but this, I'm sorry to say, is not really the case. Yes, his hairstyle looks good and his teeth are shiny, but it is the car that matters most of all.

In a Williams, Damon Hill was world champion. In an Arrows, he is an amusing sideshow. Think of it in terms of cooking. Give me the freshest ingredients and I'll knock up a supper that will cause your taste buds to die of a broken heart. Give Gary Rhodes a tin of pilchards and you'll get a tin of pilchards.

At Le Mans, I was part of the Panoz team that had been put together by his Noeliness, Mr Edmonds, and for five days I hung around in their pit, wearing a serious expression and pointing at things.

It all began on Wednesday evening when the cars were

sent out to qualify – nail-biting stuff because the two slowest entries would not be allowed to race. A year's work would be bundled onto the back of a lorry and sent home.

On his first lap one of our drivers, Andy Wallace, reported over the ship-to-shore radio that his Panoz was 'absolutely f★★★★★★ undriveable'. Under braking it was bouncing all over the track but, worse than that, the 6.0 litre V8 Ford engine had no power coming out of the corners.

The news was bad, but the other car was in even worse shape. Its oil repository had seemingly been modelled on a colander.

So when the session was terminated at 12.30 it seemed likely that neither of our cars would make the grade. We had one more chance on Thursday night to try and make those cars fly.

That morning, I asked by far and away our tweediest driver, James Weaver, to explain the problem, expecting him to lift the bonnet and point at a wonky part. But no. I was taken into a back room where men in spectacles were hunched over a bank of laptops, staring at graphs.

'There,' he said. I peered at the read-outs for some time, my face scrunched up like I was trying to read a sign from a long way away, but could make neither head nor tail of them. So he explained that the blue wheel-speed trace did not match the red rev trace; that there was a glitch, and that the men in spectacles were interrogating the engine's electronic management system to find out why.

I'd noticed this the day before. Whenever the car came back to the pits no one ever went near the engine. They simply plugged computers into it and banged away at keyboards in a Rick Wakemanesque frenzy.

And they kept on banging away all through Thursday and all through the vital evening practice session. And still the engine wouldn't work properly.

The mechanics had sorted the handling problems and the drivers were giving it their absolute best, but none of this really mattered because somewhere deep in the bowels of that multi-million-dollar carbon-fibre race car there was a morsel of silicone having a genetic tantrum

Right at the end of the second practice session one of our cars had made it – just – but the other faced being eliminated by the number 50 car from Lotus, which was out there on the track doing its stuff.

It's funny, but for three-and-a-half minutes – one lap of Le Mans – my whole life was focused on the timing screen, waiting for that Lotus to finish its do-or-die run. Every second seemed like an hour ... because every second was an hour. Lotus would have been better off with an Orion diesel.

I celebrated by goose-stepping through the Porsche pit, which was considered poor form, but still those men in specs pumped away at their laptops, desperately trying to turn our engine from Aled Jones into Pavarotti.

They were monitoring everything as the race began, but even so, just two hours down the line, one of our cars ran out of juice. The driver got it back to the pits using the starter motor but this technique wore out the battery, which had to be replaced, costing even more time.

The mechanics worked like ants, the drivers gave it their best shot and everyone agreed that the British-designed chassis was superb. But at 3 a.m. the engine on one car blew up and it was out. The other engine gave up the ghost seven hours later.

So, what have I learned? Well, if our cars had been equipped with a bank of Holley carburettors instead of stupid electronic fuel injection, there's a better than evens chance both would have finished.

When a carburettor goes wrong – and they almost never do – you rip it off and fit a new one. It costs a tenner and takes about five minutes. But when an electronic pulse goes bonkers, even Bill Gates would be flummoxed.

In motor racing you need an engine that can shovel fuel into the cylinders, a hot chassis and a team of mechanics who'll build the parts properly. A driver whose eyes work is not a bad idea either. We had all that.

But computer geeks have now shambled into this high-octane world and I just don't know why. They have a place, of course, but it should be in a loft somewhere, searching the Internet for photographs of naked ladies.

The Skyline's the limit for Gameboys on steroids

The Japanese car makers should take a long hard look at Linford Christie and Barbara Cartland. One does not attempt to win 100 metre races and the other does not try to look like a big, pink crow. They should say to themselves, 'All the best-looking cars in the world are European or American and if we try to copy them, we end up with hopeless facsimiles like the Supra.' And they should go further: 'Boys, we do not understand "soul", so let's not try to replicate it.' 'Soul' is what you get when you've won the Formula One world championship and Le Mans 99 times. You can't design 'soul' or 'character'. You can earn it.

Cars are like friends. I have many, many acquaintances, but friends are people whom I've known for years and years. 'Soulful' friendships are forged when you've been drunk together, arrested together. That said, there are short cuts. I'd be pretty matey with someone who gave me a million pounds. And I wouldn't slam the phone down if Princess Diana rang, feeling a bit horny. The Nissan Skyline GT-R is just such a short cut. Nissan accepted they could never match European finesse and style, so decided to go where Europe can't follow – into the auto cyber zone where silicone is God and Mr Pininfarina is the doormat. It worked. The Skyline is not a facsimile of something European. It is as Japanese as my Nintendo Gameboy, only more fun. I was smitten by the old model, but now there is a new version which, after a week-long orgy of big numbers and lurid tailslides, has left me in no doubt. Forget the Ferrari 355. Forget the Lotus Elise. For people who want their car to be the last word in ball-breaking ability and to hell with style and comfort, the Skyline is Mr Emperor Penguin. King of the hill. The biggest cheese in Stiltonshire. Whether its ability is down to the four-wheel drive system or the four-wheel steering or the peculiar diffs and electronic whiz-bangs, I don't know, I don't care.

The Skyline goes around corners faster than anything else. And when it does get a bit skew-whiff, it's a doddle to rein in again. Unfortunately, the price tag has gone right above the skyline: from £25,000 for the old model to a stratospheric £50,000 for this one. But the biggest problem is not the price, it's bloody Nissan GB. As before, they won't import the Skyline officially, saying it would cost a million quid to make it Euro-legal; they add that if a hundred people show real interest, they may take the plunge. A miserable hundred people. For heaven's sake, thousands spend a fortune every year on golfing trousers and thousands more spend every surplus penny in their bank account on model aeroplanes. Surely, there are a paltry hundred people out there who would make the very sensible decision to buy a Skyline instead of a Porsche, or an M3 or even a Ferrari.

I fully understand that the Nissan badge is a turn-off, but the Volvo badge wasn't something you shouted about until the T5 came along. Once a few people have a Skyline and word gets out, you will be seen as a wise and thoughtful person with immense driving skill. Women, almost certainly, will want to spend the night with you. At the same time, your customers will see you as a restrained person with no need for frills. They will double their orders, enabling you to spend even more money with Andy Middlehurst, taking the motor up to perhaps 420bhp. Including the cost of replacement turbos – the ceramic ones can't cope – this will set you back £3200 – beer money in Porsche land.

As far as reliability is concerned, I understand that there are no real problems. The Marquess of Blandford says that his old model with 390bhp never went wrong in 40,000 miles. He points out that there is no other comparable car that can handle the snow in Verbier, a family and the need to maintain a low profile. All that and a top speed of 180mph.

I know I go on about this car, but every time I drive it

I can't wait to get to a computer to write about it. Wordsworth was moved by flowers, I get all foamy about the Nissan.

Henry Ford in stockings and suspenders

It's a glorious summer's evening and what started out as a quick tincture after work now looks set to become a drinking marathon that'll last until your liver explodes.

I used to be able to cope with this quite well. There'd be a hangover the following morning, of course, and maybe a little chat with God on the great white telephone, but by lunch time the next day, all would be well again.

However, today, hangovers arrive like a tropical storm and for days afterwards send regular 4000 volt lightning bolts to every far-flung outpost of my body.

I have therefore learned to spot the moment when a quick drink after work becomes the start of a rock 'n' roll frenzy. When one of the party says, 'Oh dear, I seem to have my drinking trousers on tonight,' I get up from the table and go in search of a burger.

Fast-food joints exist for this purpose – to let a potential drunk line his stomach with something spongy before the next round is delivered. A fast-food burger is therefore not food as such. It is preventative medicine.

I'm a Big Mac man myself, but I can, at a pinch, wolf down a Whopper. I have, however, always tried to steer clear of a Wimpy. The name 'Wimpy' is all wrong – it smacks of nasty little houses with purple up-and-over garage doors. It says Avon Lady. It says you'd be better off eating the carton.

So, of course, I fully understand why middle England chooses a 3 Series BMW or an Audi A4 instead of a Ford Mondeo. The name 'Ford' is all wrong. It smacks of DIY superstores and salesmen in cardigans chatting over the garden fence.

But look. I actually had a Wimpy burger the other day and it was jolly nice – well, as nice as medicine can be – and that made me start thinking . . .

Here's the deal. Give BMW £20,000 and you get meat and bread. The 318i may have a great badge but it's drearily slow and equipped by the prison service.

Give Ford £20,000 and you'll be going home in a top spec Mondeo which comes with electrically adjustable leather sports seats, an electric sun roof, four electric windows, central locking, traction control, a sophisticated stereo and air conditioning.

Under the bonnet of a £20,000 BMW there's a four-cylinder, 1.8 litre engine while the £20,000 Ford has a 24 valve, 2.5 litre V6 with cheese and pickles. So having gone from 0 to 60 in seven seconds, Mondeo man is at home in front of the television, after a lovely dinner, before Bee Em man is into third.

To ram the message home the Ford is available in Super Touring guise, which means the car is bedecked in a party frock. There are skirts, big fat alloy wheels and the sort of wire mesh grille you might find fronting a rabbit hutch. Or a Bentley.

In fact, it couldn't look more menacing even if it had turned up carrying a Thomson sub-machine gun, which is why my first trip was something of a disappointment.

Oh no, I thought. This is going to be like a Robert De Niro and Meryl Streep film. The ingredients are all there but the end result, somehow, is weapons-grade drivel.

Because this was a Super Touring – named after the Touring Car race series – I was expecting a hard ride and twitchy steering, but what I had was a pinstripe suit and table manners to shame the Queen. It was sensible and civilized . . . right up to the moment when I decided to go stark staring mad.

Then it ripped off its Saville Row garb to reveal it was wearing stockings, suspenders and, if I'm not very much mistaken, split-crotch panties. And, boy oh boy, did I have fun with it.

I'd driven the old Mondeo V6 before and was impressed, but this one rode more quietly, leaving me free to enjoy the power and the grip – both of which were delivered by the bucketload. The only fly in the ointment was a tendency to pull to the left, which was cured by adopting the time-honoured fashion of keeping my hands on the wheel.

Without any question or shadow of doubt, this car is vastly superior to any similarly priced offering from Germany, but before rushing out there with your hair on fire you must be made aware of the downsides.

I know it looks very good and I know it's fast and exceptional value for money, but it is a Ford, and the blue oval does not cut much mustard at the golf club. Another reason why I like it.

Furthermore, it isn't just your colleagues and neighbours who'll sneer. Whereas BMW and Audi dealerships are quite happy to provide a heart donor should your own ticker give up the ghost, most Ford salesmen think private customers are only one step up the evolutionary ladder from dog dirt.

There are good Ford garages, of course, but I have a catalogue of letters from people saying that the vast majority are a complete and utter waste of everybody's afternoon.

So, as you wander into the showroom, brandishing a banker's draft for £20,000, don't expect the red carpet treatment. Expect a punch in the mouth and an apple-pie bed that night and you won't be too wide of the mark. But if you can live with that, and further punches in the mouth each time the car goes in for a service, you will have what I consider to be the best mid-range saloon on the market today.

NSX – the invisible supercar

If you want to know whether a car is going to be popular or not, ask Kylie Minogue, who, I feel sure, has more of a clue than me.

In 1992, I described the Ford Escort as a dog and it went on to become Britain's best-selling car. A year later, I reached out into 95 million homes around the planet and said the Toyota Corolla was so dull it should be supplied with a cardigan, and ever since it's been the *world*'s best-selling car.

Undaunted, I went out there again and argued vehemently that the Renault A610 was a masterpiece and that it represented truly unmatched value for money. In its first year in Britain, they sold six.

But the biggest puzzler to date has been the Honda NSX. In 1994, I showered it with literary rose petals saying that Jesus had come among us once more. They sold 19.

Things were a little better in 1995 when 55 found homes in Britain, but in 1996 a new targa-roofed version came along which could be specified with push-button gear changing. The future looked so good for Japan's first supercar that I took the corporate shilling and sang its praises in a showroom video. Sales fell to 38. And they're still falling.

These numbers are seriously small, but the picture becomes even more bleak when you remember that some of these cars must have been registered to Honda themselves as demonstrators. If you could peek inside the computer in Swansea you might come up with something startling – in 1996, not one single person in the whole of Britain actually bought a new Honda NSX.

And I bet Honda simply can't understand what on earth they've done wrong. They gave the world an all-aluminium supercar with one of the most technically advanced engines seen outside a sci-fi movie. They made it reliable and no harder to drive than a pram. They kept the price in BMW land and placed one with Mr Wolf in *Pulp Fiction*. And they were rewarded by people staying away in droves.

Well, to try and put some zest into what was already a vindaloo, they've beefed up the engine, added electric power steering and garnished the finished product with a six-speed gearbox. And now I'm going to ensure it's a spectacular failure by telling you that it's one seriously impressive motor car.

I spent a day with it at the Mallory Park race track in Leicestershire, and can safely say that round the fearsome Gerard's Corner it is a match for even the Ferrari 550.

This is a truly nasty bend: a long, long 180 degree right-hander that tightens up right at the very end. You need to lift off the power a bit but you can't, because at the very same point there's a slight crest which causes the car to go light.

Back off and you'll go backwards into the crash barrier. Keep going and you'll go forwards into the crash barrier. Be in an NSX and you'll make it, sweating a bit and promising you'll go to church next Sunday, but you'll make it and that's all that matters.

The electric steering is a bit of a gimmick but the grip and the 'feel' is awesome. And the grunt is capable of making you best mates with the horizon in ten seconds flat.

You still have a V6 with variable valve timing – whatever the hell that means – but it now displaces 3.2 litres so you get from 0 to 60 in a whisker over five seconds, on your way to a maximum of 170mph.

Not that you'll ever want to get there. What you'll want to do is go through the gears endlessly, because from inside the snuggy cabin that engine makes a noise that could curdle mud. After five laps my soul was so stirred you could have served it up as soup. I never thought it was possible to be in love with a noise, but take an NSX up to 8000rpm and you'll be heading for the registry office.

It would be a good partner too, because unlike a Ferrari, it is a perfectly serviceable everyday car. And it is so damn easy to drive. Even my granny could manage it, excepting the fact that she's dead of course.

My only real worry is the styling. Even Honda would admit in a quiet moment that they copied Ferrari, but that's like asking a nine-year-old boy to copy the *Haywain*. It won't really work, and it especially won't work if he tries to improve on the original.

Honda thought it would be a good idea to give their supercar a boot, so the rear overhang is rather larger than it should be. And they felt it should have headlamp washers, which means the smooth front end is sullied with plastic protuberances, like Claudia Schiffer with blackheads.

Now I've always subscribed to the theory that you should judge a book by its cover. I will, for instance, never buy any novel unless it has a fighter plane or a submarine on the front, but I do urge you to ignore the Honda's skin and study its meat.

It is not a match for the Ferrari 355, but then it's £20,000 less expensive. And if you scour the secondhand columns of this paper you'll probably be able to find one for £40,000, which, for a machine like this, is car-boot sale money.

I bet you're going to have a look right now, aren't you? And you'll keep looking right up to the moment when you buy a Porsche.

Corvette lacks the Right Stuff

So, underachiever, how do you feel today? Let me guess: you got up, went to work, flirted with the secretaries, came home and watched telly. Now, *Newsnight* is on and you're reading this, yawning and wondering why you've got nipples. It's OK, I do pretty much the same sort of thing most days and that's why I know Hoot Gibson will gall you as much as he galled me.

Here is an all-American dude with Paul Newman eyes who learned his art in Vietnam, flying F-4 Phantoms and shooting down MiGs which may, or may not, have been piloted by top-flight Russians. He was so adept at blowing things out of the sky, they sent him to the Top Gun Academy, where he became a better instructor than Kelly McGillis. And after that he found himself stationed at Pax River, flying all the new, experimental fast jets. When his navy flying career was over, instead of a desk, the services gave him a space shuttle – something he's used to visit space on no fewer than five occasions.

So what then, does Mr All-American Hero choose to drive when he's back on Texan earth, and restricted to 55mph? A Viper? A Jag? A Bimmer? Er, no. Mr Gibson has a Toyota Camry, finished in aubergine with a matching interior. I pointed out that this was a terrible car, and he agreed but said it was, at least, reliable – 'something that's important to me'. OK, I can understand that, but in *The Right Stuff* – the best book in the world, incidentally – Tom Wolfe says all the early test pilots and astronauts hurtled into town in Corvettes – the first American sports car. Why, I suggested, do you not have one of those? 'Because,' he said, 'it is a piece of junk.'

Whoa there, boy. Mr Pumping Pecs calling his auto equivalent 'junk'? This needed exploring and so, two days later, in Nevada, I hired myself an egg-yellow convertible with a slushmatic box. I slithered elegantly into the vibrantly shiny cockpit, the 5.7 litre V8 burbled into life and the sleek nose edged its way onto Las Vegas Boulevard. I felt good. The Corvette is dangerously handsome and my views on US V8s are well documented. The steering was quick, the stereo was sound and I began to suspect Hoot should stick to sounding off about planes. But then I ran over a piece of chewing gum. Jesus H. Christ, did you know the 'Vette has no suspension travel at all? The wheels are connected directly to your buttocks. I suspected that there was something wrong with it, and then, that night, it broke down altogether. But the red replacement was just as bad.

OK, I'll let you in on a secret. The Corvette is a slow motor car which does not handle at all. Because there's no suspension to absorb the roll the car just slides, which must be why it has traction control. But this comes in so viciously and so early that I decided to turn it off and . . . whoops eek and wahay, guys and gals, we're going backwards. It was fun right up to the moment when I saw the guardrail approaching. Here's another secret. Anti-lock brakes don't work when you're going sideways. But it was OK – I ground to a halt with a good 5 inches to spare. I was doing that post-trauma bit where you breathe out and lower your shoulders by five yards when an officer of the law arrived. The guy knew his cars and, pretty quickly, conversation turned to the Corvette that had nearly killed me. 'You know the big problem with the 'Vette?' he said. 'It's the worst goddamn car in the whole world.' He hadn't actually seen my spin but said he wouldn't even think of writing out a ticket for speeding because he knew just how easy it is to lose control of Detroit's biggest balls-up. 'Goddamn 'Vette spins so easy, you can park one outside a store and when you come out, it'll be facing the other way,' he added. As he climbed back into his cruiser, he gave me some advice. 'Tonight, leave the roof down and the keys in. With luck, someone'll steal it.'

I've always liked the Corvette, and once toyed with the idea of buying one. But I'm better now. It's simple, really. The Americans are good at space shuttles. And we're good at cars.

Footballers check in to Room 101

Making a living from writing about cars may seem like the Holy Grail to anyone who's intrigued by the niceties of internal combustion.

But there are downsides, chief among which is a constant need to reassure people that I won't waste their entire evening by talking about the track rod ends on a Triumph TR5. When I walk into a room, non-car people dive behind the sofa or, if I catch them by surprise, pretend to be deaf and mad. I'm always prejudged to the point where women will jump through the French windows, screaming, rather than talk to me.

Last year, I had to sit and watch Nick Hancock take me to the cleaners on *Room 101*, a television programme where guests consign life's little irritations to the flames of eternal hell.

Being assassinated by someone you've never even met is terribly disappointing, but rather than sulk I leapt at the opportunity to take part in a new series.

As is the way with programmes like this, you don't get

the chance to meet the host beforehand, which meant I only had the hour-long recording session to convince a man I've always liked and admired that my head is not entirely full of acceleration figures and comparative rear-seat legroom dimensions.

It obviously didn't work because, while promoting the show, Hancock said he was 'allergic' to people who like cars, and that my choices had been 'boringly obvious'.

Fine. I tried to be reasonable, but now it's payback time because Hancock, I know, is a huge football fan. And I loathe football fans.

I have just finished Nick Hornby's *Fever Pitch,* and found the whole sorry saga more sad than *Born Free.*

While I hero-worship Ian McCullum and Tommy Lee Jones, whose talents are boundless, these people drool over footballers who, almost without exception, are so stupid I'm amazed they can put their shorts on the right way round.

Last week I found myself sharing a hotel with a team that, thanks to the libel laws, shall have to remain nameless. But honestly, the lobby was like Darwin's waiting room.

They had got it into their tiny, tiny minds that I was Jeremy Beadle and, for two hours, could only say 'Watch out. Beadle's about.' At least, that's what I think they were saying, because speech was an art form they hadn't mastered properly.

I noticed that they didn't have conversations like normal people. One would walk up to a huddle of others and make some kind of farmyard noise. This would prompt the others to cluck or moo and then everyone would disperse.

Now, we're dealing here with a bunch of young men who, because they can kick an inflated sheep's pancreas some considerable distance, get paid anything up to £40,000 a week. And if you give young males that sort of money, they will be tempted to spend large chunks of it on a flash set of wheels.

Apparently, this worries Manchester United boss Alex Ferguson, who tries to veto some of the more extreme vehicular demands from his players. It seems that having paid squillions of pounds for a new player, he doesn't sleep well at night if he thinks the guy is charging around the city centre at 180mph in a Vantage. Driving while under the influence of the Spice Girls is not illegal, but it is dangerous. And so is not being able to read STOP signs.

Obviously no such ban is enforced at Liverpool FC, where a spokesman said the car park is 'incredible'. Apparently, there are several of those 'upmarket Land Rover things' (Range Rovers I presume) and the rest are all 'sporty Porsches' – as opposed, I guess, to the much rarer non-sporty variety.

This would indicate that the successful footballer is something of a petrol-head. And further research has proved this to be true. Alan Shearer has a Jaguar XK8, while Les Ferdinand has a sporty Porsche 911.

David Seaman (a rather unfortunate name) and Ryan Giggs (who's Welsh) have Aston Martin DB7s, and Teddy Sheringham pootles around in a Ferrari 355 Spider. John Barnes has just picked up a Mercedes SL.

David Beckham, a man so bright he's able to date Posh Spice, is to be found behind the wheel of a BMW M3 convertible, and Jason McAteer has a Porsche Boxster.

Now this lot would bring Hancock out in a rash, so are

there any football teams out there whose players are not interested in cars?

It was not easy finding out because I either had to ring up clubs, and speak to people who can't or I had to graze the Internet, a process that takes so long I'd have been better off using a carrier pigeon.

However, I think I've found one – Stoke City. A delightful receptionist there called Lizzie, who speaks coherent English, furnished me with a list of players' cars and it's just horrific. Kevin Keen has a VW Polo while Fofi Nyamah has a Vauxhall Astra. Other cars that litter that Potters' car park include a Ford Escort 1.8 LX, a Citroen, a Mazda 323 and a Mercedes C180 – by far and away the slowest car in the entire world.

Obviously, this disregard for automotive niceties endears the players enormously to their number one fan – a chap called Nick Hancock. And how do I know he supports Stoke? Because he talks about little else.

Big fun at Top Gun

If you're one of our more level-headed readers, you might think that when it comes to no-go areas of office conversation, cars top the list here at *Top Gear* magazine. I mean, for 16 hours a day these guys drive cars, and in the remaining eight, write about them. The last thing they want to do over a beer or in sub-zero fag breaks is to discuss the merits of a Proton over an Escort.

Well I'm going to tell you a little secret. They don't talk about cars very much, but it has nothing to do with overkill. They don't talk about cars because they are too busy talking about bloody motorbikes. The Editor rides bikes. The Assistant Editor rides bikes. The Art Director rides bikes. So does the Art Editor – and she's a girl. I've just been to Barbados with the Road Test Editor, and he sat on the beach every day reading *Bike* magazine. I've given up calling in because if I do, I always forget the rules and mention the 'c' word. I mean, it is a car magazine; maybe the people who work on it would be interested to hear that I've just driven a turbocharged Ferrari F50. So I'll say, 'Hey everyone, I drove a turbocharged F50 yesterday,' and, guess what . . . nothing happens. So I'll tell them again, and if I'm very lucky, one will stick his head up and mumble something about it not being as fast as the Triumph T595. Then they're off. 'Yeah, but the chassis on a 'Blade is better.' 'Oh sure, but I prefer the 43mm Showa usd teles on a 916.' And me, I'm the pork chop in a synagogue. I've given up arguing. Yes, yes, yes, bikes are cheaper than cars, more fun and, providing you never encounter a corner, they're faster too. I've tried pointing out that round a track, where there are bends, a car will set faster lap times, but a deathly hush descends over the office as everyone sets to work with slide rules and calculators. Three minutes later, the Managing Editor will announce that, at Thruxton, his calculations have shown a T595 would, in fact, be faster than an F50.

Well, I can now shut them up for good because I've just flown an F-15E, and no bike on Earth even gets close. Oh, and you'll note I said 'flown' and not 'flown in'. Even though I've never even held the stick in a Cessna, the US Air Force let me take the controls of a plane which cost $50 million and, in 90 minutes, used $7000-worth of fuel.

You might guess that once you're airborne there is no real sensation of speed – but this is simply not the case, a point the pilot was keen to prove. So, at 1000 feet he hit everything to slow the plane down to something like 150mph. And then, after asking me if I was ready, he lit the afterburners. And let me tell you this, Mr Sheene and Mr Fogarty: you know nothing. I wasn't timing it, but would guess that in ten seconds we were nudging 700mph. And then, just to show what an F-15 is all about, he stuck the plane on its tail and did a vertical climb from 1000 to 18,000 feet in exactly 11 seconds. You've all been in lifts which make you feel funny if they're fast, but just think what it feels like to do a 17,000-ft vertical climb in the time it takes a Mondeo to get from 0 to 60.

There was no let-up, either, because having shown me how fast an F-15 accelerates, I was then introduced to its manoeuvrability. Put it like this – in a gentle Sunday afternoon turn it'll dole out 10 g, and I don't know of any bike which can do that. And nor can a bike post a 1000lb bomb through your letterbox. What's more, in a battle between a MiG-29 and a Ducati 916, the Italian motorcycle would lose. Whereas no one has ever shot an F-15 down. Ever. But the best bit was when the pilot said, 'You have the plane.' I did a roll and a loop, flew in tight formation with another F-15, went for a peek at BMW's new factory, flew over Kitty Hawk and got within a fraction of going supersonic. The plane can do Mach Two, but only over water, and my ejection training had not covered survival in such conditions.

I really didn't mind, though. I honestly believe I've now experienced the ultimate; from this point on, everything will be a little bit tame.

As I see it, a bike only has one advantage over a fighter-bomber. On a bike, you don't get sick. In the plane, you do. Twice.

Traction control loses grip on reality

I am a patient man but Vodafone should be advised that it's run out. Either they build more of those relay towers or I'm coming down to their head office with a pickaxe handle and some friends.

My mobile phone has worked 100 miles from Alice Springs in Australia and on a glacier in Iceland. It was fine on an oil tanker off South Africa, and just last week in Italy – Italy for God's sake – I used it for an hour while driving down the autostrada and it never fizzled out once.

But it doesn't work in Fulham, or on the Oxford ring road, or on large chunks of the M40, or near Coventry. Which means Vodafone are charging me for a service that they are simply not providing. And that, I'm afraid, means they're going to need some new office furniture. And some teeth.

It's the same story with fax machines. My first simply tore any paper that came near it into very small pieces. And my new one just does alternate sheets until it gets bored. Then it starts screwing them up and throwing them on the floor so the dog can eat them.

It's all a marketing thing. I have to have a fax machine because the hype says you're a nobody if you don't. Having a fax that doesn't work is fine, but not having one at all is social herpes. And can you imagine going to a meeting and telling someone you don't have a mobile? It would be worse than not having genitals.

And now this phenomenon is creeping into the world of cars as well, in the shape of traction control.

There are a number of different systems, but each, effectively, does the same job. If you apply too much power, sensors detect the moment when the driven wheels are about to lose traction, and issue warnings to the engine management system. It then reduces the power being despatched to the overloaded wheel, and as a result you don't crash. The trouble is that, like mobile phones and faxes, traction control doesn't work.

If I put my foot down on a wet road in the Jag, it senses that something is wrong and does what we all do when we're in a quandary. It goes for a long walk round the garden, where, after much chin-scratching, it decides that, yes, it ought to warn the bridge.

But way before the central computer pushes the throttle pedal back where it belongs, the car is going backwards through a hedge. Electrons are fast, but once the pendulum effect of a tailslide has gotten its teeth into the equation, the result is a sure-fire certainty.

And anyway, the usual cause of a tailslide has nothing to do with excess power. It's when the driver realizes he's turned into a corner too fast and backs off. This causes the weight of the car to pitch forwards, lightening the rear end and causing a spin. No power is involved and, as a result, the traction overlord is about as useful as a picnic basket.

It can only sit there feeling dizzy as the car spins round and round. Unless, of course, the driver is a talented and brave young soul who knows how to react when the rear end makes a break for the border.

He knows he's going to need power to sort the problem out, but the traction control will have none of it. Any attempt to press the throttle down will be met with a metaphorical slap in the face.

This means that good drivers tend to hurtle around with the traction control turned off. And that's the biggest problem of them all, because *everyone* is a good driver. Everyone thinks they can beat the system, so everyone turns it off. Driving around with your traction control on is the same as walking down the High Street telling passers-by that you're impotent. It is deeply, deeply uncool.

And that's staggering. We're all gladly paying for something that doesn't work, and then we're turning it off. Why?

Simple. Any car maker knows that traction control sounds good. It implies that the car to which it's fitted is such an untamed monster that ordinary drivers couldn't possibly be trusted with all the power.

Wow. The makers themselves admit that the car is too fast. I must have one, and then I shall turn off the device meant for *ordinary* drivers. Men, remember, are egos covered in skin, and the car makers know this.

But unfortunately, the boffins in the back rooms with the beards and the taped-up spectacles do not. Such has been the demand for traction control in recent months that they've started to improve its reaction time, believing this is what we want.

Every time you put your foot down in the new Jag, or

the Ferrari 550 for that matter, the electrons go bonkers and it feels like you're low on petrol. The engine stutters. The ABS system cuts in and even sane people begin to wonder why on earth the damn computer won't unleash the full potential of the car.

So they turn it off too, and then ring up the dealer to express their concern. But unfortunately, they do so on a mobile, and the dealer is left wondering why his phone keeps ringing but there's no one on the other end.

This article was first published on August 10th 1997 and refers to levels of service at this time.

Driving at the limit

If you'd followed me around this week, you might have suspected that from time to time I was driving while under the influence of a blindfold.

But it's OK. I was in a Range Rover, and the damn thing just wouldn't go in a straight line, unless, of course, I wanted to go round a corner. By normal saloon car standards, it really is absolutely hopeless and so pedestrian that I kept being overtaken by continental drift.

Throughout August, Chipping Norton has been host-ing a championship to find the World's Slowest Driver, which is no big deal when I'm in the Jag – I just press the noisy pedal and surge past – but in the Range Rover I came home with the trophy.

In London, things were even worse. In the cotton-thin residential streets of Fulham, where, for some extraordinary reason, everyone has an off-roader, it felt as wieldy as Pooh after a honey binge.

And you can't park it anywhere either. I tried to go out for dinner at the Mao Tai on the New Kings Road, but no space within a mile was even nearly big enough so I ended up in the Blue Elephant on Fulham Broadway which, as usual, fielded the rudest waiters I have ever met.

I should have driven the Range Rover through their indoor flowerbeds, instead of a tip, but you can't really take it off-road in case it gets all dirty.

Strangely, I still love this enormous great brute of a car, and that's mainly because of the driving position – you really do feel like you're bouncing along in an automotive penthouse flat, looking down on the riff-raff.

You should be warned, though, that they are not looking up at you. They hate you on a cellular level. They would like to feed you, and everyone you've ever met, into a lawnmower. In just one day, two people suggested for absolutely no reason whatsoever, that I worshipped at the altar of Onanism.

They hated me even more than if I'd been drunk, and finally I get to the thrust of this week's rant – drinking and driving. And specifically, this ludicrous idea of reducing the limit from 80 to 50 milligrams of alcohol in 100 millilitres of blood. In English, that works out at a pint.

Now look. It really isn't fair to take away someone's licence and therefore their job just because they had an extra big helping of sherry trifle at lunch time.

I've never met anyone who is pissed at the current limit – only relaxed, and surely that's a good thing. Certainly, I score better times on my Sega Rally Machine after a calming drink than I do after a row, or when I've got hay fever.

Baroness Hayman, who is Labour's minister for road safety, says that the decline in drink-related accident casualties has levelled off – but decreasing the limit to the point where a pipette of ginger beer makes you Myra Hindley will only *increase* the figures.

Think about it. If every driver who crashes is psycho-analysed to see if they've ever had a beer, just about every accident will become 'drink-related'.

And anyway, the figures have only tailed off because so few people drink and drive these days. In 1996, the police breathalysed more people than ever before – 780,000 – and only 13 per cent were over the limit.

This means that 87 per cent of people who were seen driving in an erratic fashion were stone-cold sober. So, if the baroness wants to do something about road safety, this lot would surely be a better target.

Certainly, there is no point fiddling about with the limit because this won't give old people better eyesight, and nor will it mend the ways of the so-called 'hardcore' drink driver. It won't temper youthful exuberance either.

And to be perfectly honest, another round of tear-jerking advertisements to ruin the feeling of good cheer as we run up to Christmas will also be a huge waste of money because, frankly, most of us think the drink drive rules are a damn nuisance.

We don't do it because the punishment is horrific – a year or more on the bus. And on this front, I can see a big problem just around the corner. Buses are getting nicer.

The pro-public-transport people should remember, as they campaign for more trains and comfy, air-conditioned double deckers with Jacuzzis and satellite television, that if buses suddenly become a viable alternative to the car, drink driving will go through the roof.

We need to go the other way. Buses should come with luggage and chickens on the roof. The suspension should be replaced with scaffolding poles, and passengers should be encouraged to cook in the aisles on Primus stoves.

And as for the trains: make them late on purpose. Even if the Fat Controller reckons one is going to reach the station on time, he should order the driver to slow down . . . as jerkily as possible.

And instead of forcing a drunken driver to use public transport for a year, it should be five years for a first offence and life thereafter. If you want to keep them off the road, hit them with a stick the size of a giant redwood.

And use a cattle prod on anyone caught driving badly while sober, unless of course they have the perfect excuse: 'Your Honour. I was in a Range Rover at the time.'

Global Posting systems

People say that the world is a smaller place these days. Well, having just been to South Africa via western Canada, I can only assume that it used to be absolutely bloody enormous.

The first leg of the journey, from Heathrow to Calgary, was undertaken in a Boeing 767 which only has two engines. Thus, if one should develop a fault you have to run around the cabin screaming.

But even when both are working, it's a winged

Volkswagen Polo diesel. Point it at a stiff breeze and all attempts to fly forwards are thwarted. You end up landing in reverse, six hours later, in Helsinki.

Happily, we had the wind so nine hours after setting off I was cruising towards the tumbleweedy town of Red Deer in Alberta, which was playing host to a Jehovah's Witnesses convention.

By pretending to be a blood transfusion specialist, I managed to keep them quiet in the lift on the way to breakfast. And even more amazingly, I managed to win a trophy later in the day for taking part in a combined harvester V. banger race, which put me in good spirits as I boarded an Airbus for the trip home.

Now the Airbus is great. Even though it had four engines, which is about half as many as I like while over the North Atlantic, it was as quiet as a lift full of Jehovah's Witnesses when a 16 stone man is glowering at them.

Certainly, it was much quieter than Heathrow, which, these days is twinned with Brent Cross. I sat next to Sir John Egan at a dinner the other night and thought he was looking a bit pleased with himself.

No wonder: he's worked out that as chairman of British Airports Authority, he can get men to do what a billion women can't – shop. In my six-hour stopover, I went mad.

Burdened with four new pairs of sunglasses, some Pink shirts and a watch I don't need, I set off for South Africa and my appointment with the *Jahre Viking*.

This is the world's biggest supertanker, and could swallow St Paul's Cathedral – four times over. However, as there's little demand for ecclesiastical removals in the southern seas it was, in fact, carrying 137 million gallons of crude – enough to power every Jehovah's Witness in all of Canada to Mars. But not quite enough to bring them back again.

After a day on board, mostly looking for somewhere to smoke, we had a bit of bother with the weather and had to be rescued at four in the morning by a tug which was exactly the same size as an ashtray. This meant that in a raging storm I had to climb down the side of the hull on a rope ladder which had been wrested from the ship's mascot – a hamster.

There was no sleep that night, and none the next either, because South African Airways models the seats in its 747s on those found in rural Vietnamese buses.

So, in nine days, I'd slept in a bed just three times. I'd done 24,000 miles. I'd crashed a combine and had been through the most dangerous seaway in the world on a floating bomb.

But travel does broaden the mind, which is why I can now impart two nuggets. First, Air Canada's business class is very good, and second, you shouldn't buy a Japanese or Korean car.

Here's why. In America, fuel is cheap and people are fat so American cars tend to be large with a voracious appetite for gas. In Europe, the streets are narrow and fuel costs a bomb, so Renault and Fiat give us little cars with pipettes for petrol tanks.

That leaves the cars that come at us, like a blizzard, from the Far East, cars that are sold in Milton Keynes, Montreal and, because I loathe alliteration, Agadez.

Now look, they can't have it all ways. They can't tell a Canadian that it's a full five-seat sedan, an Italian that it's a nifty little pocket rocket, an Australian bushman that it's tough and the American safety lobby that it's soft.

Cars like the Hyundai Accent must be aimed at some-one, and now I know who – African taxi drivers.

In the Third World, people have grown up with an acquired immune deficiency syndrome towards the notion of cars being, in some way, linked to social standing. Alfa Romeo is currently promoting its 146 by saying that 'everyone in the office will think you've been promoted' – a slogan that wouldn't work at all well in Angola.

African taxi drivers are not bothered about a car company's past racing successes, or styling or whether it can generate 4 g while parking. They want total reliability at a nice price, and that's what Japan and Korea are giving them.

Go to any African state and you won't find a single new Fiat or Chrysler. It's just row after row of anonymous saloons.

Now, if this were the business section of the paper, there'd be a temptation to castigate Europe's car makers for failing to exploit the emerging world, but I find business about as exciting as fish.

I'm really only bothered about cars and, in the same way that you wouldn't drive a Chevrolet Caprice because it's unsuitable in pub car parks, you shouldn't drive a box that was designed for people who put plastic gold crowns on the dashboard.

Europe has a car industry which makes cars for European conditions. You should remember that when deciding what to buy in the run-up to 1 August and the R-plate madness.

Fight for your right to party

Later this summer, Ferrari is celebrating its 50th birthday in Rome with a party that will make Elton's half-century look like an old people's whist drive. They say that Rome will be brought to a standstill by 10,000 Ferraris and that even the Pope will be there. The Pope, for Christ's sake. The Pope is going to a car firm's birthday party.

Check out Q magazine's gig guide and I doubt you'll find a single rock 'n' roller on stage that night. Eric Clapton, Chris Rea, Jay Kay and Rod Stewart have each bought a 550, and the word is they'll all be in Rome, talking Armani and quad-cam motors. Me though, I'm not going. I have decided that I shall be at the Coventry British Legion that night, where Jaguar is celebrating – not its 50th – but its 75th anniversary. That's not fair. The ball, in fact, is being held at the Brown Lane factory and 1000 people will be there, including er . . . David Platt . . . possibly. The Queen – our equivalent of the Pope – is sadly unavailable because she's opening a computer park in Telford that day. Or is it a dog food factory in Cwmbran? Honestly, it's pathetic and it isn't Jaguar's fault. In fact, they've done bloody well to scrape up 1000 people who are prepared to get out there and celebrate the birth of what we're told is a bunch of wires, some Zyklon B and a slab or two of metal. It's amazing. Since British Aerospace handed Rover over to the Germans, I've had hundreds of letters from retired majors in Bognor Regis, saying that it's all deplorable, hardly worth fighting the war. . . etc. . . . etc. . . . But people in the UK are told cars are dirty and that we're no good at making anything, and we shrug and accept it. We accept almost anything.

Some years ago, the European Community, as it was called at the time, decided that all beaches must achieve a certain standard of cleanliness, which was not one of their more idiotic ideas. Naturally, the British delegation dispatched beardy types in parkas to our sandier bits, where, to their horror, none met the new requirements. Cue the *Daily Mail* with all sorts of headlines deriding Britain as the dirty man of Europe. But, according to my sources, this isn't an entirely fair picture because the other countries had simply gone home and done ... precisely nothing. No beardy types had been sent out to check; they just said, 'Our beaches are all clean.' So hey, it turns out that the unspoiled wilderness in northern Scotland is filthy while that turd-infested expanse of litter-strewn shingle called Greece is dew fresh.

Continental types treat rules with exactly the right amount of disdain. Because Italy has had so many rulers this millennium and so many governments since the war, they've learned to treat authority as though it's something they've trodden in. What's the point of obeying one new rule when next week Hannibal is coming over the mountains with an elephant and an entirely new set? Over there, you can run around waving your arms in the air, telling anyone who'll listen that Ferrari is a symbol of the unacceptable face of capitalism, and that cars are killing children. No one will give a damn. The same happens in France. When the government tried to impose new taxes on truckers they didn't have a puny strike. No. They blockaded motorways and stood around smoking Gitanes, until sense prevailed. Even the Belgians are out and about throwing rocks as I write because Renault is closing a factory down. But here, apart from a bunch of long-haired ne'er-do-wells with suspicious stains on their trousers, no one ever complains. This is why, in Italy, the whole country will be out on the streets celebrating Ferraris, while in Britain, Jaguar's birthday will be marked by one person in every 56,000. We shouldn't expect more really, because if you went into the street and put up bunting, a council official would tell you to take it down again. And if you held a street party, number 54 would ring the police, who'd ask you to turn it down a bit.

The only consolation is that things are worse in America. I'm told that in Los Angeles nowadays, it is illegal to consume alcohol after 2 a.m. . . . even in your home. I bet General Motors' big birthday party will be a real wow.

Gravy train hits the old buffers

This week, I had the most fantastic night of my entire life, accompanying A.A. Gill to a restaurant he was reviewing.

Even though he'd booked under a false name, hoping they wouldn't realize he was from the *Sunday Times*, the head waiter clocked him immediately and began a bout of Herculean fawning. We could have poured custard down the man's trousers and he'd have laughed the laugh of a man whose daughter's life depended on it.

Now Gill is probably used to meeting people who have a degree in advanced grovelling, but it made a refreshing change for someone who's entrenched in the motor industry.

I dislike being anecdotal in print, but this one bears repeating. Many years ago, the entire public relations staff at Land Rover left very suddenly and were replaced by anyone they could find who knew which end of a telephone to speak into. Not easy in Birmingham.

Anyway, the next day, I rang saying that I was a free-lance journalist and that I needed a Range Rover for a story that I was writing. The new girl – and you need to read this in a big, big Birmingham accent – said that she was very sorry but she had 'specific instructions not to lend any cars to freelancers'.

Puzzled, I asked what would have happened if Stuart Marshall had made the request – Stuart being the motor-ing writer for the *Financial Times*. 'Oh,' she said, 'he could have one.'

'But he's freelance,' I replied. This confused the poor girl, who thought for a moment before the lights came on. 'Whoops, I'm sorry,' she said, 'I meant we won't lend our cars to free*loaders*.'

This, however, isn't true. Freeloaders can have as many cars as they like from whichever manufacturer they choose. And when I say freeloaders, I mean you.

What I'm going to do now is explain how, for no out-lay whatsoever, you can spend the rest of your life driving a brand-new car every week. They will be delivered to your house, clean, fully insured and with a full tank of fuel, and then collected when the ashtray is full.

The only trouble you'll have is finding time to drive them, because twice a week you'll be flown to exotic locations all over the world, and housed in the sort of hotel that would chuck Dodi Fayed out for being poor.

Sounds appealing? Well here's what you do. Call up

the editor of your local freesheet newspaper and ask if you can write a gushing little piece about cars every week. It's OK, you don't need to be a terribly good writer – no one expects car buffs to know how to put a story together.

So now you have an outlet which means you can telephone, say, BMW who, once they know you really do have a column, will be duty-bound to lend you a car. And then it's yours for a whole week and you may go wherever you wish in it.

The next week, another car will be delivered and the week after that, another. Then you'll start getting clever, ensuring you have the right wheels for the right occasion. You've been invited shooting, so you'll have a Ford Explorer. It's your daughter's wedding, so you'll have a Mercedes S Class.

By this stage, you will have been noticed by the industry's public relations people, who will start inviting you on their infamous car launches. Now you really are in the big time.

Every new car – and there are about two a week – is introduced to the press at some far-flung ivory tower, which means your life will become a hectic blur as you ricochet round the globe. Nissan launched its new Primera in South Africa. Mercedes let everyone sample their M Class in Alabama. Jaguar takes people to France.

It's an orgy of champagne from the moment you climb on the plane until you're sent home again clutching a little gift – a computer, perhaps, or a piece of luggage.

Now remember, all you've done to earn your spot on the gravy train is write a couple of hundred words for a local newspaper every week. But quite frankly, this is getting to be a bore. You like the five-star life but you hate the bottom-drawer wages.

You want the free cars and the global travel without having to write the column – no problem.

On the launches, suck up to the public relations people as though they hold your life in their hands. Tell their bosses how good they are at their jobs. Eulogize about the new car, even if it is a Nissan Almera, and ensure you are the life and soul of the party. Over dinner, regale everyone with amusing anecdotes and be prepared to stay up till 4 a.m., drinking the bar dry.

So, when you give up the writing, the public relations people – who need bums on seats to justify their existence – will keep the invites coming. And the cars. They know you. You're a mate. They will still make sure you have a nice big diesel estate when holiday time comes around.

And therein lies the reason why motor industry people don't fawn on journalists. They're in the hot seat, deciding who gets to drive what and who gets to go where. Why should they grovel when they know that without their assistance the motoring journalist is up the creek without a boat, never mind a paddle?

Weird world of Saab Man

By computing the position of various stars on 11 April 1960, an astrologer would be able to deduce that I'm selfish, arrogant and thoughtless.

But this seems like an unnecessarily complicated palaver. I mean why bother with reference books and slide rules and telescopes when you can simply ask what sort of car I drive. See the car. Know the man.

Kind, gentle people do not drive Ferraris in the same way that Sylvester Stallone does not have a Peugeot 306 diesel. Or a Subaru Justy. Or a Skoda Felicia.

Spot a Lada bumbling down the road and there's no point peering inside to see if Richard Branson is behind the wheel. He won't be. It'll be a bloke wearing one of those suits that's neither green nor grey nor brown, but a curious cocktail of all three. It's a colour worn only by old people in Ladas. It's a colour that should be called 'old'.

Lada Man votes Labour, likes pies and checks the price of things before putting them in his supermarket basket. He is usually called Derek and he's 53. And you won't get that kind of detail from his star sign.

I can do this sort of thing with any type of car. Show me someone, in an Audi A4 and I'll show you someone with a mistress. Show me someone in a BMW 316 and I'll show you an idiot, a man who would wear Ralph Lauren shirts that had been made in Hillingdon by someone called Singh.

There is, however, one car that's much harder to pigeonhole. If you drive a Saab, all I know is that you have made one of the oddest buying decisions in the entire history of shopping. You've looked at a feast and chosen instead to eat your own shoes. You've considered a job as chief polisher to Sandra Bullock's nipples but decided that you'd rather do a milk round.

I've just spent the last few days driving around in Saab's new 9-5, which is a large four-door saloon that costs, depending on engine and trim levels, between £21,000 and £28,000. It is therefore a direct rival to the BMW 5 series.

Now, the 5 series is nigh on perfect in every way, but the Saab . . . isn't. Sure, it has powerful headlights and a remarkably comfy ride, but this simply isn't enough in a package that also has average styling, average handling, average performance, a poor gear change and a wonky driving position.

Yes, it is well priced and yes, it is generously equipped but overall this car is beaten mercilessly, not only by the BMW but also the Audi A6, the Mercedes E200, the Volvo 850 and, quite frankly, the Ford Mondeo V6. Saab has served up a good car in a world that expects excellence.

Now ordinarily that would be the end of the story, but people are going to buy it. They're going to notice the jerky gear change, the roly-poly cornering and the way you need to bend your foot back to get it on the throttle. They may even discover it shares a chassis with the Vauxhall Vectra, but they're still going to reach for the cheque book.

Why? And why for that matter do people buy the Saab 900, which is also beaten by the competition, and the Saab 9000, which isn't just beaten; it's bent over the sofa and subjected to cruel and unusual torture by every other car in its class . . . except the Nissan QX perhaps?

I mean, it isn't as though the Saab badge stands for anything particularly dramatic. This jet fighter thing seems a bit weak somehow, and anyway it wasn't that long ago when Saab were selling their cars on the safety ticket. And before that, they were doing rallies. The result of all this haphazard marketing is that, today, the cars are almost completely image-free.

And that, I suspect, is where their appeal lies. They are sold to people who don't wish to use their car as a style statement, people who simply need four wheels and a comfortable seat so that they may get to work as easily as possible.

I think, therefore, we're probably talking about fastidious, meticulous people for whom slaphappiness is the eighth deadly sin. It's the sort of car that would suit an architect, or an astrologer.

We're getting somewhere here, because if this is true it explains something else – no one has ever been carved up by a Saab. Think about it: has a Saab ever jumped a red light or tailgated you on the motorway? Have you ever seen a Saab being driven in anything other than a considerate and stealthy fashion? No, and neither have I.

This is because the sort of people who are drawn to this image-free environment are the sort of people who don't use their subconscious to drive. They know that to do it properly they have to concentrate, absolutely, on the job in hand. So they do. And that's why they never carve us up.

Eureka. We can learn something about Saab Man after all. He is, without doubt, the safest driver on the road today. Insurance companies pay him to have a car. He is never harassed by the police. He has no points on his licence. Without any doubt at all, he is a Virgo.

And I do mean 'he' because I can't recall a single time when I've seen a woman at the wheel of a Saab. Weird.

Freemasons need coning off

I've just driven from Milan to Avignon via Pisa, Bologna and Monte Carlo and in not one of the 1500 miles did I see a single motorway lane closure. There were no roadworks at all. There were no cones. It was a high-speed highway to heaven. Even though the Lancia Dedra Estate I'd rented was terminally backward, and any assault on the car's upper rev limit caused my ears to explode, I could do 90 for hour after hour after hour. On one downhill stretch I hit 100, but the doors fell off. Then I came back to England, where on the simple 85 mile journey from Gatwick to Oxford there were three major sets of roadworks.

Now the M25 I can understand. They screwed up and built it too narrow. Fine. God made a mess, remember, when he did the flamingo, which is an idiotic bird with legs that are far too long. Then he did the totally purposeless nettle. We all make mistakes.

So ever since the M25 opened they've been widening it. Then there's the roadworks on the M40, which, again, are understandable. The road has worn out and needs replacing.

Mind you, I don't understand how they intend to do this by coning off the offending few miles and employing guys in hard hats to stare at it. I've driven down the single lane they've left a lot recently, and I have yet to see a single person doing anything. Still, they're experienced roadworks johnnies so we can rest assured they know what they're doing. But I do not understand what is going on where the M40 meets the M25. The signs say it's being widened, which is nonsense. It was already wide enough, by miles. The M6 needs widening. The M5 needs widening. The M1 needs widening. But they've decided that none of these real problems will be addressed until they've had some practice on the under-used M40.

And boy are they going to take their time – two years, to be precise. Now look, a road is some stones covered with sticky stuff that sets. In two years, I could build a road from here to Sofia. In two years, they could close the M40, plant crops, allow them to grow, harvest them and then build a new motorway. And there'd still be time to stand around in hard hats, pointing at things. When an earthquake devastated Los Angeles, I don't recall signs saying the freeways would be open again in two years. No, I saw teams of worker bees shovelling ruined bridges away and building new ones so the entire network was up and running again in less than 12 months. In Japan once I saw them replace an entire Tokyo highway before sun-up.

Now at this point, some people will be reaching for the notepaper, eager to point out that we don't pay tolls to use our motorways and that we can't expect better service as a result. Well, that's crap. Britain's motorists pay £25 billion a year through vehicle excise duty and petrol tax and car tax and VAT on tax, etc. etc. etc. That's a lot of money. In fact, we're paying so much, the government simply doesn't know what to do with it all. This can surely be the only reason why they're spending two years widening an already wide road. Either that or it's the bloody freemasons again. In the past, I've blamed free-masonry for the destruction of our car industry, arguing that a component buyer from British Leyland wouldn't fire a company for sending dodgy parts if its managing director had a weird handshake. Week in and week out, lorry loads of crappy speedos, or whatever, were delivered and no one did a damn thing about it because of some barbaric ceremony every Tuesday night where a bunch of grown men run around throwing salt at one another. Well now they're at it in the construction industry, taking ten times too long to do a job that didn't need doing anyway, in exchange for a new apron and an oddball boater.

Britain stands no chance of becoming a driving force in Europe unless we build roads properly and get urgent repairs done quickly. I suspect things will be better under 'call me Tony'; he is a village idiot and his backbenchers are teachers with beards, but they haven't yet been exposed to white-collar Britain.

So when Mr Motorway Builder walks in and shakes hands while doing a handstand, they will ask him to leave or they will call security.

The curse of the Swedish smogasbord

Oh deary me. It seems that every five days air pollution exceeds harmful levels somewhere in Britain, and that as a result we're all going to be dead by tea time.

That's if we don't choke to death first. According to the National Asthma Campaign, Britain's 3.5 million sufferers are fed up. 'People should not have to make the choice between their health and being able to go outdoors and live a normal life,' said a spokesman.

Absolutely. I want to see a ban on the production of bread too. I am sick and tired of being struck down by asthma every time I wander through a cornfield – and I hold the Hovis board entirely responsible.

Grass is nasty too. Ask me to stroll down Jermyn Street on a hot day and I'll suffer no ill effects whatsoever, but let me loose in our paddocks on a summer's afternoon, and after a minute or so I'm a dribbling vegetable.

I was therefore impressed by Indonesia's attempt to help asthma sufferers by burning the countryside, though I see it's all got a bit out of hand now and that entire villages are being wiped out. You can see the smog from space.

You can also see an equally large and gaseous cloud over Britain, but this time it's coming from the British Medical Association. It says that traffic levels, diesel emissions and vehicle noise should be reduced, and that to help, we must all hop to work. Or use a bicycle.

Now this is odd. I'm used to vegetarians running around pointing their organic fingers at the car, but now a bunch of doctors has also decided that motoring is bad for your health.

Well now I'm sorry, but I suspect that this is nothing more than sour grapes. I mean, really, can it be a coincidence that the BMA report came out in the same week that Volvo announced it was to terminate production of its horrible 900 series?

This is a bad car with a power delivery that beggared belief. In most cars the throttle is connected to the fuel injection system by a cable or, increasingly, by an electronic fly-by-wire pulse.

But this obviously wasn't the case in Volvo's old barge.

Put your foot down in a 900 and it simply telegraphed a message to the engine room, where a fat man in an oily vest reluctantly put down his copy of *Razzle* and, after a bout of anal scratching, chucked a few more lumps of coal on the boiler. Then there was the handling. Or rather, there wasn't.

It was safe though. I've often wondered why Middle Eastern suicide bombers bother to load their cars with difficult and complicated explosives when they could achieve exactly the same level of destruction by driving an old Volvo into the building. The added bonus, of course, is that in the Volvo they'd survive.

The 900 series wasn't so much a car as a statement. By driving around in this wheeled house-brick you were telling people that you had no interest in motoring – though of course we all knew that simply by looking at the way you drove.

When we saw a Volvo 900 coming the other way or lumbering up a side road we took nothing for granted.

Just because it was in the left-hand lane with the left-hand indicator flashing didn't necessarily mean it was actually going to turn left. It may have gone right, or straight on, or stopped very suddenly for no obvious reason.

All the country's bad drivers were in Volvos, and we could therefore prepare ourselves. We would give them a wide berth because we *knew* the driver's reaction times could be measured with a calendar.

Now, because the old warhorse is gone, some may worry that these bad drivers will no longer be so easy to identify. Some may disguise themselves and buy sturdy Mercs, while others could go for a Rover 800. There may be a few who stick with Volvo . . . buying one of the new superfast C70s, for instance.

But I urge you not to be too concerned. This won't happen. The people who bought the evil-handling, sloth-like 900 – and a lot of them were doctors – will not be scattered to the four winds. They will deduce that there is no alternative and simply replace their car with a bus pass.

And then they'll insist that we follow suit. That's why the BMA wants us to hop to work – it's because Volvo has killed their beloved car.

The solution as I see it is simple. The last of the 900s are being fitted with 2.3 litre light-pressure turbo engines which should make them sing a bit. And remember, they are rear-wheel drive, which is what enthusiasts want.

Inside, you'll get heated, leather seats, air conditioning, a sophisticated stereo and electric windows all round. There's a three-year warranty too, and an airbag, all for £18,500.

My advice is to buy one. Other road users will think you're an idiot and flee from your path, thus ensuring you're never held up in a jam again. You can drive like a fool and people will expect it, and if you do crash, it won't hurt. What more could you possibly want?

Well there is one thing I suppose. What I'd really like, even more than a Volvo 900, is for Britain's doctors to stick to mending people and stop trying to shape the way we live. And anyway, every doctor I've ever met *does* smoke.

Pin-prick for the Welsh windbag

Bad news, I'm afraid. Kinnock's back. After we decided it was a bad idea to let someone who's Welsh represent our interests on the world stage, he disappeared into the Euro-abyss, where, it turns out, the Man of Harlech has been biding his time, waiting to wreak his revenge on the people who snubbed both him and his nuclear-free wife.

In his role as European Union Transport Commissioner, he has decided to turn Christmas into an orgy of orange juice and church by harmonizing drink driving laws across the Continent. This will mean bringing the British level down from a couple of pints to one wine gum.

This idea was mooted a couple of months ago, but as is the way with New Labour, all 'proposals' become 'discussion documents' if there's the slightest hint of an outcry. And there was – so much so that I thought the monumentally stupid plan had gone away. But now, thanks to Captain Kinnock, it's almost certain to become reality.

Taffy told a meeting of European transport ministers that there is a fivefold increase in the risk of an accident when a driver's blood alcohol is 80 mg compared with his proposed limit of 50 mg.

Sadly, he was unable to verify that with actual crash statistics. He just says we're five times more likely to run into a bus queue if we've had two wine gums, rather than one, and we're supposed to take his word for it.

Well hold on a minute. What if we lower the limit to nothing at all? This would surely remove the risk of an accident altogether. So let's go the whole hog. And let's all slow down to 4mph. And let's ban cars from towns and villages. And while we're at it, let's really nail the companies that actually make the damn cars in the first place.

I was horrified to see that Chrysler, the smallest of the US car makers, has been ordered to pay £164 million in damages after a South Carolina jury decided the company had sold its people-carriers with rear-door locks that it apparently knew to be faulty.

It seems that a 12-year-old boy was killed when the back door on his father's Dodge allegedly flew open. Never mind that the van had jumped a red light and was hit by another car, and never mind also that Chrysler had offered to change all its door locks free of charge.

With this settlement made, and it's more than twice the size of any previously doled out by a car maker, the floodgates are set to open with 37 other cases being lined up to punch Chrysler on the nose. Experts are saying it could eventually cost the company £5 billion, which would pretty well finish them off.

Now you have to remember that this is America, where there are two types of people: dim ones and lawyers. If a lawyer can ham it up in court, the dim people in the jury will think they're watching *Oprah* and vote to finish the big, nasty, child-killing corporation.

Apparently, the jury in Chrysler's case were peeved that a car maker had seemingly put economy in front of safety, but look: having seen the Mercedes in which Princess Diana was killed, and noted that the front seat passenger survived, I am more convinced than ever that the S Class is about as safe as cars get.

But Mercedes could do more. They could limit the top speed to 10mph and fit a device that would prevent the engine from starting if the driver had eaten some sherry trifle. They could fit airbags in the ashtray. All the technology exists to do this, but it is so expensive that no one would buy the end product.

Even Mercedes could therefore be accused of putting economy before safety, but come on, if money wasn't important we'd spend all day under the bed, refusing to work in case a tree fell on our heads.

America seems to have forgotten that while life is precious, it isn't much fun without at least some risk.

So what's to be done? Well, there's talk that Clinton is going to limit the awards made by a jury, but this move would be fraught with danger. You must remember that in the bad old days Ford was alleged to have sold the Pinto knowing full well that in a rear-end collision it could catch fire. It was claimed they did nothing because the cost of changing the design outweighed the odd death.

If this had been proved, and it wasn't, obviously it's only right and proper that a jury should have been free to beat Ford about the head and neck with a chain saw. If a company wilfully exposes its customers to an early grave, hit them with a fine that would wipe them from the face of the earth. And to hell with the thousands of workers who'll be thrown out of a job through no fault of their own.

And who cares about the towns and cities that depend on the auto maker for life itself? Close Ford down and in Britain alone you close Coventry, Essex, Newport Pagnell and big bits of Liverpool.

Frankly, big awards aren't the answer. We must find the individuals – accountants usually – who decided to carry on making a car they knew to be dangerous and sentence them to life, in a cell, with Neil Kinnock.

Showdown at the G6 summit

You know how Greenpeace is prone to charging around the sea in small boats, trying to stop perfectly harmless oil rigs from being sunk. Well once – just once – they came up with a cunning plan. They argued that the earth is 46 million years old, a number that's hard to handle. So they asked us to think of it as being 46 years old – middle-aged in other words.

A leaflet explained that almost nothing is known about the first 42 years and that dinosaurs didn't appear until just last year. Mammals came along eight months ago and it wasn't until the middle of last week that apes began to walk on their hind legs. This was an amazing read, but it was all complete mumbo jumbo because their claim that the earth is 46 million years old is simply not true. It's actually 4600 million years old, which makes their idea even more mind-boggling. The last Ice Age didn't happen at the weekend. It happened half an hour ago!

However, I don't want to get into an environmental debate here. What I want to talk about, in fact, is the puniness of Nelson Mandela. If you divide time by a thousand million, to make the planet 46 years old, it means that 70 years passes in four-hundredths of a second. So, as far as the Earth is concerned, Nelson is simply not relevant at all. And nor was Hitler. And nor was Jimi Hendrix. Truth is, in four-hundredths of a second absolutely nothing that you do or say will make the slightest bit of difference. For 4600 million years you weren't born, and you'll be dead for even longer so it is therefore vital that you explode out of the womb like your hair is on fire. In real time, you've only got 600,000 hours and then you'll wind up on the wrong side of the flowerbed.

So what's the best course of action? Well you could watch *Pride and Prejudice* which manages to make an hour seem like a day, but prolonging a boring life is worse than not starting it in the first place. That's why you must also not drive one of the new Toyota Corollas. Certainly, it is not exciting to behold. Yes, it has a bobby-dazzler of a radiator grille and the sort of eyes that only exist deep in the ocean where light is at a premium. But from this point backwards, there is a styling vacuum whether you're talking about the saloon, the estate, the liftback or the hatch. However, this time round there is a sporty figurehead – the G6. (I always thought it was G7, but perhaps Japan got lobbed out for making dull cars.) Anyway, this has some definite sporting overtones, in the shape of alloy wheels, red instrumentation and a leather steering wheel. There is a nifty little six-speed 'box too, which beeps when you put it into reverse.

Excited? Thinking of getting one? Well whoa there, because it is powered by a 1300cc engine, the smallest of all the new Corolla's power plants. This means old people in their not-at-all-sporty 1.6 litre liftback will be able to blow you away at the lights. Toyota argue that by putting a small engine in the G6 they've kept insurance costs down. But that's like choosing a mild curry in case your arse hurts in the morning. Life's too short to be bothered about insurance premiums. Or a fiery ring-piece. The G6 Corolla amazed me, time and again. No matter what I threw in its direction, it behaved like the school swat and refused to join in the fun. The engine is actually quite sweet and the gear change utterly delightful, but to take it through the gears is about as rewarding as eating flour.

One night, I sneaked it into a stubble field, knowing that any form of motorized transport is a laugh when there's 100 acres and a surface slippery enough to be an East End geezer. I did some handbrake turns and generally looned about and came home suffering from acute stupefaction. Honestly, I'd have been better off reading a book with an orange spine.

The G6 is, far and away, the most idiotic way of blowing £14,000. This is a car for people who see life as a chore to be undertaken, rather than as an experience to be milked. It is for a cardigan-wearing, non-smoking gardening fanatic who thinks 'E' is a vowel. It is for people who think that living to be 75, rather than 70, really matters. It is therefore not for you, and it sure as hell is not for me.

Spelling out the danger from Brussels

Last week I had to make the annual trudge to Germany, where I spent two days living on a diet of beer that tastes like chlorine and sausages that get up and walk home if you push them to the side of the plate.

The biggest trouble with Germany though, is that you feel duty-bound when on a derestricted and quiet piece of autobahn to travel as fast as the car will go.

This was a huge worry last week because I was in a 7.3 litre Brabus-tuned V12 Mercedes that had wormed its way into *The Guinness Book of Records* by doing 206mph and thus becoming the fastest saloon in the world. Incidentally, 206mph is classified by scientists as f★★★★★★ fast.

Now call me a wetty if you like, but I chickened out when the clock wound its way round to 300kph, which works out, in English, at 186.

At this speed you see a truck and wham, you're in its cab, bleeding. You're covering ground at the rate of 272 feet a second, so that if you sneeze you can miss an entire country.

Everyone who reckons the 70mph speed limit in this country is silly and old-fashioned should be made to do 186 because I feel sure most would sing a different song afterwards. 'Radar Love' would be replaced with some happy-clappy gospel. 186mph puts you on the next table to God. 186mph is seriously scary.

But in Germany it is also legal. Now that's interesting in these days of Euro unity, because at the exact moment I was chanting Hail Marys in my supersonic Brabus Benz, a friend of mine was rubbing his rosary in a Norfolk courtroom.

He'd been caught doing 107mph in a county where people still point at aeroplanes. Astonished magistrates who had only read of such speeds in Isaac Asimov books took away his licence for three weeks and fined him £600 plus costs.

They're right, of course. We can't have people doing 107mph on dual carriageways, and the punishment needs to be severe. The whole of Western Europe is clear on that, but what would happen, I wonder, if a pan-European speed limit were to be mooted by the European Union? For once, I suspect, Mr Kohl's Helmut really would turn purple.

The Germans like the idea of ultra-high-speed travel. It means they can get home faster and therefore have more time to eat sausages. They don't want to be told by a bunch of meddlers that they must slow down, and that's fair enough too.

There are age-old customs in each European country and we can't bulldoze them away in a pointless quest for uniformity. That's why I'm so pathological about this drink driving business. As regular readers of this column know, Kinnock wants our limit brought down from 80 mg in a vat of blood to just 50, so that we stand alongside the French.

Thus, if you are caught driving home after drinking a pint, you will lose your licence for a year and be fined until you're urinating lemon juice. You will then lose your job and your wife will run off with a fitness instructor who has a Porsche.

But in France things are somewhat different. If you're over the 50 mg limit, you get three points on your licence and an on-the-spot fine of 900FF. If you break the 80 mg barrier – the current British limit – you get six points and a slightly bigger fine. You need to be hog-whimperingly drunk before they'll take your licence and, even then, you can get it back if you go on a two-day road safety course.

So, we may end up with the same limit as France but

the punishments could not be further apart, and this is just one more example of Britain being kept in the dark and kicked around by the Continental bullies.

The only shred of dignity Britain will have left after Europe becomes an amorphous blob is the English language, which most experts agree should become the official Euro-tongue.

However, a secret document allegedly found in a BMW communiqué to Rover suggests that even this might be tweaked a bit.

It says that English spelling does leave room for improvement and that a five-year plan has been drawn up to develop EuroEnglish. In the first year, 's' will be used instead of the soft 'c' and 'k' will replace the hard 'c'.

Not only will this klear up konfusion and make the life of sivil servants easier, but also komputer keyboards will need one less key.

There will be growing publik enthusiasm in the sekond year, when the troublesome 'ph' will be replased with an 'f'. This will make words like 'fotograf' 20 per sent shorter.

In the third year, publik akseptanse of the new spelling kan be expekted to get to a stage where more komp-likated alterations are possible. So double letters will be removed to inkrease the likelihod of akurate speling. And the horrible mess of the silent 'e' wil be banished.

By the fourth yar, peopl will be reseptiv to steps like replasing 'th' by 'z' and 'w' by 'v'.

During ze fifz yar, ze unecesary 'o' kan be dropd from vords kontaining 'ou', and similar modifikations vud of kors be aplid to ozer kombinations of leters.

After zis fifz yar, we wil hav a sensibl riten styl. Zer vil be no mor trubls or difikultis and evrivun vil find it ezi tu understand ech ozer.

Ze drem vil finali kum tru.

Dog's dinner from Korea

All week, I've been watching newsreel footage from South Korea of International Monetary Fund bankers trying to sort out what economists call a big financial mess.

It seems that most of the banks are technically insolvent, having been forced by the government to finance massive growth in the industrial sector – growth that just didn't translate into sales.

Now of course, it would be easy for fat Westerners to sit back over a glass of port and laugh, saying they grew too fast and now they've fallen over. *Filthy little yellow nouveaus. Got what was coming.* But when the people of a country are having to fill a van with money every time they want a pound of rice, that country is weak. And sitting right on South Korea's border is North Korea, a country that spends all *its* money on plutonium and mad German scientists. If the West does nothing, the Far East could become mushroom city.

And then you've got that oriental dignity to deal with. Analysts seem to be saying South Korea really needs a loan of $40 billion yet they've only asked for £2.50.

So, all things considered, it can't have been much fun this week for the IMF Shylocks. All that political and economic turmoil to worry about, and nothing to look forward to at night except another plate of roast dog.

However, every time I saw them arriving at yet another meeting in a blizzard of flashbulbs they seemed to have bemused grins on their faces, like there was something warm and comfortable in their trousers.

It took me a while to figure it out, but now I understand. They were being chauffeured around in Korea's answer to America's Cadillac. It's called the Kia Enterprise.

It's priced at the equivalent of £40,000, which seems like rather a lot for a car that's the size of a Scorpio. Certainly, you aren't paying for much in the way of styling.

What they appear to have done is taken an old Toyota Corolla and blown it up with a bicycle pump. They should have blown it up with Semtex, but never mind.

To ensure, however, that no one is in any doubt that this is a serious player, it comes with a gaudy bonnet mascot fashioned to look like a golden dog turd. Clever stuff this – you eat the animal and use its excrement to enliven the look of your car.

From the back, you've got a sign in the rear window which says 'intelligent control' and, of course, the word 'Enterprise' picked out in gold on the boot lid.

Do not, however, expect much in the way of warp speed. The engine compartment may house a 3.6-litre V6 which is said to be capable of propelling the car to 144mph, but acceleration is not so much *Star Trek* as Star Stroll.

I blame the gearbox, which inevitably, is an automatic. Now, the whole point of an auto is that you just get in and steer; you don't have to worry about gears, but in the Enterprise you can think of little else. The lever is festooned with buttons that make it do all sorts of things you don't need.

And that really sets the tone for the whole car. Switch on the engine, which is commendably quiet, by the way, and the dashboard doesn't simply come to life. It explodes into a technicolour blaze that can detach retinas at 400 paces.

There's a digital read-out for every single feature of the car, and this car comes with the lot. Adaptive damping, traction control, mirrors that fold away, a fridge on the rear parcel shelf, parking proximity sensors. I mean it; the lot.

There is a television too, but instead of simply shutting down when you set off, a message flashes on the screen, saying, 'Attention on Driving'. Well, it's hard to comply when you're driving into what looks like a forest of lasers.

It's in the back though, that things really go bonkers, and this I guess is where the IMF boys have been seated.

First of all, there's almost no legroom whatsoever, but you can remove the backrest from the passenger seat and use the squab as a leather footrest. Nice.

You can also move your seat around electrically, change television channels and adjust the temperature from a wood-look console in the centre armrest. But I've saved the best bit till last. The reason why all those IMF chaps are wearing bemused grins is because the back seat vibrates.

All over the world, car manufacturers spend an absolute fortune making their cars quiet and relaxing. Kia too must have blown millions dealing with what's called NVH – noise, vibration and harshness. Yet, having eradicated it, they allowed their engineers to put it back.

No wonder they nearly went to the wall last summer. If people want a car that vibrates they'll spend £100 on a secondhand Morris Marina, not £40,000 on a style-free wasteland with dog dirt on its bonnet.

At present, Kia's British importers have no plans to import the Enterprise, preferring to stick with whatever it is they are already bringing over. There's a very cheap hatchback with a warranty, a four-wheel drive thing and a saloon of such enormous tedium I can't remember its name or what it looks like.

Let's ensure we keep it that way. Send the people of South Korea food parcels and emails wishing them well. Send money in brown envelopes, but make them promise that the Enterprise boldly stays at home.

New Labour, new Jezza

Well it's been a lovely, long hot summer and frankly, right now is a good time to be British. The economy is booming. House prices are back where they belong and unemployment is at its lowest levels since 1981. By pulling all the right faces and not actually doing anything, tony@numberten.co.uk seems to be popular, and even when his fat sidekick, John Prescott, made some silly noises about two-car families they were drowned out by reports that half a million people had bought a new set of wheels in August.

The trouble is, of course, that columns like this thrive on bad news. I need to stand on a rake or fall in a vat of sheep excrement for there to be something to write about each month. Good news, frankly, is dull. I haven't even had the privilege of driving any spectacularly awful cars in recent weeks. There was the Toyota Corolla, of course, which is motorized mud, but it's not 'bad' by any means. And the same goes for Saab's 9-5, on which you light the blue touch-paper and then hang around – nothing at all exciting will happen. In a world of ceremonial fireworks, this new Swede is a damp sparkler. And anyway, this dreary twosome are more than outweighed by some of the most exciting stuff we've seen in years. There's the Puma, of course, and the new 911. But what can I say about that? It's very reliable? Whoa Jezza – incisive stuff.

In the spring we were treated to an onslaught of new convertibles like the SLK and the Boxster, and now they're tickling our erogenous zones again with a welter of coupés. Alfa has announced that it will be importing the 220bhp, six-speed three-litre GTV, but it'll find life tough out there as it competes with the Mercedes CLK, the Peugeot 406 and, of course, that rocket ship Volvo C70.

The next big deal will be the advent of the serious niche car. There's the Land Rover Freelander of course – a car that's making our nanny almost moist with anticipation. Then there's the BMW Z3 coupé, the VW Beetle and the Audi TT. I'm starting to swell just thinking about them. Obviously, what's happening here is that platform-sharing is starting to pay dividends. If you can bolt any body onto any chassis, you can make new cars more quickly and cheaply than ever before. In the past Ford could never have given us a Ka, a Fiesta and a Puma, but seeing as they're basically the same, nowadays they can. And this means more choice for you and I, which makes picking your ideal five-car garage harder than ever before.

Obviously, I'm a fifth of the way there because already I have a 355. But in La-La Land it would be a Berlinetta, and not a GTS. This would leave space for my convertible to be a big fat barge of a car – and that leads me straight to the door of the Mercedes SL. Also, now that I've started to shoot anything that moves, I'll need a four wheel drive and, much as I respect the Land Cruiser and the Grand Cherokee, I'd have to have a Range Rover. It would come in new 'Autobiography' trim where you get to select whatever colour and interior appointments take your fancy. I'd demand wood from that 2000-year-old tree in California – just to annoy the Americans – and then I'd fit television screens in the back of the front headrests. These will be visible to following traffic to make for all sorts of fun as I drive up and down the motorway with *Debbie Does Dallas* on the video. As far as an everyday car is concerned, I'd have the new Jaguar XJR V8 for all the reasons I outlined last month, which leaves me with the need for a family estate car. I've considered, obviously, the Volvo V70 T5 and its V8 rival from BMW. The Mercedes 300E is a contender too, but I've decided the kids should walk and that dogs don't really need to go on outings. My final car would be one of the 100 Nissan Skylines. I don't care that it got trounced in our Nurburgring feature last month or that it failed to do well in this month's handling test.

We need cars like this because, pretty soon, tony@ numberten.co.uk will stop pulling faces and let Fatty Prescott loose. Time is running out. Winter is almost upon us. For God's sake, get out there and live.

Sad old Surrey

Careful and studious readers may know that A.A. Gill is being hauled in front of the Commission for Racial Equality after describing the Welsh as being 'pugnacious little trolls'.

Well, though we write for the same newspaper, I wish to distance myself from these attacks. Wales is a pretty and charming part of the country and the Welsh have a rounded range of abilities – singing and er . . . setting fire to things.

I think if we're going to single out a part of Britain for ridicule and hatred, Wales comes a very distant second to that jumped-up lump of suburbia called Surrey. If I may be permitted to liken the British Isles to a beautiful woman, Surrey is her most stubborn dingleberry.

In the past three years I have travelled to many countries and seen traffic to frustrate even the most dedicated petrol-head, but on Monday morning Guildford made Tokyo look like the Brecon Beacons. To get from one side to the other took two hours, at an average speed of 6mph.

All around, people were sitting in their horrid neo-Georgian houses congratulating themselves on having moved out of London to the country, obviously unaware that they have not left London at all. They're as much a part of the metropolitan sprawl as Tottenham.

Except that in London, if a main thoroughfare is full locals can use any number of rabbit runs whereas in Surrey this is not possible.

Sure, there are a few open spaces and, given Surrey Man's tendency to drive a large four-wheel drive car, none would present much of a problem, technically speaking. But to drive off-road in Surrey is to invite a confrontation with one of its rangers.

Now, a friend of mine once signed on at the Kensington dole office saying he was a shepherd, and I dare say an investment banker would find life hard in Swaledale, but a ranger? In Surrey? Why?

What they do, apparently, is drive around the much coiffeured heathland in Land Rovers telling other people in Land Rovers not to drive off-road, and to get back to central London where they belong.

So everyone sits on the roads, not moving for hour after hour after hour. Every Laburnum Close and Orchid Drive is full. Every B road is full. Every dual carriageway is full. And there's no way in hell that Fatty Prescott is going to get this lot onto a bus.

For these people, image is everything. They won't even admit to living in Surrey, saying instead they live on the Surrey/Hampshire borders or, for those in the know, that they live in GU4 – which, the postman will tell you, is a ritzy suburb called Shalford.

Here, I saw mothers depositing their children at school from cars that were several miles long. One had an American off-roader that was easily bigger than an Intercity 125. Why should she use a train when she's already got one?

And these people don't park their cars neatly outside the school gates. They simply abandon them nearby and stand around with the other mothers, who've abandoned their space shuttles and coaches, arranging bloody coffee mornings. 'Actually I can't make it today. I'm having sex with the gardener.'

That, of course, is after the gardener in question has helped the ranger to chop down a few more trees. Trees need to be murdered here because, to convince themselves that Surrey is not simply London SW37, the locals demand that the open spaces be kept as such. They call them beauty spots, and that's exactly what they are – spots, tiny little pinpricks of manicured green in a sea of fake marble pillars and Mitsubishi Shoguns.

When the rush hour has subsided and Surrey Woman is at home watching the gardener pant over her panties, old people come out of their houses and climb into their Chevettes and Rover 600s and head for the hills – where the ranger has ripped up some more trees to make car parks.

I spent two days in such a car park this week, and have rarely felt so depressed. The view was undoubtedly pretty, but you know that it's stage-managed and that just over the next hill lies Esher, which isn't pretty at all.

And you know that you must not let your dog off its lead or pick a flower. This is countryside in the same way that the Spice Girls is a rock band, that is, it isn't countryside at all. If it were cheese, it would be Primula.

And the visitors know it. They sit in their cars, not daring to get out in case they break one of the ranger's rules, and they stare at that pitiful facsimile of nature for hours on end. They don't talk. They don't eat. They don't read.

They're sitting in a bloody car park, surrounded by hundreds of other people in cars, listening to lorries lumbering up the A25, watching a tree being chopped down by nature conservationists.

One man turned up in a brand-new Bentley Turbo R

and sat in his car facing, not the view, but the café which sells chips.

And the staff there explained that Paul Weller is a regular visitor. Small wonder the poor bloke has such a strange view of the world when he's forced to sit in a traffic jam for two hours just to get one.

Surrey is more awful, I suspect, than hell. If that's the future for commuting then, my God, you can have my keys right now.

A frightening discovery

I've been sitting at my computer now for two hours, unsure about how this week's column should begin. You see, after years of Biro-sucking, I've finally decided the Land Rover Discovery is absolute rubbish.

But we're talking here about a national institution – an automotive Prince Philip – and you can't just launch into attack mode saying it's a completely useless waste of everyone's afternoon.

But it is, that's the trouble. It's ugly; really, really ugly and I have no idea why this has never occurred to me before. It's been around for years but only this morning did I start to ask the important questions.

Why does it have that raised bit at the back? No dog I've ever seen is 15 feet tall and not once, ever, have I heard of someone keeping a pet giraffe. The Discovery doesn't need that rear end lump.

And why's the back window cockeyed? And have you

seen the panel gaps, for God's sake? I reckon you could get into a Discovery without opening the door. And the windscreen's too flat, and the wheels are lost in those huge arches. They're like Polo mints mounted at the entrance to Fingal's Cave.

Seriously, next time you're down in Guildford have a look. You'll see that the Discovery is even uglier than a Ford Scorpio.

It is also dangerous. Now that's contentious stuff. You can say a car manufacturer's new product is a waste of the world's resources and they'll do nothing. You can liken it to a cup of cold sick and refuse to test it, saying it's more boring than dying, and still they won't react. But call a car dangerous and whoa, what's this? A writ? Blimey.

Well, here's the defence. I've always felt that all cars are capable of stopping in roughly the same distance but this, it turns out, is just not true. I tested a handful of cars last week and was simply amazed by the results.

A Lexus GS300 took just 139.8 feet to haul itself from 70mph to a standstill whereas the aforementioned Land Rover Discovery came to a halt in an almost unbelievable 224.1 feet. And that, to save you the bother of working it out, is a difference of 84 feet. I'll say it again: 84 feet, 28 yards, five car lengths.

Think about that. You crest a brow on the motorway to discover the traffic ahead is stopped. If you're in a Lexus you'll pull up just in time, but if you're in a Discovery you'll still be going at a fair old lick when you have the smash.

Now I want to make it plain that the Discovery is not the only car to perform badly in this test. The Toyota

Rav4 is awful and the Ford Explorer is horrific, but whereas the other two have many strings to their bow, the Disco does not.

Yes, it is a fine off-road car, as well it should be with those Range Rover underpinnings and a lusty V8. There's a diesel too, but quite frankly, I'd rather take my own appendix out.

The only good thing about the diesel is that it's not terribly powerful. Thus, you'll never get up enough speed to turn it over, which is something that I suspect could happen very easily indeed in a V8. A top-heavy, 2 ton car simply cannot be as wieldy as a low-slung saloon.

Of course, the big safety device fitted to all Discoveries is the build quality. As they spend most of their time on the back of low loaders, all the braking and cornering problems are cured at a stroke.

Now, I'm machine-gunning the Discovery because I've recently spent some more time with the new Freelander, whose praises, you may recall, I sang a few weeks ago in a deep and lusty baritone.

Well, after several thousand miles I can report those initial findings were just about right. On the road, the Freelander stops and corners like a normal car, even if it is perhaps a little slow. On a long uphill motorway gradient, you sometimes need fourth gear to maintain a 70mph cruise.

Off-road, however, it's even better than I first thought. On one shoot, mud that stopped both a normal Land Rover and a Toyota Landcruiser proved no problem at all for the Freelander's traction control. I simply adore this little car which, in every way, knocks spots off its bigger brother.

What Rover must do, and now, is stop making the Discovery. It is so far past its sell-by date it should really only be sold in one colour – mould.

But even if they do, there is still the problem of used Discoveries, sitting on the secondhand market looking all innocent and tempting. Pop down to the auctions and you'll find J-registered diesels going for less than £8000. And more worrying still, a P-registered 25,000 miler is available now for just £19,000, making it seem like a large and sensible alternative to the Freelander.

It isn't. It's a huge, salivating dog that, at best, will sit around in your drive wetting itself. Worst case scenario? It'll tear your leg off and beat you to death with the soggy end.

The choice is easy. Buy the puppy instead, the dog that you know has been bred properly by a registered member of the BMW kennel club. Buy the Freelander.

Hannibal Hector the Vector

Atlanta is one of the world's most peculiar cities. It has the requisite pointy skyscrapers and if you ask for a small Coke in a Taco Bell, it still comes in a bucket. This is America.

And yet somehow, it isn't. The people, largely, are slim, and regularly you'll see a well-dressed, pretty girl in an Alfa Romeo Spider.

And then you've got the valet parkists at the Ritz Carlton. They're efficient for sure, but they don't crawl across the driveway on their stomachs, clutching at your legs like you're the only person in the world who shares the same type of bone marrow.

I read recently that America's business travellers had voted Atlanta the rudest city in the world . . . and that's it. That's why I like the place so much. Ask for a pail of Sprite in a restaurant and you'll be ignored. Summon a man to fix the television in your bedroom and he'll stomp around, prodding the remote control and swearing at you for breaking it. It's fantastic. It's just like Britain.

And it's very like Britain if you head north to the town of Braselton, which was recently bought by Kim Basinger. Here you will find Road Atlanta, which isn't a road at all. It's a swooping, Spa-like race track where the girl on reception greets you with the distinctly un-American 'Hello love.'

Men drift around in the background, being English, and then you're introduced to the boss who, it turns out, lives in Field Assarts – a small village just outside Chipping Norton.

I was there to drive the Vector, an American supercar about which I had serious doubts. When I first heard of it, 20 years ago, it was being made in California by a man whose mouth was so big you could park a lorry in it.

He used to claim that his car, which had a twin-turbocharged Corvette V8, could do more than 200mph, but I saw no test results to back this up. Indeed, the only time I ever even saw a Vector was in the film *Rising Sun*.

Anyway, he went bust and the company was relocated to Florida by the same Malaysians who own Lamborghini. But not long afterwards, they simply locked the factory doors and walked away.

However, despite this chequered past, there, at Road

Atlanta, was an enormous American lorry which housed a brand-new Vector, and alongside it there was a huge black limo which had been driven overnight from Jacksonville, six hours to the south.

In the back was Vector's new boss. Now I was expecting a ten-gallon hat to stumble from the back door, followed several hours later by a stomach. But no, a cheeky chappie in a two-piece pinstripe bounded over and introduced himself in a Thames Estuary accent as Tim.

Turns out, he served his time at Lotus, where there is only one mantra. The car must be light.

But the £100,000 monster being poured from the back of that gigantic truck appeared to have been fashioned from a cocktail of lead and mercury. It was huge: 6 inches wider than a Diablo and 10 times more striking.

As is the way with supercars, getting in is like potholing. You crawl under the Kevlar gull-wing door and burrow over the sill to find an interior which is shockingly cramped. Put a veal in there and Dover docks would be closed down for a month.

Still, you turn the key and behind your head a 5.7 litre, 500bhp Lamborghini V12 explodes into life. This is what supercars are all about. Deep discomfort, allied to unspeakable noise and fear. If you feel like a veal with a rocket strapped to its back, you're in a supercar where Nessun Dorma. If it's all comfy and quiet, you're in a Nissan Dormobile.

It was odd then, to discover that when I shoved the throttle into the carpet the car merely went a little faster. Only when the revs crawled past 4000 did it really wake up, but by then I was already tired. The steering is power-assisted, but only a little bit. And you don't press the brake pedal to slow down; you have to climb in the footwell and use a jack hammer on the damn thing. It absolutely will not oversteer either.

Now at this point, I'd like to say that the first man ever to buy one of these cars had the right idea. He took down the wall of his house, put the car in his sitting room and built the wall again.

And yet I suspect that somewhere in the package there is a good car. It reminded me in many ways of an early '80s TVR which we could, so easily, have written off as kit-car junk. With a bit of careful development, mainly to make the engine work at low revs, the Vector could pick up the baton that Lamborghini, I understand, is soon to drop. Rumours coming from the factory in Bologna talk of an empty order book and even emptier pockets.

It is, of course, pretty damn hard to take on Ferrari and Porsche but there's no doubt in my mind that Vector is using the right recipe – a British chassis, Italian power, American prices and Buck Rogers styling. They just want to make sure they don't get that all the wrong way round.

If they succeed, they'll be selling the nicest American surprise since Atlanta. If they don't, they'll be selling a crap car.

F1 running rings round the viewers

Every year, I predict who will win the Formula One world championship. And every year I am completely and utterly wrong. This year, I said it would be Jacques Villeneuve . . . but don't worry, I'm not losing my touch.

Martin Brundle's job is safe, because I was wrong again.

I may have been right with the outcome but, as with examinations, you must be able to show how you worked it out. And on that front, I was all over the place. You see, I said Jacques would win every single race, have it wrapped up by Silverstone, and that we were in for the dullest year of racing since the drivers' strike.

And I wasn't alone. Everyone who knows which way up a helmet goes agreed with me. So what went wrong? Well I'm not big on conspiracy theories. I don't, for instance, believe that Princess Diana was murdered by one of the Queen's corgis. My hair was not cut this morning by Elvis Presley. And I think Neil Armstrong did make his giant leap on the moon, not on a soundstage in Nevada. But at the end of qualifying for the European Grand Prix last month, one of my eyebrows was raised just a little higher than normal. And at the end of the race, the other one had joined it. With hindsight, you can see things starting to go awry in Austria. Schumacher was romping away with the title when he was hit with a 10 second penalty after passing Heinz-Harald Hopeless under a yellow flag. Result: Villeneuve closed the gap.

Then there was Japan, when Jacques could have sewn it all up. But no. He didn't slow down for a waved flag while qualifying, he was under a one-race suspended ban and that was it. He was out. Result: Schumacher closed the gap.

And just in case Jacques thought about appealing, he was warned that Eddie Irvine had done this before and had seen his ban extended from one to three races. Result: a bunch of promoters with Blair-style grins. These penalties had been imposed for clear misdemeanours, but

I find it odd that the only two drivers to have fallen foul of the law this year were the two fighting for the title, and that both did so in the championship's dying hours. Anyway, when the circus arrived in Spain for the big showdown, Villeneuve and Schumacher were one point apart, and I had buttocks you couldn't have prised apart with a blow torch. However, during qualifying we were asked to believe that Michael and Jacques on this, the greatest day in motor racing, had driven round the circuit at exactly the same speed – something that had never, ever happened before. Far-fetched? Not if you think *Star Wars* is a true story.

Then came the race. Over the year I've come to respect Schumacher, who seemed to be genuinely pleased when he won. He undoubtedly had an inferior car – one of my beloved Ferraris. He had proved himself a truly great driver, and after his praise for Eddie Irvine in Japan, a gentleman. But in Spain he proved that, when all is said and done, he is still a German. So he was out and Villeneuve was on his way to victory, not only in the race but in the championship too. Hip hip hooray and so on.

But wait. What's this? Team managers dash about in the pits, and look what's happening. Hakkinen has overtaken Coulthard. On the straight. Fisichella has been blue-flagged, and Villeneuve's car seems to be suffering some damage after all. Now, obviously it would be improper for me to suggest even for one moment that there had been some behind-the-scenes jiggery-pokery going on, but did you see Coulthard on the podium? He looked like a man whose dog had just died. Even Hakkinen, who I expected to burst with pride when he finally won a race, looked like he'd just failed all his A-levels.

There's talk that Sylvester Stallone is working on a Hollywood blockbuster about Formula One, but if someone presented him with a script based on the 1997 championship, he'd dismiss it as completely implausible.

Bernie Ecclestone has done a magnificent job with Formula One and he needs these last-minute showdowns. But we, the keen viewers, need to be assured that it is still motorsport, with young men going wheel-to-wheel in a life-or-death struggle for glory. And not panto.

Big cat needs its tummy tickled

I'd only been driving the new Jaguar for 20 minutes when, inevitably, it happened. On a rain-streaked M42 my rear-view mirror filled to overflowing with the menacing sight of a steel-wheeled BMW 316. Inside, the driver was barking into a mobile phone, his face contorted with rage that I should be in *his* way.

Now there was a time when I'd have eased the Jag into a lower gear and floored the throttle, but next weekend I shall be 38 and I just can't be bothered any more. So, as soon as a space appeared, I moved politely into the middle lane and smiled as Mr Neo-Georgian screamed past, on his way, no doubt, to yet another crisis at the photocopier shop.

It was, I'm afraid, a rather patronizing smile because matey could have brought any BMW to the battle and he'd have lost. I was driving the new supercharged XK8 you see, and no German production car can even get close. Not even the new Porsche 911.

The standard XK8 has already been voted the most beautiful two-seater sports car in the world by a bunch of Italian designers, but now Jaguar has added some teeth to create what's called the XKR. Basically, it's propelled along by a supercharged 4.0 litre V8 engine which produces a simply staggering 370bhp – roughly the same as a Ferrari 355.

This is a natural successor to the old Jaguar XK120 which, 50 years ago, was also voted the most beautiful car in the world. And then, on a deserted stretch of Belgian motorway, it achieved 139mph, making it the fastest too.

The new XKR goes further. Even though it is burdened by various pieces of Prescottery in the exhaust and an automatic gearbox, it will get from 0 to 60 in 5.2 seconds and onwards to a top speed in excess of 175mph. Well it would, but an electronic referee blows the whistle at 155mph.

These are impressive statistics but it is the quality and the relentlessness of the power delivery that leaves a more lasting impression – that and the top-end clout. From 140 to 155mph, when the supercharger is eating steroids by the handful, it is an almost unbelievable 50 seconds faster than a standard, unsupercharged XK8.

So that passers-by in their weedy BMWs are able to tell what they're dealing with, the XKR has different wheels, twin bonnet louvres and a microscopic boot-lid spoiler. And that's it. There isn't even much of a difference with the price – the coupé is £60,000 and the droptop I tested is £66,000.

So it's well-priced, pretty and very, very fast. What more could you possibly want?

Well actually, quite a lot. You see, the new car is

smooth and wafty and will, I don't doubt, find a great many friends at golf clubs all over Houston. I can see Los Angeles dentists arriving to buy this car by the bus load. They're going to just lurrrve that wood and leather interior, and the comfort, and the silence.

However, there are some changes I'd like to see before I'd buy one, and first of all they need to get it down. A real jaguar, the furry Attenborough type, prances about on tiptoe when it's cruising, but as things get serious it'll hunker down on its haunches – and the XKR doesn't.

It sits too high, with a huge gap between the wheels and the wheel arches – you could erect a tent in there. One designer I know described the XKR as an off-road sports car, but he was wrong. It isn't a sports car at all because the suspension, though beefy, is sadly boneless. My six-cylinder XJR saloon has a much meatier feel, both in terms of the ride and the steering. It feels like it's in attack mode, which for a car of this type is how things should be.

Then there's the seats. If you are broad-shouldered – like the chairman of Jaguar, incidentally – you will find them so narrow that driving through high-speed corners, you fall out of them. BMW offer sporty, heavily bolstered alternatives and I really can't see why Ford's luxury division does not.

This, I guess, is my point here. BMW make lovely cars for a certain type of person but they know there will always be a small number of people who want *bigger better more*. So they have the tiny and separate Motorsport 'M' division who take the best . . . and make it better.

Mercedes does the same sort of thing. You can buy a car off the peg, or if you want some snarly, bone-snapping get up and go, you can buy one that's been tweaked by AMG.

What Jaguar needs is the same sort of thing – a wholly owned, but separate, engineering-based outfit that could take a car designed for everyman and turn it into a road-going rocket for the few.

They wouldn't even need to touch the engine – it's easily good enough already – but that said, if by fiddling with camshafts and fuel flow they eked out a few more horses, I wouldn't complain. Some exhausts that offer a muted V8 beat wouldn't go amiss either.

The standard XKR is a superb dish that's been well thought out and beautifully prepared. It is an excellent grand tourer, but with some slightly different seasoning it could be turned into something else. It could be turned into a sports car.

Elk test makes monkeys of us

Imagine the horror. You're a cameraman with the BBC's natural history department and you've been dispatched to Tierra del Fuego in South America where, once every ten years, a strange frog comes out of the mud, mates, and then dies.

You've been sitting there for the best part of a decade when the need for a crap becomes utterly overwhelming. So you scoop up the bog roll, a copy of *Viz* and disappear behind a rock. And while you're gone, Froggy comes out of his muddy home and struts his stuff with Mrs Frog.

Well last month, Britain's motoring journalists were on

the bog while all hell was breaking loose all around. Some had driven the Mercedes A Class and glowing reports were appearing in magazines all over the land. *Autocar* said it could dive through tight bends with agility. *Car* magazine said much the same thing, while *Auto Express* praised its responsive chassis. Now I don't want to sound smug about this, but after half a mile in the new baby Benz it became very obvious that its handling was not agile and it certainly wasn't responsive. It was utterly and completely crap. Contrary to what many may think, we road testers do get swayed by the opinions of colleagues and I found myself in a quandary. Here was a car from one of the world's most ruthlessly efficient manufacturers, a car that my colleagues liked very much.

It takes a very special kind of bombastic arrogance to be that little boy in 'The Emperor's New Clothes' – to stand up and say: 'Actually, its handling is appalling.' But thank God I did, because just a week later a Swedish magazine found to its cost that, while performing what's become known as the 'Elk Test', the A Class rolled over and put its occupants in hospital. A German magazine then repeated the procedure and subsequent examination of the film showed that what we had here was A Class One Disaster. Experts immediately dismissed the Elk Test as unrepresentative, but I disagree. Swerving one way, then the other, to miss an obstacle is worthwhile in any environment. Sure, we don't have elks in Britain but we do have children and dogs, and debris in the outside lane of the motorway. And Mercedes agreed because first they said they'd change the tyres, then they said stability control would be fitted as standard, then they stopped the production lines. The fact is, Mercedes screwed up and our journos missed the biggest story since the Ford Pinto.

Well now it's time to wake up and smell the coffee. Drive into a roundabout at a sporty, rather than aggressive, pace and understeer is colossal. Switch direction and massive body roll attaches itself like a 2 ton barnacle to a problem that shouldn't have been there in the first place. It's impossible to miss. I only feel guilty that I majored on the car's good points. But at least I spotted the flaw. In the early days of car journalism, it was important to be on the ball because rotten and dangerous cars lumbered onto the market every week. But in recent years, the whole game has shifted. We assume a car is safe and reliable and make our judgements instead on what the badge says about the driver, value for money and so on. Sure, not many magazines can afford to do crash testing like car companies do, but most have forgotten how to do any testing at all other than zooming up and down with stopwatches. Thank God a little Swedish magazine still does things properly.

Everyone else took it for granted that the A Class would be safe and steady, and talked instead about the space inside and the fact you could park a three-pointed star on your driveway for just £14,000. No one actually stopped to think, hold on, this is Mercedes' first ever attempt at a front-wheel drive car. Let's assume nothing. Let's do a lane-change manoeuvre. Let's wiggle the wheel a bit to see what's what. It doesn't matter what Mercedes do with the design of their new car now. The A Class is dead. And with it has gone the reputation of Britain's motoring journalists.

We should all be sent to Iraq, but I fear that as the

F-15s sweep in from Turkey to post bombs through Saddam's letterbox, we'd all be on the beach, filing copy about fine wines and nice cheese.

At the core of the Cuore

In Britain, the great European debate centres on two issues – tradition and trade.

One group says if the Queen's head is removed from our banknotes, fire and pestilence will rain down from the heavens and a plague of locusts will infest Gordon Brown's underpants. The other says this is jingoistic nonsense and concentrates on the implications for business, pensions and immigration. Frankly, it's all so dull I'd rather eat card-board.

But over dinner the other night with Wolfgang Reitzler, who is a significant oberlieutenant at BMW, I discovered that in Germany things are rather different. He said, 'I am in favour of the EU because it would prevent another war,' and I damn nearly fell off my chair.

I mean, my God, this is something I'd never even thought of. In Britain it is considered inconceivable that any two Western European nations could open hostilities with one another, but obviously the Germans are still looking wistfully at Poland.

So, frankly, I was delighted this week when I heard that BMW has added Rolls-Royce to its portfolio of British investments. The more they own over here, the less likely they are to drop bombs on us.

I was, however, a little upset by the £340 million that

Vickers seem to have accepted for what everyone seems to agree is the ultimate British brand. When you consider Rolls-Royce has just spent £200 million developing the new Seraph and its sister car the Bentley Arnage, it could be argued that the actual sale price is just £140 million.

But in fact it's even less than that because remember, much of the £200 million went to BMW, who are supplying engines and various ancillary parts for the new cars. In real terms, BMW has bought Rolls-Royce for about 18p, which seems rather low.

Still, it's all irrelevant because who's going to buy a Rolls-Royce now that Mr Prescott is offering a 50 quid cashback offer on the road tax of a Daihatsu Cuore+?

This is a remarkable offer which I just know will have all of you perched on the edge of your seats. Secretly, I suspect, you've always wanted such a car, and now that it comes with New Labour's seal of approval the temptation is almost too much to bear. So come on Clarkson, tell us. What is it like?

Well, it is 20 inches shorter than a Ford Fiesta and, amazingly, 7.5 inches narrower, which means you don't ever have to worry about parking. You just put it in your briefcase. If you don't have a briefcase, don't worry, it comes with a carrying handle that's been cunningly disguised as a rear spoiler.

However, despite the diminutive dimensions the + model I tested comes with five doors and enough space, even for 17 stones of me, behind the wheel. Indeed, it needs a big heavy driver or it would simply blow away in the breeze and you'd spend your entire life looking for it up trees.

I, however, tied it to some stones and spent the best

part of a day trying to find the engine. With the bonnet raised I climbed into the engine bay, initially dismissing what appeared to be a small matchbox nailed to the inner wing. This, however, was a mistake. This was the engine – all 850cc of it.

It only has three cylinders and produces a catastrophically miserable 42bhp, making it 30 per cent less powerful than a Mini. I suspect it is also 10 per cent less powerful than my Moulinex Magimix.

So, though I could fit inside, there was some question about the car's ability to actually move me around. But it did. Obviously, I wasn't going to build up much of a supersonic shock wave, but even though the 0 to 60 time of 16 seconds looks feeble on paper, it felt quite sprightly.

With the missing cylinder causing an imbalance in the matchbox, it sounds almost exactly like an air-cooled Porsche 911 which is rather endearing. But just as Porsche Man is shifting into second, Daihatsu Man is out of puff. I saw 85 on the clock once, but as I was being overtaken by continental drift at the time it's possible the speedo may have been lying somewhat.

Of course, a car like this isn't supposed to be fast, and nor is it supposed to go round corners very well. So it doesn't. What did come as a surprise was the fitment in the luxury Cuore+ of electric windows, a stereo and, particularly, central locking. Why do you need central locking, pray, in a car where you can reach all four doors from the driver's seat – with your eyelashes? I think I'd prefer the standard three-door Cuore which is £700 less expensive and comes with four wheels and a seat.

But either way, you are in for one very special treat. Even though I drove this little car with verve and aplomb

I managed to go 53 miles on one gallon of petrol, making it by far and away the most economical car ever to come under the command of my size nines.

It is also one of the cheapest. The Cuore+ is £7200 while the standard model is just £6500.

The only way you can do better is by fitting wheels to your rabbit hutch and attaching the motor from that juicer you never use. Or jogging.

Last 911 is full of hot air

Reviewing music has to be the hardest, most pointless job since Twinkletoe-Winkletoe Fffiennes walked to the North Pole wearing nothing but a dressing gown and slippers. Or something.

Imagine, please, being instructed to write about the latest All Saints album. You'd listen, hate it and say so. And a week later, all the 14-year-olds who took it to number one would burn your house down.

I could sit my mother in front of the stereo and play her 'Life Through a Lens' by Robbie Williams and she'd look like someone was using a staple gun on her nose. Ask me to listen to Joni Mitchell, and I have to put my finger in my ears and sing 'Baa Baa Black Sheep' at the top of my voice. Tell me to review one of her albums and I'd say it was like chewing on polystyrene.

Cars are so much simpler. They're either fast or slow, spacious or cramped, expensive or not so expensive. And reviewing them is a black-and-white science right up to the moment when you surge past, say, 60,000.

At this point, if you're buying a car you're not buying it for practical reasons. You're buying a brand that you've dreamed about since you were two. And it really doesn't matter if the brand in question has, in the meantime, become a joke. For some reason, the name 'Bristol' has just sprung to mind. And here comes another one: 'Maserati'.

And then there's the Porsche 911, which, of course, is emphatically not a joke. I have just spent a week with the very last of the old air-cooled monsters and it failed to raise a smile even once – a bit like Rory Bremner's BAFTA presentation last weekend.

Now for some, the passing of the old 911 is right up there with the passing of Princess Diana. I have seen people weeping in the streets and threatening to hurl themselves off a tall building unless Porsche bring it back.

I, however, couldn't care less because, at two, I dreamt about Ferraris and in the playground I would fight people from Planet Porsche. I would push their heads down lavatories until they admitted the 911 was smelly. Say it. SAY IT or I'll give you a Chinese burn.

And as a result I now find it rather difficult to review this, the last of a breed, sensibly. I can tell you that it's got a twin-turbocharged 3.6 litre engine which develops 450bhp. I can tell you that only 33 examples of this so-called Turbo S have been made and all are sold. I can tell you, too, that this car is supposed to be the ultimate 911, that it brings together everything Porsche has learnt from 35 years' practice.

These are the facts ... and now, here come the opinions. It is, without doubt, the scariest, nastiest, ugliest piece of donkey dirt that has ever graced my drive. The only possible way you could have fun with this car is by dropping a lighted match into the petrol tank.

On the motorway, the ride is so firm it blurs your vision and in town, the ground clearance is such that some of the more vigorous speed bumps in the Socialist Republic of Hammersmith and Fulham brought it to a dead halt.

Sure, the technologically sophisticated four-wheel drive system would enable the Porsche to keep going in a ploughed field, but not with that nose – it's so low it snorts the white lines off the road.

Then you've got a mixed bag of turbo lag, fearsome acceleration and brakes which inspire no confidence at all. Hit them gently and nothing happens – hit them hard and your nose slams into the windscreen, which is too close at the best of times. You don't so much look through it as wear it like a pair of spectacles.

I truly hated this car and am glad it's no longer in production. Indeed, I believe that the devil himself would drive such a thing.

So I was horrified to hear a man from Porsche say the other day that so long as there is a Ferrari, there will be a Porsche Turbo . . . which means that in the year 2000 we can expect a blown version of the new 911 – a car I quite like as it is.

Anyhow, that's then and this is now and I'm still stuck, trying to review a car that's about as much use as a CB radio in a Vodafone world.

So, in the interests of balanced journalism, I managed to find a five-minute gap in the Marquess of Blandford's diary when he actually had a driving licence, and asked him to have a go. He is a huge 911 fan and approached the egg-yellow monstrosity as though it were the Turin Shroud. 'To drive the last of the air-cooled 911s,' he whispered, 'is a real privilege.'

One hour later he was back, beaming the smile of a man who'd been taken a little closer to Godhood. 'You've got to understand a 911's little foibles. If you really understand these cars, you will know that this is just the best of them all,' he added bouncily.

He's sitting behind me now, tied to the chair with a bar of soap in his mouth. And I won't let him go until he stops calling my Ferrari a Fiat, and admits the worst car in the world is not, as we'd suspected, the Vauxhall Vectra.

False economies of scale

For the first time in years you can now buy a brand-new car for less than £5000. It's from Malaysia and it's called the Perodua Nippa.

Now I want to make it plain from the outset that I have not driven this car, and there's a strong possibility that I never will. I mean, if we live to be 70, we only have 600,000 hours to play with and I'm not prepared to spend even a tiny fraction of that in a car that has no radio.

Besides, it's not that long since I spent a whole day playing with its rivals – a bunch of so-called microcars to delight Crasher Prescott and his team of weird beards.

After the budget in which Golden Brown said he'd provide a £50 cashback offer to those who bought a small, clean car, I thought I'd better wise up on this new breed of city car, to see if £50 was enough of an incentive to entice people out of their BMWs and into a wheeled shoe-box.

Let's start with a car that's ghastly – the Suzuki Wagon R. Powered by a 996cc engine, you don't expect it to be the fastest car in the world, and it doesn't disappoint. It isn't. With a top speed of 87mph, it is exactly half as fast as the Porsche turbo I wrote about last week – and, unbelievably, it looks twice as daft.

It's very narrow, very short and for some extraordinary reason, very tall. I, for instance, enjoyed a good foot of headroom, but why? The only people who could possibly need such an enormous amount of space above their heads are people who wear bearskins, and let's be honest here – a guardsman spends all day standing in a Wendy house so why should he want to drive home in one?

So, unless you enjoy being laughed at, avoid this stupid Suzuki like you would avoid unprotected sex with an Ethiopian transvestite.

I wouldn't mind, but it isn't even desperately economical, which, surely, is the whole point of a small car. The cheapest model costs £7400 and only does 47mpg.

Which is why I was led to the door of the 1-litre SEAT Arosa, which costs £6995 and does, as near as makes no difference, 50mpg. Now this looks like a normal car and goes like one too. With determination and a hill, you could do 100mph.

I still think the Daihatsu Cuore+ is a better bet, for all the reasons I outlined a month ago – four doors, low price, greater economy, etc.

But since then we've heard news of a new small Fiat called the Seicento, that Perodua, a Volkswagen version of the Arosa and, at long last, a replacement for the Mini. Having failed to make cars run on electricity, hydrogen, runner beans or any other fuel of the future, car makers have obviously decided to give us half what we're used to.

And this is bound to have an effect. You're bound to be impressed with the promise of halved fuel bills, halved insurance and easier parking in a car which can now be had for less than five grand.

But there is one significant drawback to these cars. Basically, if you crash a microcar there's a greater chance of dying than if you crash something large.

It's all very well saying that these are city cars, and that you're only likely to be doing 7mph at the time of impact, but come on; from time to time you will take them on the motorway, where their miserable acceleration will put them in wheel-to-wheel combat with Scania Man.

And while you may marvel in a showroom at how the bodywork of these cars seems to have been moulded to fit your body, consider this. Your feet are no more than 2 feet from the front bumper. Your shoulder is jammed against the B pillar and your children's heads are perilously close to the rear tailgate.

Last week, the chairman of Jaguar climbed unscathed from a Daimler that had barrel-rolled several times up a motorway embankment. And I feel sure that if he'd been the chairman of Perodua, or Suzuki or any of the others, he'd have been drinking through a straw for several months. Or playing a harp.

America is currently awash with statistics, one of which suggests that you're four times more likely to die in a small car than a Range Rover. Another says that as downsizing takes hold, up to 3900 more people will be killed on the roads each year.

But statistics can be moulded to say anything you want, so let's ignore them and concentrate instead on simple laws of physics. I think it was Isaac Newton who said that if you crash a big car, you'll be better off than if you crash a little one.

Actually, I might have made that up but I'm sure he would have said it, if he'd been good at sound-bites and aware of the car.

So, here's the deal. Don't swallow the government's line on small cars. The whole point of environmentalism is the preservation of life – but what's the point of helping to save the trees if you wind up dead after running into one?

If you're limited by budget, don't buy a new small car. Buy a used big one.

Blowing the whistle on Ford and Vauxhall

Since Rover fitted chrome kickplates to its mid-range saloons, the British fleet market has been left to Ford and Vauxhall. Vauxhall made the early running with all sorts of new and exciting products – the Frontera, the Calibra and the Tigra, which, we were told, went from concept stage to production in 14 minutes. Ford, meanwhile, were fast asleep. They'd hit us with the new Escort in 1992, but it was such a dog people were surprised it didn't have a tail. Then there was the Probe. A nice enough coupé, but we all knew it cost £28.50 in America and couldn't see why it was more than 20 grand here. To fight the Frontera, they teamed up with Nissan – always a mistake – and launched the Maverick, which was ugly and hopeless.

But, all of a sudden, things began to change. It turned out the Frontera was built like an Airfix kit and so was less reliable than Israeli politics. And when the launch-time brouhaha around the Calibra and Tigra died down, people began to notice that, as driver's cars, they fell some way short of the mark. Indeed, the mark was in Latvia and these things were just outside Leamington Spa. Things went really pear-shaped, though, when the Cavalier won the British Touring Car Championship. Great – except it walked off with the laurels at exactly the same time as it went out of production. Its replacement brought the whole house of cards tumbling down. The Vectra was unpleasant to drive, uninspiring to behold, not especially cheap, and took tedium to unprecedented heights. If Vauxhall can be likened to Manchester United, the Vectra was a performance that would have guaranteed a 4–0 defeat at the hands of Doncaster Rovers. Using football analogies, actually, is a highly dangerous game because I know nothing about it, but I'll give it a go. Just like Vauxhall 'gave it a go' with the Vectra. Right now, watching Ford and Vauxhall slug it out is like watching a game of soccer. Both teams field 11 players, with some geared for attack and some for defence.

Ford's new star of the front row is the Puma, which can run rings round the Tigra – bad news for Vauxhall, whose Calibra has been sent off just as Ford substitute the tolerable Probe with the amazing Cougar.

Out on the wings the Frontera is still falling to pieces while the Maverick spins its wheels in wet grass, so we'll call that one a draw. But there's no doubt the Explorer is a damn sight more able than that Japanese player, the Monterey. It's the same story on the other side of the pitch, where the Sintra is made to look wooden by the Galaxy – even though identical players are on offer for much less money elsewhere. Vauxhall gain a little ground in the midfield because while the Omega and Scorpio are equally talented, there's no way you could raise additional funds by flogging posters of the Ford to teenagers. It doesn't matter, though, because in the midfield Ford scores its biggest trump of the lot – the Mondeo. Even if Vauxhall's front row could break past Ford's, the Mondeo would stop them dead. The Corsa, too, is no match for the Fiesta, and Ford even provide the ball, in the shape of the Ka.

And now we come to the defence. Ford has the Escort, and Vauxhall has the Astra. And both are utter crap. They just bumble about, earning both teams a poor reputation for shoddy, unimaginative thinking. However, both are about to be pensioned off to run bars in the East End, allowing new, and apparently fresher, players to take over. For Vauxhall, the new Astra is critical. At the moment, their entire team is out of date or useless, or both. Ford's army of fresh-faced attackers has a clear run of the whole pitch. If the Astra works, however, the blue-and-white team from Essex will be in trouble. They have a great front row and a stronger midfield but they'll be up against a great defence.

One of these days, Vauxhall is bound to wake up. Two years ago, Man United would have beaten Derby without trying. But today?

Ford has to remember they're up against General

Motors, which has slightly more money than God. And in the end, as Fulham are about to prove, money is what matters.

Hell below decks – Clarkson puts das boot in

As I see it, there are three possibilities after death – heaven, which should be very nice; nothingness, which will be just like sleep; or hell, which no longer concerns me.

I don't care what foul vat of sewage has been dreamt up by Lucifer because it cannot possibly be any worse than life on board an American aircraft carrier.

As you were heading for work on Monday morning, I was on board what looked like a winged washing machine, outbound from Norfolk, Virginia, to the nuclear-powered, 100,000 ton USS *Dwight D Eisenhower*, the biggest, fastest warship the world has ever seen.

Now I've talked before on these pages about the braking ability of a Porsche turbo, which goes from 70mph to 0 in 2.8 seconds. But that's nothing. My plane, which was called a 'cod' and flew like one, hit the deck at 175mph and was stationary two seconds later.

Picking bits of spleen from the inside of my float-coat, a loud and hectoring sailor ushered me from what he called 'the most dangerous place on earth' – the flight deck of a carrier – into what I now know to be the *worst* place on earth – the bowels of a carrier.

A crew of 5000 live down there, spread over 17 decks which are interconnected by 17 miles of corridor, 66 ladders and a thousand watertight doors on which you bang your head.

There are no open spaces to sit and chill out. You aren't allowed to have sex with the 500 women. Everything is fashioned from steel. And there are no windows. Then there's the total lack of privacy, even when you're on the lavatory, and the constant, deafening noise, 24 hours a day, for month after interminable month.

To get even the slightest idea what life is like for these sailors, imagine being locked into the back of a steel container and driven around on the back of an articulated lorry for six months. For company you have two eight-year-olds, one of whom is learning the recorder, and the other the violin. Then there are 30 young men with spots who shout at one another all day, and night, and for good measure one bloke who follows you around blowing a hairdryer in your face.

On the *Eisenhower* I was allocated a guide who walked like Herman Munster and talked like Barney Rubble. He was not a bright man. If I asked him a question he would repeat it, very slowly, and then, before answering in navy gobbledegook, say, 'Now let's see.'

On the second day, the ship's tannoy announced that an F-18 had suffered an engine fire and was limping back to the carrier on just one of its General Electric turbofans. Barney, however, couldn't care less because I'd just given him $10 to settle an $8 bill and this was confusing the hell out of him. 'Now let's see,' he said.

It all ended well though, and a booming voice rose above the violin practice to say the plane had landed safely and that everyone on the ship, including Barney, I presume, had done, 'an outstanding job'. This happens a lot.

At four in the morning, you will be woken by the

booming tannoy: 'Someone you have never met has done something you don't care about in a part of the ship that you will never go to. Outstanding job.'

All I wanted was a drink and a cigarette, but there was one small hurdle that stood in the way – my fish/washing machine had to be attached to a steam catapult and hurled over the front of the ship.

Now this catapult is quite a thing. If it could be angled properly it would throw a Volkswagen Beetle 12 miles. But it was not angled and my cod weighed much more than a car. I just knew the plane would flop into the sea and tumble under the ship until it reached the stern, where the four 30 foot propellers would shred it, and me, into bite-sized chunks.

Here's what happens. They attach the front wheel of your plane to what looks like a half-brick, which is then fired down 100 yards of track by a steam-powered piston. At the far end the wheel detaches itself and, hey presto, you're airborne.

So in 100 yards, and just 1.5 seconds, I would accelerate from 0 to 175mph, and at long last we have a motoring flavour.

Apparently, there was some discussion at the very highest levels within BMW about the new M5. Burnt Fish Trousers himself was quoted recently, saying that there's only a small gap between the current 540i and the physical limits of acceleration.

The Bee Em boffins were seriously concerned that they'd expend a great deal of time, energy and money designing an M5 which, in the end, would not be that much faster than a standard 540i.

But I'm delighted to say they went ahead anyway, and

really, they needn't have worried about this acceleration business. Even if the new 400bhp car does 0 to 60 in a single second, which it won't, it'll be nothing compared to the power delivered by that steam catapult. Only by stepping into the path of an Intercity train could you have even the vaguest inkling of what it's like. You'd die, of course, but don't worry; that's much better than having to spend time on a carrier, believe me.

Country Life

You're fast approaching middle age. You have a child. You live in London. All your friends live in London. You love London but those earnest men and women from the BBC's Newsroom South East have planted the seeds of doubt.

Every night, they tell you that traffic has reached crisis point and that teams of scientists from the World Health Organization have found enough air pollution in Camden alone to kill every man, woman and child within a week. There's anthrax in the Serpentine and a mugger in your wardrobe.

This has an effect, and you're starting to wonder if maybe it's time to move out. And then, one day, you pick up a copy of *Country Life* – the most dangerous magazine published today – and there, between the story about handkerchief makers and some bird with straw up her backside, you note that for the price of your four-bed-roomed house in Fulham you can buy Oxfordshire.

Instead of a backyard, you can have six buttercuppy

acres, an Aga, a barn, a brook and a wide and varied selection of something called reception rooms.

I know how this feels because it happened to me. After 18 unswervingly happy years in Fulham I was exposed, for no more than 10 minutes, to a copy of *Country Life,* and within a month I was on my way to Hackett for some tweed. I was off to a new life in the Cotswolds.

We'd found a magnificent house and, even after we'd festooned it with satellite dishes to keep us amused on long, dark winter evenings, it was still magnificent as the removal trucks disgorged our entire belongings into a cupboard under the stairs. Understand, please, that furniture which fills a house in London isn't going to fill a lavatory in the country.

Happily, we only had to watch buggy racing from Finland and the Dubai racing results for six nights before a cousin of someone who once sold a dog to a mild acquaintance in London rang, and we were off to a bottom-sniffing, getting-to-meet-the-locals drinks party.

This was just like the drinks parties we used to throw in our early London days, but it quickly became apparent that one vital ingredient was missing – there was drink, and bonhomie and an inglenook. But no one was flirting. Aged 20, you only want to meet people so that you can get them into bed. Aged 40, you only want to get to bed.

Samuel Johnson, it seems, was right. When a man is tired of London, he is tired of life. You go to London to live. And then, when you've served your biological purpose and had children, you move out. To die.

Children are always the excuse . . . but permit me to blow away a few myths on that score. My eldest is now nearly four and she hates – and I mean really hates – wind, rain, mud, trees, fields, tractors and snow. Once a wet leaf attached itself to her shoe and she wailed like a stuck pig for two hours.

Even though we have six pastoral acres to play with, she confines her activities largely to the playroom and her colouring set. And you should see the look on her younger brother's face when I start the ride-on mower. You know how Robert Shaw looked when he was being eaten by that shark in *Jaws*? Well it's like that, only the boy majors a little more strongly on the fear and terror.

Most mornings both children ask if we can go to the London flat so that they may swim at the Harbour Club, meet up with friends in Tootsies and take tea at Hurlingham. Emily is known among friends as Tara Palmer Clarkson.

And don't for one minute be taken in by this health business. Mothers say that children in London suffer from asthma and streaming eyes, but out here the hay fever is simply appalling and I heard last week of someone with diphtheria. Consumption is commonplace, and all I've got to look forward to is gout.

Then there's the noise. Hear a sound at 4 a.m. in London and you'll turn over and go back to sleep. Hear a sound out here at 4 a.m. and you'll jump half-way out of your bloody skin. Twice a month, at least, I'm to be found in the middle of the night stomping round the house in my dressing gown, convinced that the scuffling sound my wife heard is a one-eyed Jethro who has broken in for a spot of under-age rumpy-pumpy. Usually, though, it's a muntjac, which is a sort of big rat.

It's nonsense to say the countryside is quiet. No one in London is troubled by wisteria tapping on their bedroom

window, or crow-scarers. You get the rhythmic and distant rumble of jets on their final approach into Heathrow
– we get the nasal drawl of model aeroplanes. You get
the odd burglar alarm or party and you moan like hell.
But it could be worse. You don't, for instance, get combine harvesters working through the night, do you? Or
badgers tripping your security lighting. Or campanology
every bloody Sunday morning.

Sure, you have a constant background traffic roar,
but we have born-again bikers who can be heard from
40 miles away. And if your child runs into the road he'll
be hit by a car doing 10. Out here, it'll be doing 100.

In London, children can learn to ride their bicycles
on Clapham Common in almost perfect safety. Mine
will have to take their chances on a road that makes
Silverstone's Hangar Straight look like a farm track.

Mind you, cars are the only things that do move
quickly. In London, you can pop to the corner shop for a
packet of fags and be home in 30 seconds flat. You ask
the shopkeeper to give you 20 Marlboro. He says £3.38.
You pay. And that's it. Me? I have to drive to the shop,
and when I get there, it's like I've got the lead in a soap
opera. You can be there for hours.

First time I went, the bloke in front was holding a
white fiver. The woman in front of him had a purse full
of half-crowns. Nice thrupennies though. You can say
that round here. No one knows what it means.

In the bank, the cashier always studies the cheques we
pay in and exclaims in a loud enough voice to knock
down barley, 'Ooh that's a lot of money.' I'm not kidding
here. It happens all the time, even when it's a BBC repeat
fee for £18.

I think half the problem is that in the countryside you can't actually spend money. Go to the pub and people are playing shove ha'penny, so you leave. Go to the cinema and, although you can park outside, it is showing *Lethal Weapon* and everyone is coming out of the matinée saying they should make a sequel. Out here, Marc Bolan has not yet been supplanted by Mel Gibson as a sex symbol and Leonardo di Caprio has not yet been born.

However, I must confess that a world removed from that Norman Lamont lookalike in short trousers cannot be all bad.

It isn't, really, and the chief reason is parking. In London, I became used to spending the last half hour of my day roaming the streets looking for somewhere to stop. And I became used to this, like you can become used to white noise or pain, only really noticing it when it goes away.

And believe me, I've noticed that nowadays my car is parked right outside the back door every single morning. Think about that. Think about never worrying about looking for a meter ever again.

And think, too, about pulling out of your gate in the morning, knowing that you can be doing 100mph in as long as it takes your car to get there. Oh sure, provincial towns have horrendous rush-hour jams but they're short-lived, lasting from five-to until five-past nine, and then in the evening from 5.29 until 5.35.

People, after all, are in no hurry to have a drink after work with their new secretary, or that girl they met at a drinks party last night. They have their vegetable gardens to weed and things to shoot.

Seriously, everyone in the countryside shoots everything. I've become so caught up in this that last night I went into the kitchen garden and shot all the thistles from their moorings. In London, all I ever wanted to do was meet Kate Moss. Out here, all I want to do is shoot a muntjac. I also talk about moles a lot.

And this is a worry. I do get up to London regularly, and I do meet up for lunch with bright-eyed urbanites, but when it's my turn to talk I have to pause and gather my thoughts before speaking. When they've been on about Mogens Tholstrup for half an hour, it's important to avoid the mole-hill tangent.

And anyway, all I ever want to do is get lunch over so that I can wander up and down Jermyn Street. I could walk for miles in London, breathing in the abuse from taxi drivers and checking out the hemlines. The shop windows are full of mysteries: motorized pepper grinders and compass cuff links. There's a bustle. The people have a sense of purpose.

Contrast that to a walk in the countryside. It's an aimless amble with just one goal – to get back home again, to your Aga and your noisy plumbing. On a walk in the countryside you'll see trees and brambles, but where's the fun in that? You've seen them before, and you'll see them again. A bramble is not, and never will be, even remotely interesting. And nor is a fern. And nor is a woodpecker – not that we see too many of those out here. They've all been shot.

So why then, really, am I here? Well it's simple, actually. If you live in London, you can't have a Ferrari.

Beetle mania

Launching the new Beetle to quite the largest gathering of motoring journalists I've ever seen could not have been easy for Dr Ferdinand Piech, head of Volkswagen. Obviously, he had to make reference to the old Beetle – which, rather inconveniently, was inspired by Adolf Hitler. This is not a big selling point. Hitler told his motor industry to design a little car so people could enjoy the new autobahns. It should cost less than 900 Marks and it would be called the 'Strength Through Joy'. Again, not a big selling point. Only after the war, when a British major got the old Wolfsburg factory up and running again, did the rear-engined tool with its unusual faired-in headlamps come to be known as the Beetle. And who came up with that? Step forward Gordon Wilkins – one of the first *Top Gear* presenters. Does this mean that in future the Vectra will be called the Dungheap?

None of this war stuff was mentioned in the press conference. Instead, we got Janis Joplin singing, rather cleverly, 'Oh Lord, won't you buy me a brand-new Beetle'. And afterwards, in one of the most lavish corporate videos I've ever seen, we saw hippies and flower-power people, at Woodstock and in San Francisco, naked and stoned. Earlier, we had been to a huge party in the old Roxy Theatre in Atlanta, Georgia, where, to the accompaniment of the worst Hendrix tribute artist in the world, waitresses in miniskirts and waiters in tie-dye T-shirts offered us free love and beer. But why, for heaven's sake? The Beetle has been around for seven decades. Why should it have come to symbolize the '60s?

The video could have shown SS stormtroopers burning books in Poland, or vast hordes of underpaid Mexican peasants, or my mum using her Beetle to jump-start yet another of my dad's ailing Fords. And it would have been just as relevant. I mean, the Queen Mother was around in the 1960s too, but she's hardly an icon of free love is she?

Anyway, when the rather clever video, which had been set to The Who's 'My Generation' and the Stones' 'Under My Thumb', finished, the lights in that vast auditorium were turned back on and there on the stage were . . . seven Germans in suits. They'd been hammering away all evening about what fun the old Beetle had been and how much fun the new one was, and yet . . . and yet. Fun. German. German. Fun. These two words do not sit well together. Dr Piech, notorious in the car world as easily the least funny man alive, tried to smile, but I suspect there was a public relations man under his desk tickling him. It was more of a grimace.

I suppose that now is a good time to explain that I was never a fan of the old Beetle. I mean the engine was air-cooled – why? And located at the back, behind the rear axle – why? It had a crappy suspension design too, so anyone trying to corner with any verve would end up facing the other way, or dead. The heater didn't work, the six-volt power supply was disingenuous, and if weathermen even thought it might drizzle later, the sills would oxidize. It was a poor design, badly built and horrid to drive. And that's exactly why it did so well in the 1960s. It was bought by a bunch of tree-huggers precisely because it was crappy. Ideally, they would like to have driven around in a bush, but as this was not possible they chose the worst car available. Like now. Visit any road protesters' hide-out and you'll find the car park awash with 2CVs. Another anti-car car.

At this point, fans of the Beetle will doubtless point out that 21 million have been sold, many to people like my mum, who has never felt tempted to hug a silver birch. Quite right, and nor do the vast army of South American Beetle drivers have much to do with trees – except for chopping a lot of them down, that is. Sure, but, you see, the Beetle's greatest strength has always been its cheapness. It was designed to be cheap, and in Mexico, where it lives on, it still is. My mum had one because it was cheap. Tree-huggers had them because they were cheap. Students buy them even today because they're cheap. But they are not, and never have been, fun. Whereas with the new car, it's the other way round.

Football is an A Class drug

As it's the British Grand Prix today, you're probably expecting this column to focus on the battle between the talent of Michael Schumacher and the technical supremacy of McLaren. Well sorry, but I just don't care any more.

Along with most of the country I've recently been introduced to football, and I've seen the light. That match between England and Argentina was, without any question or shadow of doubt, the most gut-wrenching two hours of my entire life.

I genuinely do not know how football fans live though this torment every weekend. During the penalty shoot-out I was, medically speaking, in the throes of a massive coronary. My heart had stopped, and even if a Bengal tiger had started to savage my wife, I'd have been unable to move. Football has introduced me to the true meaning of passion.

If someone overtakes a Ferrari during a motor race, I'll tut and wander into the kitchen to see if by some miracle there's a cold sausage in the fridge. But when Argentina were awarded that penalty, I found myself sprawled on the floor demanding that a gunboat be sent immediately to Copenhagen harbour. I wanted to rip the little mermaid to shreds. I sobbed to my wife that we would never, EVER have Danish bacon in the house again. EVER, do you understand? And then David Beckham was sent off.

I thought I hated the referee, but this was something else. My brain concocted a whole new chemical for this one. Had you cut me at that moment, I'd have bled concentrated sulphuric acid. When they showed that slow-motion replay of Beckham's ill-tempered foul, I really and truly wanted to smash him in the face with a tyre iron. And even now, two weeks later, I still lie awake at night dreaming up new and imaginative ways of making him pay. I'm told he sometimes goes out dressed as a woman. Well good, because when I'm finished with him he will be one.

It wasn't all hatred though. As the game unfolded I began to fall in love with Tony Adams, fantasizing about moving with him to a little cottage in Devon and rearing geese. And when Michael Owen scored that goal to put England ahead, I experienced a euphoria way beyond the ken of mortal man.

So when England were knocked out I couldn't simply

stop watching. I needed more of these incredible highs and they came in spades when Germany lost 3–0 to a country that didn't even exist five years ago.

I suspect all of us like to see the Germans fail from time to time, and that's why I've rather enjoyed this whole Mercedes A Class saga. This tiny little hatchback, with its extraordinarily large interior, was going to take over the world. But half-way through Poland it fell over.

A journalist discovered that if you performed a sudden lane-change manoeuvre at anything above 37mph, the A Class would flip onto its roof. On the whole, this was a 'bad thing'.

So bad, in fact, that Mercedes withdrew their new wonder-car from sale and set about making some changes. Clever traction control was added, the suspension was modified, and now the A Class is back.

So, have the changes worked? Well this week I waved a tearful goodbye to the children, checked my life insurance policy, took the baby Benz to an airfield, and went completely bonkers. I built it up to 90mph and went from full left lock to full right lock. I braked in the middle of corners, I did handbrake turns. I completely wrecked the tyres. And I'm sitting here now, writing this, so all is well . . .

. . . nearly. Unfortunately, the changes Mercedes made to the suspension have endowed the A Class with quite the most unforgiving ride you could possibly imagine. You would have David Beckham round for tea before you'd deliberately run over a cat's-eye in this car.

I was therefore staggered to read that, for £180, you can fit firm, sports suspension. Really, you mustn't. It's quite firm enough already, and anyway, this most definitely is not a sports car. Try any sort of speedy driving and in half a nanosecond the traction control comes down on you like a silicone ton of bricks.

I really do believe that what we have here is a bad chassis with a Band-Aid on it. It doesn't work and, anyway, the A Class is ridiculously expensive. I know it's a Mercedes but I find it hard to accept that a car which is shorter than a Ford Fiesta should cost, depending on the engine and trim you choose, between £14,490 and £17,890.

This is a great deal of money, especially when I tell you that Renault will sell you a larger, more comfortable and more practical Scenic for less than £13,000.

And yet. To drive a Scenic is to advertise the fact that you've had it. You have children and a gut. Your life is ruled, not by a need to be attractive and funny, but by the prices at Ikea. I've seen you in the supermarket, buying washing-up bowls.

The Mercedes may be horrid to drive and stupidly expensive, but in St Tropez this year it is *the* car. It may come with a boot, space inside for five and an engine that, in a head-on crash, slides under your legs. But it is also cool and funky. For years we've been eating lettuce and now Mercedes has given us some rocket.

Yank tank flattens Prestbury

I suspect that last year some corporate bigwig at General Motors was given an atlas for Christmas. And I suppose it must have been quite a shock for the poor chap to find that his teachers, the newspapers and the television news had all been lying.

Imagine. For 50 years he had known without any doubt that God was called Hank and that the world stopped at Los Angeles. He knew the Americans had tried without success to find new civilizations – the launch pad at Cape Canaveral was proof of that. But here on his lap was this atlas – a book which spoke of strange and exotic new places where people breathed air and had central heating and Corby trouser presses. And yes, even cars.

Back at work after the Christmas break he would have been treated as something of a lunatic, as he rushed around telling his colleagues that there were life forms outside the USA. 'What? In the ocean you mean? Fish? Whales? Sea cucumbers?'

'No no. There are bipeds. In places like Japan and South Africa and The United Britain. And we can sell them cars. All we need do is put the steering wheel on the other side. We'll be rich.'

And this did it, because now there's an armada of General motors heading for the UK. There's the Chevrolet Camaro, the big four-wheel drive Blazer, the Corvette and, most amazingly of all, the £40,000 Cadillac Seville STS.

Oh dear. I appear to have put Prestbury into a state of cataclysmic shock. For years, people in the neo-Georgian suburbs of Manchester have been on the look-out for something a little more vulgar and ostentatious than a Rolls-Royce, and now it's coming. Not since Parker Knoll brought out their last recliner has Cheshire been in quite such a heightened state of expectation. The people there need to know what this new car is like.

Well now, I drove a Cadillac Seville last year and it was simply incredible. You could stop, get out, go shopping, have dinner and when you got back to the car three hours later it would still be rocking back and forth.

It may have looked a little more restrained than the finned, chromed monsters from the late 1950s but it was still as soft as a puppy, with the directional control of Bambi. In Arizona it was, of course, very comfortable, but for trips into Wilmslow it would have been utterly hopeless.

Cadillac, however, has not only moved the steering wheel but they've also changed the suspension. Indeed, I drove one this week and can report it doesn't really have any.

They've noted that while American footballers take to the field in an all-over body tampon, rugby players protect their bones with nothing more than a shirt. So they obviously figure we don't need springs or dampers – just four bloody great RSJs. And the he-man steering is so macho the wheel has a full beard.

There is, however, terrible disappointment elsewhere in the interior. Cheshire, I'm sure, was hoping for pearlescent vulgalour upholstery, Las Vegas lighting and button-backed, white carpets. But no. You get black leather and exactly the same sort of wood they used on Garrard turntables in 1975. You don't even get a back-lit gold Cadillac motif in the middle of the steering wheel.

But do not think for one minute that this is a low-key sports saloon like the BMW 5 series. It has front-wheel drive for a start, and the automatic gearbox works in geological time. Put your foot down and several aeons will slide by before it kicks down.

And nor is it an executive cruiser. First of all, it is stupidly noisy thanks to absurd tyre roar and second, the driver's seat is modelled on an electric chair. It features a device called auto lumbar support, which moves in tandem with your body.

Unfortunately, it was designed to support the average American back, which is a task every bit as difficult as propping up the Leaning Tower of Pisa. I suspect that slender, lettuce-fed Prestbury backs will tire very quickly of being asked to rest on what is basically a piece of heavy engineering.

But you will never tire of the engine. This 4.6 litre V8 produces 300 horsepower, and that's enough, despite the best efforts of the Darwinian gearbox, to get the Seville from 0 to 60 in 6.8 seconds and on to 150. Apparently, 19mpg should also be possible, and it is, if you are towed everywhere.

So what's the big deal then? Well, not only does it sound utterly wonderful at high revs but it only needs servicing once every 100,000 miles. And, thanks to clever cylinder management, it can even run for 50 miles with no oil and no water. This will be a boon to the mink coat and no-knickers set, who are forever laughing over a lettuce lunch about how, that morning, they filled up the washer bottle with diesel.

An amazing engine, however, isn't enough. If only they'd sent the Seville here with pleblon seats and Fablon decals down the side, Prestbury would gave gone, to coin a local phrase, 'mad for it'.

But instead, they've toned it down, hoping to pick up a few BMW and Jaguar drivers. And they've failed on that one too, because the Seville just isn't good enough. It was a brave effort from our man with the atlas, but then it was a brave effort too the last time an American picked up a map of the world . . . and the army got sent to Vietnam.

Supercar suicide

Tiff doesn't want you to know this, and after telling you I'm probably going to need another boyfriend, but last week, at the Pembury race track in Wales, he stuffed a Honda NSX. When Quentin and I heard, we exchanged glances and immediately guessed what had happened. Tiff, we reckoned, was too vain to wear his glasses on television, but without them he's something of a mole. He was just trundling along, flashing his boyish smile at the camera and quite simply, never saw the corner.

In fact, the truth is somewhat different. You see, I've now seen the footage, and Tiff saw the corner just fine. He was sailing round it with a fair bit of understeer which he tried to correct with a little flick – a little *soupçon* to upset the back end. It worked too, but the rear just snapped round, lunging Tiff and £70,000-worth of supercar towards the end of the pit wall. Now, if you hit the end of a wall sideways, at 80mph, you're dead. It would have been Goodnight Tiff. But, amazingly, the car slid into the pit lane and had scrubbed nearly all its speed off when it hit a bank of tyres at 30mph or so. Tiff says the accident happened in slow motion and that he had time to sit there wondering what on earth had gone wrong. I mean, being a racing driver, the accident obviously wasn't his fault . . . And much as it pains me to admit this, I think he's right. It isn't that he's old and blind. It's the NSX. I think there's something wrong with it.

You see, back in the dim and distant past, Derek Warwick tested one of these cars for us ... and spun off. In the Nurburgring story we ran, back in the autumn, the only car to leave the track with Barry 'Whizzo' Williams at the helm was the NSX. And now I'm hearing rumours that Mark Hales, another seriously good race driver, recently stuffed one.

So that's the NSX off the list then. You see, here's my problem. Last night, while my wife and I snuggled up in front of *Kavanagh QC,* she leant across and, out of the blue, said: 'So, are you going to keep the Ferrari then?' This is the equivalent of a salesman with his foot in the door just before he barges into the living room and spreads brochures all over the rug. Right now, she's wondering why I keep a car that I hardly ever use. Pretty soon the wondering stops. And the recriminations start. To be honest, I've toyed with the idea of changing it, but for what? Certainly not an NSX and, much as I am impressed by the sports-exhausted Diablo, I'm not a has-been rock star. I have to admit I've been going through a Lancia Stratos phase, but I fear I'd use that even less than the 355. (He strokes his chin . . .) I've also thought about the Jaguar XJ220.

I'm sure you know by now that the new left-hand drives are up for sale at an Essex dealer for the sum of £150,000 each. You can, however, acquire a lightly used right-hooker, I understand, for a mere £85,000. At 17 feet long and 7 feet wide, you have to admit that the 220 is an awful lot of car for £85,000. I had to admit it too, which is why last week I found myself in Wales driving an XJ220 for the first time, in anger. I'd been told it was heavy and cumbersome but when you bury the throttle, the power is sensational. This is noticeably more accelerative than a Ferrari F50 and, as we all know, much faster at the top end. It's also a stable high-speed cruiser. You may have noticed in recent years that in race trim the McLarens have sprouted elongated tails – which makes them more steady at 220mph. But the Jag comes with a lengthy back end to start with.

I have to admit, I was falling madly in love with this previously unloved hypercar – until I needed to brake hard, that is. In the nick of time, I heard a little voice in my head. It was Tiff saying the XJ220 would be a lot better if only it had brakes. And so, when I stamped on the pedal, I was half ready for what happened. And what happened was nothing. Honestly, it was amazing. I had both feet on the pedal and I was still doing 100mph or more. I was still doing 100mph when I got home and saw the Ferrari just sitting there. Swap it? You must be joking. I'd rather lend it to Needell.

Bedtime stories with Hans Christian Prescott

I have a dream. I see a world with happy, rosy-cheeked children scrumping apples. When you ring to book a seat at the cinema you will talk, not to a machine, but to Ma Larkin. And there will be an interval in the film where you eat pork pies and fudge.

No one will have a mobile phone that plays 'The

Grand Old Duke of York' and Bernard Cribbins will run your local railway station. Estuary English will be spoken only in the Thames Estuary which, incidentally, will be full of cormorants. And no one will die of anything.

Now I could publish this in a White Paper but you'd all laugh. You'd know that the Thames Estuary children would shoot all the cormorants and that Bernard Cribbins is already dead.

Well it's much the same deal with the vision of Britain outlined this week by Mr Prescott in his much talked-about White Paper. I've read every one of the 160 pages and it is fantastic. No one could possibly argue with any one of the fat man's dreams but, sadly, that's what they are – dreams.

Take point 5.10. 'We need to improve the image of the bus if we are to attract people who are used to the style and comfort of modern cars.' And it goes on to say that the bus industry must respond to the challenge with a vehicle designed for the twenty-first century.

Right, well if you want me out of the car and in one of your buses by 2000, you've got 18 months to come up with a vehicle that can do this . . .

Yesterday, while making a white sauce, I found I needed some more milk and had to get to and from the shop in less than three minutes. I shall need a service that can handle this.

This afternoon, my mother is coming to Oxfordshire from Peterborough with two small children and their nanny. They don't want to go via two train stations in London. So, if this new public transport is going to be as convenient as the car, there must be a bus service from Castor to Chipping Norton, 30 times a day.

And on board the bus I want electric Recaro seats finished in the finest hide, I want television, I want air conditioning and I must be able to play whatever music I wish without disturbing any of the other passengers. Also, there must be a screen, such as you find in the first class section of a British Airways 777, so that I can pick my nose without being overlooked. I shall also wish to smoke.

The bus must also be eco-friendly, so obviously a diesel engine is out of the question. Gas might be an answer, but a big V8 is better. Certainly, I shall be looking for 0 to 60 in less than ten seconds and a top speed of 150 or so. And it must be designed by Pininfarina.

OK. Got all that? Well it gets worse because the service, I'm afraid, has to be free. You see, Mr Prescott has said it's all right for me to have a car but that I must leave it at home more often. Fine, but I'll have paid for it and road tax is applicable no matter how infrequently I use it. I therefore can't afford to spend even *more* money on a bus fare.

To address this, the White Paper says that I will have to pay to use motorways and that I will be charged if I drive into a city centre. I see, and how will this be done then?

Will there be toll booths on every single road into London, all 10,000 of them? Or will I be forced to fit my car with an electronic device that can be read by roadside monitors? And if so, who will pay for this device to be fitted?

Sadly, the White Paper fails to explain this, in the same way that Enid Blyton fails to explain how Noddy, a wooden puppet, manages to converse with an elephant.

Undaunted, Mr Prescott goes on to say that by charging tolls to use roads, and taxing car-parking spaces, super-efficient, dream-world local authorities will be able to raise a billion pounds a year. They won't lose it. They won't waste it on twinning ceremonies. They'll spend it on public transport.

Oh dear. I'm afraid that in Mr Prescott's world, where everyone drinks Ovaltine and Jenny Agutter is 13, a billion pounds is a lot of money. But in fact a new double-decker costs £130,000, and as a result a billion won't even buy *one* for each town in the country.

The chances therefore, of getting a service from Castor to Chipping Norton 30 times a day are somewhat remote.

But that's not the end of the world. You see, if car travel were as bad as everyone says, no one would do it. And things are going to get better and better.

Already, we have the same number of cars on the roads as we do people with driving licences. So unless we perfect the art of driving two cars at once, the projected 30 per cent increase in traffic volume just can't happen. In fact, as people start to work at home more often, it'll probably decrease slightly.

I do, however, think that Mr Prescott's White Paper has a place. If it were illustrated with attractive drawings, it might even supersede 'Thomas the Tank Engine' as my daughter's favourite night-time story.

Clarkson soils his jeans

It was announced this week that the market for jeans has suffered a dramatic fall, and that I am completely responsible.

Marketing experts at the *Daily Mail* say that because of my fondness for the denim trouser, youngsters now see jeans as old people's clothing, and as a result, won't buy them. This has been christened the 'Jeremy Clarkson Effect'.

Wow. My very own effect. You can forget knighthoods and OBEs. They're for people. I'm to be talked about with the reverence of a star or a moon. I'm a galactic guiding light and I'm going to be very, very rich.

You see, the British jeans market shrank from £609.5 million in 1996/7 to a paltry £561.2 million in 1997/8. That's a fall of 14.3 per cent, and I don't doubt for one minute that the people at Levi's and Lee are desperate.

Well chaps, I have a solution. Pay me £40 million a year and I'll stop wearing them.

But what shall I wear instead? I can't possibly switch to trousers. Trousers are what you needed to wear to get into northern nightclubs, aged 18. Trousers are what you bought at Harry Fenton in the Arndale Centre. My father wore trousers. Jim Callaghan wore trousers.

I wore jeans in 1976 because there was no real choice. Oh sure, Levi's did a range of corduroys in rebellious, lurid colours and I once sent away to a company that advertised in the back of *New Musical Express* for a pair of velvet loons. But while bulldozing myself into them they burst, and that was that.

You must understand that I was brought up under the 'David Dundas Effect'. I told my careers master that I didn't care what I did after leaving school, so long as I didn't have to wear a suit. And to this day, I still don't own one.

My father used to sit at the kitchen table in his slacks telling me that by trying to be different I had ended up looking the same as everyone else. As far as he was concerned Led Zep sounded just the same as Rick Wakeman and jeans were jeans.

But jeans, most emphatically, were not *just* jeans. You would not, for instance, be seen dead in a pair of Wranglers. Wranglers were too dark and even after two years they still felt and looked too stiff. I once chucked a girlfriend when she came out at night with a brace of Ws on her backside.

Wranglers were for people who liked country-and-western music, so back then I wore Levi's with the forward tip of the flare hanging *exactly* a quarter of an inch over the platform sole of my shoe. And then, when I was first introduced to The Clash, I switched to straight-leg C17s from France.

And today, the badge is still important. Today, I see Levi's as a bit too Toto, a bit too middle-of-the-road American soft rock. I therefore wear British Lee Coopers and I can still spot M & S denim at 1000 paces.

Furthermore, I am not alone. Andy Wilman, my producer and co-presenter on *Top Gear Waterworld,* once trod in some human excrement while walking through an unlit Calcutta backstreet. As we sat over dinner in the Fairlawn Hotel he examined the splashes of faeces exclaiming, 'Well this is a bird-puller.'

Back at our hotel he deposited them in the laundry and was horrified, when they came back, to discover that the Indians had ironed a crease down the front. And of course, once a crease has been ironed in, that telltale faded line never goes away. So he threw them in the bin.

So, let's just sum up here. He was prepared to keep them after they'd been bathed in shit, but once they'd been stained with the indelible mark of a man who lives at home and lets his mother do the ironing they had to be binned. I understand that completely.

And I was therefore horrified to be accused in the *Independent* this week of wearing 'nasty, really nasty, stone-washed jeans'. I have never, and will never, wear stone-washed anything. Thank Christ it was the *Independent* so no one will have read this astonishing and libellous slur.

But the point has been made all week. Jeans are for old men. Ian McShane is 55. Tony Blair is 44 and I'm 38. We wear jeans and we listen to original Doobie Brothers tunes in the car, not the jungle remix versions. And it's all our fault that denim sales are in freefall. The moon does the tides. The sun does the central heating. And I do for Levi Strauss.

But sadly, this isn't entirely accurate. The real reason why today's *poulets de printemps* are not buying jeans is because they have a choice which extends beyond the impractical loon.

Today, you don't *have* to look like Roger Daltrey. You can, if you wish, decide to model yourself on a 19-year-old negro from the east wing of an American jail. Seriously, they're not allowed to have belts in Yank clink and, as a result, wear their trousers so low on the hip that their underwear is on plain view.

Upon their release they continue to dress in this fashion, to show they've been in jail and thus get some respect. And as a result, everyone now shows off the elastic of their underwear as a means of demonstrating some wayward past.

Last year, in Texas, I met a chap who had the low rider strides but no underpants. You could see the top of his penis, and I must confess I found this rather shocking. But that, I suppose, is the point.

At agreeable dinner parties round these parts, we sit around pompously wondering how on earth our children will shock us with their behaviour. We know all about drugs and we've all staggered from sweaty dives at four in the morning with perforated eardrums. I've even been in the cells. Twice.

But come on. We never had ecstasy or alcopops. My father used to listen to 'Hi Ho Silver Lining' with his fingers in his ears, telling me over and over that it was just tuneless noise. Which is pretty much how I react when *Top of the Pops* comes on these days. Modern music is supposed to scare the old. Yes it used to scare my Mum in the same way that I'm frightened half to death these days by rap artistes advising their fans to kill a pig.

Then there's television. We had *Tiswas,* where people threw custard pies at one another and Sally James wore a tank top. Now we've got *TFI Friday* and stuff on late at night that I really, really don't understand.

So of course jeans sales are falling. I bet Rick Wakeman doesn't shift many CDs either. Times move on.

And in terms of fashion, they've moved on to combat trousers. The original supply of army surplus clothing dried up years ago, so today the big names in fashion are offering a trouser with baggy side pockets and an infantry cut. If Tony Blair really does want to be seen as a man of the people this is what he needs – a pair of Gap cargo pants, as I believe they're called.

And the jeans people have to stop worrying. They've been at the top for nigh on 30 years and now The Verve and Oasis have kicked them into touch. They need to accept that the denim trouser is no longer at the cutting edge of rebellion and is now acceptable in all but the most stuffy clubs and restaurants.

I'm sorry to bring cars into the equation here but the denim industry would be well advised to talk to Honda, whose cars appeal largely to the older, more mature driver. What's wrong with that? What's the point of spending a fortune making something fashionable and cool when it just isn't. And anyway the old outnumber the young. It's a bigger market with more spending power.

Jeans, you see, will never go away. On Tuesday night I wore a pair at dinner in some desperately flash German hotel in Baden–Baden and the next day I was scrabbling around in them on an oily conveyor belt at Heathrow. You can't do both in any other sort of trouser.

Burning rubber with Tara Palmer-Tailslide

If you want to arrive at the Pearly Gates in soggy pants, you don't need to have died at the hands of a firing squad – just try climbing into a car with Tara Palmer-Tomkinson. I've been in an F-15 and I've done 0 to 60 in one second on a snowmobile. Next week, I shall land on an aircraft carrier and a day later, strapped into an F-14, I shall take off again. I know, understand and can cope with fear.

But I lost control completely after half an hour in a car with Tara. The bladder went, and round the back, I was touching cloth. That woman is easily the maddest driver the world has ever seen. I wasn't scared to start with; it was the tail end of the rush hour, we were in the middle of London and it was raining. So even though she was using a Honda NSX, which we know to be tail-happy and skittish, I felt we'd spend most of the time doing 3mph. But no. On Chelsea Bridge, she put her foot down hard in first gear and I felt the back starting to weave, the power straining the very upper echelons of the traction control's restraining bolt. Then we were into second, foot still hard down, heading for what was unquestionably a set of red lights. The next time I opened my eyes we were heading up Sloane Street at Mach one . . . and then we weren't. There was a squeal accompanied by a full-bore test of the NSX's anti-lock braking system. 'Gucci's got a new window display,' she wailed.

During the next 26 minutes we'd stop at red lights and, on each occasion, the driver alongside was scrutinized. If he was good-looking, there was some flirting then a race. If he was ugly, she'd skip straight to the race. And she never lost. Throughout London that night a trail of BMWs and Porsches were left dazed and confused at the side of the road, their drivers emerging from the tangled mess asking passers-by, quietly, for hot sweet tea. They looked like they'd been victims of a tornado, and in a sense, they had. They'd been TP-T-ed.

Now earlier in the day, I'd talked to various girls who

have finally realized that there's no point spending a fortune on clothes and hairstyles if they are going to go out at night in a crappy car. So Katy Hill from *Blue Peter* has a Porsche Boxster. Emma Noble has an MGF (while boyfriend James Major has a corduroy Rover 200). Dani Behr has a BMW 328 convertible and Julia Bradbury, from something called Channel 5, has a Mercedes SLK. I shot the breeze with all of them about how these new sports cars are very definitely for the girlies, and how men today need to spend more if they want something macho. And all of them said that the biggest problem they faced on the roads was blokes trying to take them on. And that struck me as odd. I mean, I'm a bloke and I never, ever, feel the need to race a girl at the lights just because she's a girl. I phoned all my male friends and they all said the same – we like a fit bird in a nice car. It looks good. I therefore suspect that it is not men who take on women. It is the other way round.

It's like Kuwait. After the recent unpleasantness people there walk around with their chests puffed out, telling their neighbours in the north that they have all sorts of new hardware and will take them on any day. Iraq, on the other hand, can afford to sit there doing nothing, and it's still frightening. Iraq is a past aggressor. Iraq is a man. And Iraq won't ever attack Kuwait again. Men, in recent months, have also been strangled by sanctions.

Turn on the television and you'll find *Playing the Field*, where a bunch of women bully their husbands, have affairs and play football. And then there's *Real Women*, where the women are men and the men are hopeless. In both, sex scenes are all-woman affairs.

We've now got women jet pilots, women boxers and

women like Madeleine Albright and Mo Mowlam sorting out the world's trouble spots; and in the charts Take That and Boyzone have been replaced by All Saints and the Spice Girls. I'm all for equality, but it does seem at the moment that the pendulum has swung rather too far in the other direction. So, if TP-T ever comes alongside me at the traffic lights she should be aware that I now have a supercharged V8 under the bonnet, that the Jag in question isn't mine, and that I will win.

Jag sinks its teeth in

So, what's been happening this week then? Well I've been to St Tropez for a little break, but as I rented a diesel-powered Citroen people-carrier it's hard to think of a motoring angle.

What else? Oh I know. While screaming down a test track in Wiltshire on Tuesday, I crashed a 215mph, £750,000 Jaguar XJR15. And I suppose that, with a bit of hyperbole here and a small embellishment there, this is quite good column material.

I'd been warned about this remarkable supercar before I ever squeezed my haddock-sized frame into that sardine tin they call a cockpit. Derek Warwick, the former Grand Prix driver, was very unkind about its waywardness and my colleague Tiff Needell called it 'the worst-handling car ever made'.

It seems that though it was conceived as a road-going Le Mans car and was turned back into a racing car again, the engine is located so far from the ground that it's a menace to orbiting satellites. So the car has its centre of gravity in the clouds and, to make matters worse, no downforce at all. And that, in a car of this type, is as bad as having an aeroplane with no lift.

This, I'm sure, was a big concern for the boys who raced them, but I was at Kemble Airfield surrounded by mile upon mile of runway.

I hit the starter and behind a wafer-thin fire wall a 6.0 litre V12 engine exploded into life. It's so noisy that it kills all known wildlife within a 50 mile radius – even worms – and at full chat it rocks the needles on seismographs in California.

Completely deaf, I started to grapple with the gear lever which, though this is a right-hand drive car, was located by my right knee. Move it a millimetre to the left and you're in first; 1.1 millimetres and you're in reverse. This makes changing gear a more exact science than splitting the atom.

And then there's the clutch. Equipped with the sort of spring that they use on oil rig platforms, there is nothing for the first 9 inches of travel and then, just when you think you're still in neutral, drive is fired at the back wheels and you're doing 80mph. Backwards sometimes, if you got the wrong gear.

I was really very scared, even though I knew there was nothing to hit . . . except the camera car. But you can't hit a great big blue Mondeo which is always travelling in the same direction as you are, at the same speed.

I did though. After a 15 minute bottom-sniffing, getting-to-know-you period I made sure the XJR15 was pointing in a dead straight line and, in second gear at 50mph, floored the throttle.

On the streaming wet runway the car started to slide, but it was hard, in the cockpit, to detect this loss of traction, this imperceptible drift to the left. So I kept my foot hard down and didn't really start to dial in some opposite lock until it was far too late. The back flicked the other way and bang, my front corner slammed into the camera car, the only solid object within 40 miles.

Damage to the Jaguar was remarkably light, but its owner pointed out that even a 4 inch crease in a double-skinned piece of honeycombed carbon-fibre cannot be knocked back into shape by an apprentice mechanic. It needed a whole new front-end, and guess what – Halfords don't sell nose sections for the Jaguar XJR15. No one does. Only 30 cars were ever made, and after 11 were crashed in one race alone there are no spare parts left.

Had I been equipped with a tail at this point, I'd have put it between my legs and run off into the bushes, where I would have beaten myself with twigs. But then the owner said he'd use a bit of filler and all would be well. Filler in a £750,000 car – that's like using Humbrol to touch up the *Haywain*.

Still, I am now officially a bad workman so I'm allowed to blame my tool. The XJR15 may be one of the most beautiful cars ever to see the light of day but it is bloody dangerous. And there's nothing to stop you taking this automotive psychopath on to the road, for heaven's sake.

I read recently that the area of contact between the four tyres on a car and the road would fit on a piece of A4 paper. So it isn't so much skill that you need to handle a car like the XJR15, but temperament. You must be aware that if the road is even slightly greasy, you cannot apply full power or the tyres will lose traction and you will crash.

In an ordinary car you can mash the throttle into the carpet whenever you like, but in a 200mph hypercar you have to employ the restraint of a saint. You must feed the power in gently, let the clutch in slowly, turn the wheel carefully. Basically, if you have a Jaguar XJR15, or a Ferrari F50 or a McLaren F1, you must drive slowly.

This, I know, is a bit like going to the best restaurant in town and ordering beans on toast, but that's the way it is. If you want to drive around like your hair's on fire, rent yourself a diesel-powered Citroen people-carrier. I did and I never crashed once.

Kraut carnage in an Arnage

So, our fat friend at the Department of Transport has decided there will be no more roads, and the summer's been terrible, but cheer up. Things could be worse. You could be Dr Ferdinand Piech's cat.

Last month, the steely-eyed chairman of Volkswagen surveyed the breadth of his domain and realized, like Alexander the Great, that there were no more worlds to conquer.

Volkswagen's net earnings were up 70 per cent and it owned a raft of other household names – Audi, SEAT, Skoda, Cosworth and Lamborghini. He was a leading German industrialist, one of the richest men in Europe and an heir to the Porsche family fortune.

And on top of all that, he had just paid £470 million

for Rolls-Royce, which meant he not only had the factory in Crewe but Bentley as well. This was the jewel in his crown and, with only the Sudetenland to go, things were looking pretty damn rosy.

He couldn't have known that within a week he would be ranked alongside that American chappy from Arizona who bought the wrong bridge. He lost Rolls-Royce. One minute he had it, and then it was gone. And worse, he was beaten by a fellow German – Peter Burnt Fish Trousers, the man with the face fungus from pipsqueak BMW.

Can you imagine what it must have felt like when he discovered that he would have to give – *give* – Rolls-Royce to this impudent upstart? The rage. The angst. The cat. I need to kick something. Where is the bloody cat?

The upshot is simple. In exchange for £2.50, BMW now owns the rights to make Rolls-Royce Motor Cars – something they will do at a new purpose-built plant in the UK.

Volkswagen, on the other hand, has a factory where the instruction manual for the boiler is written in Latin. And Bentley, whose new car has a BMW engine, a BMW gearbox, BMW switchgear and BMW power steering.

Piech is now saying he never wanted Rolls-Royce in the first place and that in the fullness of time they will be making 10,000 Bentleys a year. But the only way he can do that is by raiding the corporate parts bin, and I'm not 100 per cent certain that this is such a good idea.

You see, the only suitable donor car in Piech's armoury is the Audi A8, and I don't think any of its components would do in a Bentley. It would be like serving up food from your local pub in the Caprice. Audi's V8 engine is designed to propel a lightweight, aluminium supersaloon. Put it in a Bentley and the top speed would be 4mph.

I know this because I spent last weekend with the new Arnage, which weighs 2.7 tons. Sure, it has a 4.0 BMW V8 engine but this had to be enlarged to 4.4 litres, and even that wasn't enough. So they garnished it with a brace of turbochargers. And that did the trick.

But even so, sales of the Arnage have been so pitiful that Bentley won't say how many have found homes. I do know this though. Potential buyers have been worried by the company's future, and especially the threat by BMW to cut off the supply of engines.

It's easy, of course, to be magnanimous in victory, and as a result that threat has gone away. So, the Arnage then? Worth a punt or what?

Well it is stupendously fast. Nought to 60 is dealt with in 6 seconds and the top speed is governed to 150. Put your foot down in the mid-ranges and were it not for the headrest, your neck would snap like a twig.

It handles too and, unlike the old Bentley turbo, the Arnage does not ride down the road like a tea tray. It simply uses its enormous weight to crush road surface irregularities rather than riding over them, and inside all you can hear is the fuel swilling into the cylinders.

This is conspicuous consumption on a biblical scale. When they get round to televising Enid Blyton-Prescott's Transport White Paper, the baddie will be put in an Arnage. Already, people sneer at you and no one – *no one* – will ever let you out of a side turning.

That's OK though. The Arnage has gun racks in the boot so that you may carry weapons for dealing with those who won't get out of your way. Or you could simply ram them. Or you can just recline the electric seat a tad and savour the atmosphere.

It's not quite so opulent as the Rolls–Royce Silver Seraph in that the carpet pile is only 2 feet thick. Plus, the dials are finished in a sporty cream colour and my test car had a chunky two-tone wheel – but then the airbag instructions on the back of the sun visor were in Arabic.

Despite this, I truly loved it. It's not so good to drive as the Ferrari 456 which, in turn, is blown into the weeds by the Aston Martin Vantage, a car that combines the crafts-manship of the Bentley with the power of a Tornado jet. It is now, incidentally, the fastest, most powerful car you can buy.

Yet the Bentley can hold its head high as a stunningly fast, motorized drawing room. Choosing to ignore it because it's made by Volkswagen would be idiotic. On that basis, the Aston is a Ford and the Ferrari is a Fiat. Piech should leave his cat alone. He's bought himself a barnstormer.

Absorbing the shock of European Union

Every day, 650 Members of Parliament decide what new laws they are going to foist on the country. And they're not alone. We have parish councils and borough councils and county councils and the House of Lords and the European Parliament – thousands upon thousands of people whose job is very simple. They decide how we live our lives.

They dictate what we eat, what we say, where we go,

how much we're paid, how we cut our hair and how often we're allowed to pick our noses. Then, after a while, when we get bored with their proclamations, we have an election and they're replaced with thousands of new people, all of whom have new ideas. This is democracy at work. And when it comes to democracy there's only one end product: new laws. I read recently that, last year, the European Parliament passed 27,000 new directives . . . 27,000, for Christ's sake. That's 74 a day. Think what you were allowed to do 50 years ago that you aren't now. You can't drive quickly. You can't sell things on pavements without a licence. You can't build an extension. You can't buy more than 200 duty-free cigarettes. And if you want to smoke them, you have to stand outside in the rain like a leper.

You see, even before the government gets round to actually banning something, political correctness steps in. You can't employ a girl because she's got big tits. You can't fire her for not sleeping with you. And in the time it's taken you to read this, Europe has passed another law. You are no longer allowed to keep a pet badger. And here comes another. Fish must be orange. You are, however, allowed to pop over to France and fill your car to the gunwales with cheap plonk. It's one of life's small pleasures, a tiny crumb of comfort to the battered people of this continent-sized nanny state.

But whoah, what's this? – A press release from Tenneco Automotive which says that so-called booze cruising could be a false economy. It warns drivers that if we overload our cars with wine and beer, we may not only break the law but also wreck our shock absorbers, thus negating the savings we've made on the drink. Oh really? Yup, and they've sent me a handy ready-reckoner so that I can work out how many cases of wine is acceptable. And here's the news. If you have a small car, a Fiesta, say, and there are five adults on board, the safe limit is 10 cases of wine. Well, I reckon five people would pretty much fill a Fiesta, leaving you with just the boot. And I'm pretty certain you would only get four cases in there. So that leaves you able to carry six extra cases, but with nowhere to put them.

And don't think it gets any better if you leave your passengers at home. Go by yourself and Tenneco says you can carry 30 cases. I'm sorry, but if you can get 30 cases of wine in a Fiesta, you should call Norris McWhirter. It says on my ready-reckoner that the safe limit in a large car with no passengers is 31 cases but then, according to Tenneco (a maker of shock absorbers, by the way), a Mondeo is a 'large' car. So what is a Lamborghini LM002? How many cases of wine would I be allowed to put in this V12-powered, 7-foot-high, 3-ton monster? My ready-reckoner is unable to help, but if you are allowed to put 30 cases in a Fiesta, as the Lambo is 10 times bigger I could bring back 300 cases – 3600 bottles. Which is enough to make you very, very drunk.

But there is one small problem. You can't have a Lamborghini LM002. Well you could, but in the last few minutes the European Parliament has announced that all four-wheel drive cars must be powered by corned beef from a boneless German cow. And Mr Prescott says that if you buy anything larger than a Vespa, it'll cost you £200 a minute to park it. Tenneco says that if you're going to France on a booze cruise this year, you should think about the damage you're doing to your car, the laws you may foul – and, if on a long drive, you should take frequent breaks.

I say you should get over there, buy as much as you want, in whatever car you want, and if anyone stops you, remind them that we're living in a free country. And then pull out their liver with a rusty hook. This – and I've checked – is still legal.

Minicabs: the full monty

Book yourself in for an operation and you'll have no idea who'll be wielding the knife. But you'll *know* with absolute certainty that he'll have a raft of qualifications and no history of muscular spasms.

It's the same with restaurants. You may have no clue about who is preparing your food, but it's a fair bet he'll understand that you can't put Tabasco sauce on sherry trifle.

And yet, when you want a taxi you'll summon the services of a minicab which may or may not have brakes. And it will be steered by a man who may or may not have learnt to drive in Peru.

One thing is for sure though. He won't come to your door and ring on the bell. What he will do is pull up in the middle of the road and lean on his horn, signalling that it's time for you to stop whatever you're doing and run outside.

This is unfair. He may have taken five hours to get there and in the meantime you may have met the girl of your dreams. Or you may be my wife who, having said goodbye to everyone at a party, will sit down again and give everyone a blow-by-blow account of her life.

Whatever, you will climb into the back of the cab whereupon you will be overcome with a wave of nausea. 'Can I smoke?', you'll ask. To which the answer will be no, on the grounds that tobacco leaves a lingering odour, thus making the car harder to sell. WHAT? Tobacco would *improve* things. A giant fart would improve things. Smearing the entrails of a dead dog into the seats would improve things.

Minicabs have a smell all of their own, a smell that could not possibly be replicated even in a laboratory. It's a smell that doesn't even exist in a businessman's pants. It's not stale sick or even the driver's shirt. And nor is it a mixture of the two. No, it's the smell you get from those Christmas tree air fresheners. And it's obscene.

To take my mind off the problem, I usually try to guess what sort of car I'm in. Obviously, it will have beige, pleblon seats, and obviously it will have been made in the Far East. But is it a Toyota or a Nissan?

I can understand why minicab drivers buy used Jap boxes – they're reliable and cheap to run – but who buys these cars when they're new? And why do they treat them so badly?

By the time Minicab Man is falling in love down at the auction the wheels are always square, and if you look hard you'll note that whenever you're going along in a straight line the driver is having to turn left. And what is that noise coming from the stereo? Why is it that minicab drivers listen to radio stations I've never even heard of, and how can they appreciate the wailing of the sitar when the controller back at base never shuts up?

I was once picked up by the fattest man in the world in an FSO Polonez. It broke down in the tunnel at Heathrow Airport, and do you know something? I was glad.

You see, his driving was incredible. I think he'd been to see *Star Wars* and genuinely thought he could use The Force to miss oncoming traffic. As we shot through the 18th red light, I really was expecting him to turn round and ask me, in a rasping voice, about the plans for the rebel base. He even called me 'Young Luke'.

He was the worst driver in Britain, a complete madman, and he was in a car which had started out badly and had grown worse with age. This is a terrible combination, and as I lugged my suitcases into the daylight from that tunnel I swore I'd use a better class of minicab from that point on.

And I do. The firms I call today can field brand-new people-carriers and stretched Mercedes limos. The prices reflect this but, sadly, the driving doesn't. One bloke the other day swore vehemently at every other driver on the road, saving the real torrent for anyone who actually carved him up.

When one bloke failed to let him out of a side turning, my man began a stream of abuse which continued without repetition or hesitation for three minutes. And when he ran out of suitable English phrases and expressions, he switched to Turkish. It was both breathtaking and useful. I now know how to tell someone in Ankara that he's a ★★★★★★★ ★★★★ and a ★★★★★★ with a face like a dog's ★★★★.

I should have used this the other night on a driver who had never heard of Fulham, or a guy, a couple of weeks ago, who obviously believed that his Fiat Croma could, given a long enough run-up, beat Richard Noble's land speed record.

Now obviously, at this point, every licensed Archie is sitting at home beating his children with a rolled-up copy of the Style section, telling anyone who'll listen that I should use black cabs. This is true, but unfortunately he will still be sitting there, shouting, when I next need some wheels.

So we'll all keep on using minicabs. Just beware though. If you find yourself sitting behind a bloke who answers to the name of Darth, get out and run for your life. He's easy to spot because he has a long black cape, a black helmet and asthma.

Supercar crash in Stock Exchange

As I look out of the window today I can see the storm clouds gathering. Mr Blair will be torn from his mountings and the Met Bar will be forced to introduce an all-night happy hour.

The Dome will be cast into the North Sea and Rotherham will be wiped from the map. The brief sprouting of industry in Corby will be erased and easyJet will crash. Be in absolutely no doubt on this one: Recession Tony is on its way.

Now I think it wise at this point to explain that I have never read the *Financial Times*. Also, I achieved an 'unclassified' grade in my economics A-level, partly because I forgot to turn up for the exam and partly because I forgot to turn up for any of the lessons either.

I've tried, really I've tried, to understand the implications of handing over interest rate control to the Bank of England, but every time I think I've grasped it I fall into a deep and dreamless sleep. And I'm sorry, but I really cannot work out why the Brazilian balance of payments deficit will make me less likely to buy a cauliflower.

I write this column. Rupert Murdoch gives me some of his money. I spend it. And I'm sorry, but you can wipe as much as you like off the Japanese stock market but it won't make the slightest difference. No, don't argue. It won't.

Nevertheless, I am able to predict the onset of a financial holocaust because the world's car firms are getting cocky again.

Remember what happened last time. They reacted quite late to the mid-80s boom and began work on a series of new and frighteningly expensive cars which crept onto the market at exactly the same time as Recession John started to blow things over in EC4.

Jaguar were left with an unwanted stock of XJ220s. McLaren only managed to shift 47 of their preposterous F1s. Bugatti went to the bottom of the Tiber wearing lead pants and Lamborghini ended up being run by a Malaysian pop star.

I don't doubt for one minute that after the debâcle of 1992 every single car exec in the world declared that he would never again be tempted to make a megabucks supercar. But when the world is bathing in greenbacks it's hard to resist, and already word is starting to creep out that the car execs have relented.

I read in this newspaper that Mercedes and McLaren are planning to make a £150,000 supercar, but as it was in the Business section I lost the plot after that. I do understand, however, that Maserati is back in business with a new coupé and that Audi has bought Lamborghini, who are known to be working on a new one-ounce, one-million-horsepower dream-mobile. This will force Ferrari to fight back with something so light it'll need mooring ropes.

And then there's Jaguar who, at the Paris motor show this week, will unveil a car called the XK180. It is propelled by a supercharged V8 engine which is said to develop nearly 450bhp. And for those of you who understand, 445 ft/lb of torque.

I can't even begin to guess how fast it will go, but as the two-door, convertible body is made entirely from aluminium, obviously it should be able to outrun a Nissan Micra.

I can tell you, however, that I've seen this spiritual successor to the D type and it is quite simply the most beautiful car ever made. Remarkable, when you discover it was designed on the back of a cigarette packet.

Last December, Jaguar decided to mark the 50th anniversary of the XK range and talked about turning the supercharged XK8 into more of a driver's car, perhaps shortening it a bit and taking out some of the Houston dentistry in the suspension.

Work began in February. On Saturday mornings, a group of die-hard petrolheads could be found ferreting away in what has now become known as the Special Vehicle Operation. They went by instinct, deciding quite late on to make the panels from aluminium. And then, even later, that if they were using a new material they may as well have a new look too. It's this sort of ingenuity, remember, which gave us the jet engine and the hover-craft. And now it's given us the XK180.

It doesn't work, of course. Well no, that's not true. It does work, but thanks to all sorts of Euro busybodery, they could never actually sell one. However, at the show in Paris they will be bombarded with requests to turn this D type dream into a production reality, and it'll be almost impossible to resist.

I'm sure they'll make it, and I'm sure too, that the day they choose to launch it – with a price tag of £150,000? – will be the same day that the National Lottery goes bust. And no one will buy a car that costs four times more than a large and sumptuous house.

Fairly soon now, the car industry must learn to get itself in step with the economy. As soon as the 14th floor windowsills in the City are full of men in suits, crying, they should start to work on a high-price, low-volume supercar.

That way, the car would be ready to go on sale just *after* the economy has recovered. And that, really, is the key. The economy always does recover because we'll always want cauliflowers and we'll always want cars that go like stink. Galbraith? It's my middle name.

The school run

I remember with vivid clarity the moment when I began to grow up. I was 22 years old, standing in a hardware shop asking the assistant if I could buy a washing-up bowl.

Until that point I had spent my money on beer, cigarettes, rent and, begrudgingly, the occasional Christmas present. Never before had I wasted it on something useful.

I remember vividly driving home staring at the bowl, knowing with crushing certainty that pretty soon I'd be out there buying light bulbs and white goods, things from which I would derive no pleasure whatsoever.

Fifteen years later, stage two was reached and I retuned my car stereo to Radio Two, and then this week I reached stage three. I did the school run.

And from now on there is no chance of enjoying an impromptu night out with friends, lest it become a hazy stumble into the early hours. You can't go to bed at four in the morning when you have to be sober and hearty just three hours later.

And let me tell you this. Buying a washing-up bowl was dull. Turning to Radio Two was inevitable. But the school run is hell on earth, because it removes the last vestige of fun from the concept of motoring. It turns a car from a thrusting symbol of virility into a tool, a tool that goes head-to-head with a washing-up bowl. And loses.

For a kick-off, I have to forget all about Terry Wogan. Instead I'm forced to listen over and over to Aqua's 'Barbie' song, which lines up alongside the '1812 Overture' as the least appropriate morning music ever written. Should I ever meet this Swedish band, I promise you this: I will kill them.

And then there's my son, who, at the age of two, can identify every single car on the road. On its own this would be a source of pride, but what turns it to something less savoury is the fact that he *does* identify every car on the road.

We have to drive along with a Swede telling me that life in plastic is fantastic while the boy child shouts out the name of every car going the other way. He can't say 'Good morning' but seems to have no problem with 'Daihatsu'.

Obviously, I would like to get the journey over as quickly as the Mondeo's V6 engine will go, but there's a pressure valve built into every parent which activates as soon as children climb into the back of a car. Suddenly, I lose the ability to overtake.

I may be stuck behind a tractor and the road may be clear for 200 miles but I am incapable of dropping a cog and going for it. When I'm driving with the children, I become exactly the sort of person I shout at.

There's another issue too. You daren't overtake the car in front in case it turns out to be another parent taking their children to the same school. You'd be classified as a maniac and your children would be bullied.

So we all drive along at 2mph, each of us being bombarded with Aqua, until we arrive at the maelstrom itself, the school gate. And at this point the gene which controls manners and common sense is simply switched off.

You stop the car wherever it is physically possible to do so, not caring two hoots that you happen to be in a bus lane or blocking in someone else who is very obviously just about to leave. You want those children out of your car NOW, and frankly you will use battery electrodes and skewers on anyone who dares to get in your way.

And once the children are in the classroom, the craving for adult conversation becomes impossible to resist. You are desperate to talk to someone who can say more than plastic, fantastic, Suzuki, Rover and BMW.

So you will talk to anyone, completely forgetting that your car is in the middle of the road with three of its doors wide open.

In his ill-fated White Paper, John Prescott declared war on the twice-daily school run as perhaps the major cause of traffic congestion in our cities today. And he's right. I know he's trying to do us a favour. He's been there, though in his day it was probably the 'Birdie Song' and 'Agadoo', rather than 'Barbie Girl', and he wants to save the rest of us from the horror. It's a nice thought John, but the simple fact of the matter is this. I cannot put a four-year-old and a two-year-old on a bus, for one very good reason. Round here, there isn't one.

And I'm loath to fix up an organized rota either. I don't want to sound twee here, but it seems silly to entrust the two most valuable things in my life to someone who, for all I know, isn't a very good driver.

Some people aren't, you know. They get the clutch and the rear-view mirror muddled up, even when they're on their own. And it gets worse when they have a people-carrier full of chanting four-year-olds.

And you can't very well say to a parent who is offering to take your children to school, 'Yes, but first of all, let's see how you handle power oversteer.'

I'm afraid, therefore, I'm on the school run right up to the moment when life reaches stage four. And I start gardening.

Voyage to the bottom of the heap

The television reviewer for a local newspaper in London hates every molecule in my body. In recent years he's described me as Stephen Fry's older, fatter sister, he's said I'm talentless and recently he even wished me dead.

As a result, I know what it's like to be on the receiving end of a savage and vitriolic review. And therefore I have some sympathy this morning for the people at Chrysler who, just before Christmas, asked if I'd like to drive their new diesel-powered Voyager people-carrier.

Obviously, the right answer was, 'No, I would rather rip my own head off than drive something with a diesel engine' but sadly, my wife took the call and said instead, 'Yes, Jeremy would be delighted.'

Well, I'd only gone a couple of miles in it before I smelled a rat. This was a Noel Edmonds 'Gotcha'. I knew exactly what had happened – the bearded one had fitted the engine from a cement mixer, and hidden cameras were going to see how far I went before realizing.

Keen to demonstrate my prowess, I pulled into a lay-by just two miles from home and ripped the interior apart looking for the Pulnix minicameras. And there weren't any. This was not a joke. Almost unbelievably, this car was for real.

I know what's happened here. In the last few years, potential customers have told the salesmen in Chrysler showrooms that they would love to buy a Voyager but that the big, 3.3 litre petrol engine is too thirsty. 'If you did a diesel, we'd buy it.'

Why do people do that? When we're buying a petrol

engine we agonize over the technical data for hours, working out torque figures and analysing the brake horse-power. We look carefully at the top speed and even worry about the meaningless 0 to 60 time.

But when people want a diesel any old rubbish will do, a point that was obviously not lost on Hank the Yank from Chrysler, who simply bought 'a diesel engine' from the Italian company VM and fitted it under the bonnet of the Voyager.

The results are catastrophic. Nought to 60 takes a woeful 13 seconds and on the motorway 60 is realistically your top whack. Beyond that, the growl of the engine, allied to the whistle of the turbocharger, renders the stereo useless.

I'm told, however, that fuel economy is dramatically improved. Apparently, back in the summer, a family drove across Europe in a Voyager diesel and averaged 53mpg. Well I'm sorry but they must have pushed it because, realistically, it won't do more than 33mpg.

And nor will it go round corners properly. Even with the miserable power output which dribbles to the front wheels like it's coming out of a pipette, the Voyager diesel suffers from dramatic understeer on wet roundabouts.

I was just driving normally, and each time I let in the clutch after a down change the front wheels just skidded. It was like trying to drive to work on a halibut.

Only less comfortable. Chrysler seems to have achieved the impossible with this car, combining a bone-jarring ride with the pitch and roll characteristics of a small yacht.

So, after three long years, the Vectra has finally lost its crown. By a huge margin, this new bus from America is the worst car on the market today. And that title is earned not simply as a result of the terrible engine or the unusual handling characteristics.

I once described the Voyager as the best of the people-carriers, but for the life of me I cannot remember what possessed me to do such a thing. For a kick-off, the interior layout is all wrong, with a poky park bench in the back and only two Parker Knoll recliners in the middle. Why not three? Everyone else has three.

And why, in such a huge car, is life so cramped for the driver? You have to rattle along, hunched over the wheel, changing gear every 15 seconds to keep that useless engine in its power band. The rev counter is red-lined, for heaven's sake, at 4000rpm. What good's that?

And why is the handbrake buried under the driver's seat? And why do all the controls feel so cheap? And why is it a condition of the loan that I don't smoke while driving the vehicle? According to a letter I found in the car, future owners will be non-smokers and will not like the smell I leave.

Oh I see; they're going to buy a diesel which will pump the world's most carcinogenic substance — 3-nitroben-zothrone — into their children's frail little lungs. But they'll be put off if it smells of burned leaves. Well, they can p★★★ off.

To be honest, I can't see anyone with even half a brain buying this car. Sure, there are bound to be a few idiots who'll do so because it's a big diesel, but for the rest of us, here are the facts. The cheapest model is £19,600, but for that you only get five seats. You may as well have a Ford Focus.

To get a proper model with seven seats and a boot, you need to spend £22,000, and I simply do not have the

space here to list all the things I'd rather have instead –
venereal disease, for a kick-off.

I'm afraid I didn't even complete the test with this new
car. I eventually left it at a remote airfield in the middle of
Wiltshire and hitchhiked home instead.

Van the Man

On the face of it, motoring in India could not be easier.
The Highway Code states simply that 'might is right', and
that you must give way to anything which is larger.

At all intersections the lorry is king and then, in
descending order, you have the bus, the van, the elephant,
the car, the auto-rickshaw and finally, the mushy and
pliable pedestrian.

So why, if there is only one very simple rule, do the
Indians kill 168 people on the roads every day? Well, first
of all we must face up to the simple problem that Indians
can't drive. Think about it. The world of Formula One is
hardly littered with names from the subcontinent. And
no Indian has ever won the RAC Rally.

And then we have the question of religion. A majority
of Indians believe that their death is preordained and
that they can do nothing about it. So they arrive at the
intersection knowing full well that they *should* give way
to the truck, but they don't know which of the three
pedals is the brake, and they don't really care about the
consequences anyway.

It's a dangerous mix, and that's before we get to the
wild card, the four-legged two of clubs. If you encounter

a cow you must swerve on to the wrong side of the road irrespective of what is coming the other way, whether it's a school bus, some nuns, or Buzz Aldrin on a tractor.

Elsewhere in the world this would not be a big problem because cows tend to be kept in fields, but in India you round a bend on the equivalent of the M6 and oh no: there, right in front of you, is Ermintrude enjoying a round of gin rummy with Daisy.

Be aware, then, that if you are planning to drive in India you may not listen to the radio or chat with your passengers. If you lose concentration for a split second your head bone will become connected to your windscreen bone.

It's all so very different in Britain, but remember, we also have a wild card – White Van Man. He is our equivalent of the sacred cow. He should be in a field, with a ring in his nose, but he isn't. He's on the road with a ring in his eyebrow.

Now a report out this week tries to defend the man in a van, saying he is courteous to other road users, that he is likely to have a pet and that he is first to get out of the way should an ambulance want to come past. Of course he is. That way, he can tuck in behind the paramedics and get home even faster.

The whole point of this survey, paid for by Renault, is to demonstrate that there is no such thing as White Van Man, and that people who drive trannies for a living are as demographically disparate as the nation as a whole.

I see; so how come then, that 40 per cent of van drivers questioned said they had a satellite dish and that 28 per cent take the *Sun*? Only 4 per cent do any gardening and, here's a good one, only 4 per cent are women.

What we're dealing with here are young men who like football, beer, curry tours of Corfu and films where people get chopped up. And I'm sorry, but I don't subscribe to the report's findings, which say White Van Man drives fast because his boss has set an impossible schedule.

White Van Man drives fast because his boss will pay for repairs when he crashes. That's why he never changes gear until the valves are coming through the bonnet. That's why he lunges about with his front bumper in the small of your back. And that's why he treats red traffic lights as advisory stop signals.

The report suggests we all try driving a white van once in a while to see what it's like. Well I have, and I'll tell you. It's great.

You're big enough to mix it with the trucks, but nimble enough to get out of their way when the going gets rough. You can go head-to-head with taxi drivers, and win. And as for drivers in their precious, shiny cars. They're not people. They're targets.

You can send White Van Man on as many driving courses as you like. You can attach a 'How's My Driving?' sticker to his rear bumper, and you can fit a wireless which only plays Sade's 'Smooth Operator', but it won't make the slightest bit of difference.

Like the Indian cow, White Van Man is immune to all known forms of assault. You can carve him up and he'll hit you. You can brake-test him and he'll ram you up the backside. You can get out and remonstrate, but you'll find the back is full of navvies who will practise the ancient art of origami on your arms and legs.

The solution is obvious. Week in and week out I tell you all about the new whiz-bang GTi which will get from 0 to 60 in one second, but I appear to have been missing the point. If you really want to get around quickly, become an urban terrorist. Rent yourself a Ford Transit.

And if a market researcher asks any questions, do everyone a favour and set the record straight. You like beating people up. Preferably with chips.

'What I actually meant was . . .'

Right: think back now to the most embarrassing thing you've ever done in your whole life. Maybe it was the porn mag you shoplifted when you were 11, or maybe it was the one-night stand you had last month . . . with your bank manager. Come on. Feel the guilt. Squirm. And now, imagine what it would be like to suffer from that feeling every single morning.

Here's the problem. I get a car to test, for a week usually, and in that time I'm able to work out almost nothing. Oh sure, I can tell you how fast it goes and what it looks like. I'm even able to determine if it's noisy. But in actual fact, none of this stuff really matters. Take the Ford Puma, for instance. Having been bowled over by the styling, the performance and the promise of low, low Ford-style running costs, we made it the *Top Gear* Car of the Year.

And, impressed by our report on the programme, a friend of mine bought one. And over dinner last week he shoved his finger up my nose and explained that if you lift the tailgate up when it's raining, several gallons of water pour into the boot. I never spotted this because when I tested the car, it was dry. But in the big scheme of things, it's not the end of the world. What concerns me far more is that I can't report on the one area that really matters – reliability.

In the last series I decided that the new Alfa Romeo GTV was the best coupé you could buy. It was pretty much the fastest and, though looks are subjective, I'll come round to your house with a broken bottle if you disagree that this mini-Ferrari is a supermodel in a sea of excrement. Now I knew it would not be reliable. I knew that after six months, if I'd pressed the window switch, the boot would have opened, and that if I'd mashed the throttle into the carpet, the bonnet would have flown away. I knew all of this. But I had no proof. So I couldn't say it. And as a result Dr Lynch of Belfast bought one. And now he's written to say that it's the most unreliable piece of donkey-do ever to grace the Emerald Isle. In nine months, the car has been off the road for eight weeks. And I told him to buy one. Oh my God. The guilt. The angst. And what's this? The next morning I got a letter from Simon Saunders who, following my report, has a Land Rover Freelander. It arrived with the speedo calibrated in kilometres. And over the summer, the speaker fell out of the door, the transmission began to rattle, it ate oil like a school boiler and the air conditioning began to think it was a shower, hosing water into the cabin.

Sadly, it hasn't actually broken down so, technically speaking, under the terms of my agreement with the managing director of Rover, Dr Hasselkus, Simon is not allowed to burn anyone's house down. But he is cheesed off. And so is Andy Jones. Because he bought a Volvo

T5R, which received the Clarkson small-boy-in-toy-shop treatment on television. I loved it. I raved about it. Andy bought one, and to list all the faults he used up all the paper in my fax machine. My hair stood on end as I digested the litany of problems. Oh God, the CD stopped working; pass the razor. Oh no, it judders; where are the Disprin? And then, this morning I really did reach for the carving knife. A driving instructor wrote to say that in the last four years he has covered 130,000 miles in his Nissan Micra. It has been subjected to the worst kind of brutality from Maureen and her ilk and, apart from regular servicing, it has only needed two new brake pads.

A Nissan bloody Micra, for heaven's sake. I hate the Nissan Micra. I have joked about this lump of Japanese junk for years. It is as sensible as a sandal, with as much flair as Johnny Rotten's trousers. Yet it works, every single day, without fail. There is only one solution. Treat what I say about cars as entertainment – but under no circumstances actually go and buy anything I like.

Seriously, the guilt is killing me. Every morning, Postman Pat delivers another tale of woe from some poor sod who wanted that 150mph top speed. He wanted to generate 2 g on every roundabout. And now the car is sitting in a workshop with oil spewing out of its heater vents. Please don't write to me any more. Please. Write to Quentin. It's all his fault.

Mrs Clarkson runs off with a German

The road which passes my house is a beauty. Ten miles or so of sweeping corners, a wiggly bit, some truly mouth-watering views and a brace of long, long straights which plunge like an arrow into the heart of Cotswoldy Britain.

There is one tiny little problem though. The road's in my backyard. I don't mind one bit if you drive like a bat out of hell past someone else's house, but when you go past mine I want you to turn off the engine and coast.

I even rang the council last week and had a long chat with their Highways Department, during which the words 'rumble', 'strip', 'speed', 'camera' and 'I'll stretch cheese wire across the road if you don't do something' were used extensively.

When I put the phone down my wife was open-mouthed with disbelief. 'You bloody hypocrite,' she yelled. 'You're like one of those idiots who buy a house near Heathrow and then spend the rest of their lives complaining about the noise.'

In a temper she snatched up the keys to a Porsche 911 and roared away, saying that if I was going to be a weird beard vegetablist, I could use the Mondeo. It's hard, sometimes, living with a woman who once declared that she wouldn't drive any car unless it has 'at least 200 horse-power'.

As far as she's concerned the road outside our house is a private Nurburgring, and when she came back after her wheel-spinning foray into the night she declared the 911 was brilliant, a little jiggly at the front end perhaps, but otherwise a gem.

High praise indeed from the daughter of someone who won the VC for shooting Germans.

I figured I'd get my chance in this wondercar the next day but, oh no, by the time I was awake it was half-way to a wedding in Hampshire, where its four-wheel drive system was apparently a big boon in the muddy car park.

The next day, I was hit hard with a germ that even made my eyelashes ache, so there was still no driving. But my wife kept the information coming. You can get a child seat in the back. The noise is a bit dull. Here's a Lemsip. I'm going for another spin.

And I was left in bed reading all about this new, all-wheel drive Carrera4 which has, according to Porsche, the most advanced electronic monitoring system yet seen on a car. Called Porsche Stability Management, it can monitor the desired trajectory with the likely actual trajectory.

And then, by using the anti-lock brakezzz, and the engine management system, it makes minute alter-ationzzzz before the car becomes unstable.

It all sounds deeply impressive in a sleep-inducing kind of way, and yet, rather pointless. Here's why. When I drove the normal two-wheel drive Carrera earlier in the year, I found that it just would not misbehave at all. It's one of the most sure-footed cars on Earth, and I emerged from the experience a fan.

I said that it managed to combine the bloodcurdling excitement of a Ferrari with the loping, motorway-munching ability of a Jaguar XKR. So why, I wondered, would anyone want to spend a further £3000 buying such a car with four-wheel drive?

Five days later, and just hours before a man from

Porsche took it away, I got a chance to find out. The weather couldn't have been better. There was rain, wind, locusts and, on the road, pools of standing water deep enough to classify as boating lakes. And the 911 took everything in its stride, allowing me to concentrate on the noisy wipers and the steering wheel that creaked as you turned it. I do so love reporting faults of this nature on German cars.

And then I arrived at a 90 degree left-hander and it was time to test the PSM system. Basically, I didn't bother slowing down for the corner at all. I just turned the wheel and waited to see what the car would do.

First of all, I felt the front offside wheel being braked and then, when the nose had been brought to heel, power was unleashed to the rear, which wiggled slightly. And that was it. You get more drama from Chaucer.

But here's the deal. Who, in their right mind, would not slow down for a 90 degree bend? The electronics were working to rescue a situation that would never occur in real life.

The ordinary, £64,000 Carrera2 generates so much grip that its abilities way surpass the talent and bravery of even the most suicidal motorist. In order to make the Carrera4 work for a living you have to drive like a complete madman.

So what, then, is the point? I mean, both cars have the same 3.4 litre, six-cylinder engine, the same top speed of 165mph, the same 0 to 60 time of 5.4 seconds and the same interior. Visually too, Carrera2 and Carrera4 are identical.

However, Porsche has always said that so long as there is a Ferrari, there will be a 911 Turbo, and that we should expect a blown version of the Carrera4 some time soon.

Now to keep *that* in check, the four-wheel drive and the PSM might just come in handy. But if you buy such a car and decide to test it out on the road past my house, remember: I have a gun.

And last week, I went to the post office and spent £4 on a licence to kill.

Un-cool Britannia

I think it fair to describe snowboarding as the very embodiment of youth. It's a world where any sense of danger is masked by a constant haze of cannabis, a world of primary colours and funny hats. A world where you come down the hill at 70 – but you're over it at 21.

Now at the other end of the spectrum, we find Rover. I only need hear the word and I'm filled with an uncontrollable urge to head for the sort of pub where the fire smokes and the customers don't. It makes me want to drink sherry and snuggle down at night between tweed sheets.

Rover is an old sofa, a wingback dog with gingivitis and boils. Rover is the moleskin waistcoat worn by your doctor if you live in Arkengarthdale.

It would be easy then to wonder what on earth Rover thought it was doing when it sanctioned the recent televisual advertising blitzkrieg. The advertisement may have been set to a song that topped the hit parade in 1964, but the visual imagery was more up to date than your watch.

They were trying to tell us that Rovers are actually bought by 20-year-old girls with lacy G-strings and pierced navels. They were trying to make Aunty Rover in her big bloomers sexy.

And why? Well obviously Rover is about to launch the new 75, and they didn't want people thinking 75 was the minimum age for buying one. They wanted a youthful image for their new, youthful car.

Well I've driven one and it isn't. The 75 is wilfully and deliberately old-fashioned. If the new Ford Focus is a Canon Ixius, then the 75 is a 1950s radiogram. The advertisements have told us to expect an F-22, but the company has given us a wireless.

Naturally, I blame the Germans. They still think that in Britain, everyone is either a squadron leader or a Brontë sister. We go to work with tightly rolled umbrellas and bowler hats. We only eat food when it's charcoal and we only ever watch films about the war.

Ask a German to name something British and he'll come up with Fortnum and Holland or Holland and Royce. They like this, and that's why, when BMW bought Rover, they wanted some olde-worlde charm engineered into the cars.

So the 75 has a chrome strip down the side and chromed door handles. When you open the door there are cream dials set into a wooden dashboard, and while this may not have much to do with Conran's Britain Jurgen the German will feel like he's bought a little piece of Chester. Or York. It is like the Shambles on wheels.

No, I can do better than that. It *is* a shambles on wheels. I shall begin with the dashboard which, as I've said, features cream dials set into wood. But then, rather incongruously, there are LED read-outs and an LCD satellite navigation panel. It's a mess.

It doesn't drive well either. On challenging roads, drivers used to the lightning responses of a snowboard will find the steering ponderous and the brakes devoid of feel. They'll also find the wipers unnecessarily noisy.

And then they will arrive at a corner, where they will discover Rover's sole concession to the modern age – rock and, especially, roll. The traction control system is too eager as well, and there's nowhere for your left foot. Oh, and before I forget, the driving position is odd, the door handles feel cheap and it's hard to drive smoothly in traffic.

Then we come to the new 2.0 litre V6 engine. Well, it was out of its depth, like finding the electric motor from your daughter's peeing Barbie in the bowels of an aircraft carrier. The car feels big and heavy, like a bison, and the engine feels like it belongs in a mouse.

Obviously, the 2.5 litre V5 will be better, but then it will also be more expensive. And while we're on the subject of price, the 1.8 looks like good value at under £20,000, but I suspect it will be more of a garden ornament than a car. It won't move.

At this point, I should introduce some of the car's plus points. It is remarkably quiet and smooth on the motor-way, it is spacious and, if you're over 55, the styling is appealing.

Now I admit my test drive was short – just 70 miles, and that the car was a pre-production special. I must also add that the weather was as bad as the traffic and that I had tummy ache. But even allowing for all of this, I have to say that overall the 75 is not as good as it should be.

It would be easy, then, to say Rover has got everything wrong, that they gave up with the 'Relax, it's a Rover' campaign and went all trendy just weeks before launching something that's a lot more retro than rocket.

All true, but in Germany, France and Italy this car will sell well because the styling conjures up a tourist board vision of Britain. And for the same reason, it will sell to people in this country who have never heard of arugula; members of your local Conservative association will love the way it looks like a little Bentley.

However, the rest of us should buy either a 3 Series or, if we want more space, a 520iSE. Clever, eh, because either way BMW walks off with your cheque.

Move over Maureen

Before Quentin became an estate agent and drove around talking about people's fireplaces, he lent those dulcet tones to a programme called *Driving School*. You may remember it.

It focused on people learning to drive, and it made a star of Maureen, whose mouth was on upside down. Sadly, she never did get the hang of driving, but that didn't matter; some civil servant in beige trousers handed over a document saying that she was legally able to drive a Ferrari F40 on the Snake Pass in winter. Well that's just brilliant. And Maureen isn't alone. There was another woman in the programme who, having passed her test, had another lesson because she wasn't confident enough.

She wouldn't be, driving around with a dog the size of a wildebeest in the passenger seat.

Oh, how we laugh ... right up to the moment when someone just like dog-woman ploughs into a primary school playground, killing 30 under-fives. I'm sorry, but every day I see people in cars who were born to be on the bus. Hunched over the steering wheel, airbag an inch from those half-filled hot-water bottles they used to call breasts, they peer into the gloom, looking neither left nor right. Tom Cruise could be in the car alongside, waving his meat out of the window, but these people wouldn't dare sneak a peek. They're driving along, petrified. And petrified means 'turned into stone', by the way. They can't look in a mirror to see what's behind, they can't glance out of the side window to see what's alongside, they just plough on, oblivious to the mayhem in their wake. I found one of them yesterday doing 30 on an open, sweeping A road. The sun visor was pulled down behind her head which meant, of course, she had no idea I was overtaking when she began – with no warning whatsoever – to turn right.

We've all seen this, and we all assume the police should be more vigilant and aggressive; but be realistic. Even if they do pull someone over they'll find it impossible to charge them with 'sitting too far forward'. Or 'doing 30'. No. To attack this we have to get to the root of the problem – the driving examiner. I have some sympathy with these poor souls. Think about it. If you are scared half to death while someone is taking their test, you'll pass them. That way, there's a very small chance you'll meet them coming the other way on a dark night.

If you fail them, there's a very large chance that, in six months' time, they'll be back, ready to scare you to death all over again.

Here's the solution. First, anyone who fails their test three times is simply told that they may not apply again. They must accept that they can't drive, in the same way that I have now accepted that I'll never be an astronaut or a lesbian. Second, anyone who has not passed their test by the age of 25 shall not be allowed to do so. Let's face facts here. If you're so disinterested in driving and cars that you allow eight years to slip by without trying to get a licence, then you are just never going to make a good enough driver. Fact: if you are not interested in something, you will be no good at it. Proof: I am no good at cricket. Basically, the driving licence will become a privilege and not a right, and in order to get one I'm afraid that the test will need to be modified. You'll still be expected to brake sharply and reverse round a corner; town driving will remain to ensure you have good spatial awareness. The written test will survive too, and don't worry if you live in Norfolk or Cornwall. I have no proposals for motorways to be on the curriculum, so you won't have to come to England.

However, you will be taken to a circuit which you will be expected to negotiate in a certain time – nothing mad; just fast enough to make the tyres squeal on the corners. We need to see that the car doesn't scare you and that you're able to take it to the red line once in a while.

We don't want you to break speed limits, they're there for a good reason. But, on the A44, we want to ensure you'll go at more than 30. And if you don't, you blind, deaf, old bat, we'll come round one night and fit a turbo to your Rover 400. A turbo with the wastegate jammed shut.

Toyota gets its just deserts

Obviously, I receive a great many letters from people who are angry but this morning I've been accused of dropping metaphorical napalm on the Midlands.

A man, who wishes to remain anonymous, but gives his address as Angmering in West Sussex, says that when I reviewed the new Rover 75 a couple of weeks ago my remarks were loutish, cheap and unjustified, and that I've inflicted immense damage on Rover and its workforce.

To ensure I take these observations seriously, he points out that he has 'no connection with Rover's (sic) or any of their (sic) allied companies'. Which means, of course, that he hasn't driven the 75 yet.

Well, whoever you are, I apologize unreservedly. The new Rover is a superb car, crisp and elegant to behold and quite breathtaking to drive. The steering is a delight and the performance a credit to the brave engineers who, against all odds, battled to create something really rather wonderful.

There you are. It's not actually true, but in this business you learn very quickly that to keep the people of West Sussex happy it is important to remember that all Rovers are superb and that all Toyotas are made for people with vivid jewellery, and Africans.

I know this because I spent a large chunk of last week

driving across Namibia in an Afro-spec Toyota Camry – four wheels, a gigantic stereo and a burglar alarm.

It really was a dreadful drive, from Swakopmund on the Skeleton Coast 400 miles into the wilderness on roads that were part gravel and part dried-up riverbed. You'd be hurtling along at 90mph when there'd be a trough into which the car would crash with the sort of crunch that can loosen hair.

Stones would fly into your windscreen, creating yet more impact scars so that you'd have even less warning of the next elephant pit.

And then there were the corners. When you've been driving in a dead straight line for 50 miles you get lulled into a false sense of security so when, over a blind crest, the road suddenly executes a 90 degree left, you're not ready.

On tarmac the move would be hard enough, but on gravel you need to use the handbrake and some left-foot braking, and it's a good idea too, to flap your hands around outside the window; anything really, to make the bloody thing turn.

I'm told that 10 per cent of all hire cars in this gigantic desert state are never sold. They're rolled.

Now my Camry had done 20,000 miles as a hire car in this superheated boulder-strewn landscape and do you know something? There wasn't a squeak, there wasn't a rattle. It may be quite the dullest shape ever to have left the lead in a designer's pencil, but my God, Johnny Jap knows how to weld metal.

He also knows what's what in the world. With a miserable 2 litre engine and quite the worst automatic gearbox I've ever found, the Camry is not a fast car. But really, on gravel, cars which offer blood and guts performance usually bring out the blood and guts of those inside. A top speed of 90, I promise, is enough.

What you need from a car out there is the robustness of a lunar-rover and the reliability of a military satellite. You can't break down when the only roadside cafés are staffed by lions, and you're on the menu.

And then there's the dust, which is so persistent it could get into a nun's pants. Certainly, it managed to get through a closed boot lid and from there actually into my suitcase. If you drive an ordinary car through a dust cloud for 400 miles, Clarence will have you for elevenses. But while the Toyota lets dust in, this seems to have no effect on anything vital.

Except your hairstyle. After just 100 miles I appeared to have an anvil on my head.

Now in Britain, of course, we don't have gravel roads and our dust is converted into mud long before it can invade your bouffant. And while we have troughs in the road, called speed humps, we tend to take them at 5mph, not 70.

So it could be argued the Camry is over-engineered for our tame and temperate environment. We want more than granite door hinges and a front spoiler that's made from Kevlar.

We want restful orange dials and piped leather upholstery. We want chrome embellishments on the body and a raft of electro-techno wizardry in the dash. We want – no, we demand – traction control and an ability to get from 0 to 60 in complete silence.

Rover, for sure, has met this criterion well with the new 75 and will, as a result, sell a great many cars in West

Sussex. But then Toyota also sells cars in West Sussex, *and* West Africa *and* Western Samoa.

The simple fact of the matter is this. When Rover designs a car, it thinks only of Europe. When Toyota designs a car, it thinks of Tokyo's traffic jams, the Stelvio Pass and the vast deserts of Namibia.

And that's why Toyota is the world's third biggest car company and Rover is a small and tiresome division of BMW, which, in turn, is soon to be a small and tiresome division of General Motors.

Rover, and all the European car makers, need to think not of the European Union, or even the global village. Satellite phones and the RB211 may have made the world feel smaller. But it isn't you know.

Kristin Scott Thomas in bed with the Highway Code

Now that cruise missiles and the environment have gone away, we have something new to worry about. The police, it seems, have become racist and every night hundreds of black people die in the back of their vans.

A report in the *Sunday Times* last week showed that nearly 16 per cent of people entering the legal profession, and 23 per cent of those at medical school, are black. But even here, in society's stratosphere, they're not safe from constant police harassment. Plod, apparently, is forever barging into surgeries to look up Dr Ngomo's bottom.

So what's brought this on? Why have the heroes from an Enid Blyton world of rosy cheeks and apple scrumpers become a bunch of salivating Nazis, proven guilty of institutionalized racism?

I suspect it's because they're bored. For 30 years the police have had a purpose, a goal in life, a reason for being. They were employed, solely, to stamp out drinking and driving.

They didn't investigate the Stephen Lawrence murder properly because all the available manpower was cruising the streets looking for people who'd had a glass of sherry.

But now, the war on drink driving has been won. Britain is the safest country in Europe in which to drive. Only 14 per cent of fatalities are drink-related. And most of those are drunken pedestrians wandering into the path of perfectly sober drivers.

Now, you don't need a psychiatrist to explain what happens when all of a sudden your role in life is removed. The issue was addressed in *The Full Monty*. You go a little bit bonkers. Some people take off all their clothes. Others roam the streets at night, beating people black and blue. Well blue, anyway.

I was saddened but not in the least surprised to see that Channel Four newsreader Sheena McDonald was run over by a police van last weekend. The driver probably mistook her for Trevor McDonald.

What the police need is a new target. And now they've got one – the mobile telephone. Last year, a government-funded study showed that making a call while on the move is as much of a safety risk as driving while drunk. And now, the latest edition of the Highway Code tells drivers that phoning and driving is a no-no.

In Canada, research has shown that drivers who make a call while at the wheel are four times as likely to have a crash, and that hands-free sets are just as lethal. Talking on the phone, they say, means you're not concentrating on your driving.

So what about talking to passengers? Well, according to our Highway Code, that's fine but you must not argue with them. Even if they say the age of consent should be lowered to four, you must bite your lip or you will wind up in court charged with driving without due care and attention.

The code also says you must not eat or drink either. And it warns that navigation systems, onboard computers and even stereos can prove to be a distraction. Crikey. Who'd have believed it? Listening to Terry Wogan is now illegal.

For the country's black people, this is fantastic news. It means the police can get back in their powerful patrol cars and do what they do best – harass motorists.

They'll need expensive directional microphones, of course, and persistent offenders, I imagine, will have their cars bugged. Hidden cameras will be used to catch those using satellite navigation systems or eating a Twix.

And that's just the start. The new Highway Code also says you should not drive for an hour or more if you're feeling tired and that a break of 15 minutes every two hours is advised.

So how are they going to enforce this then? 'Sir, we've been following you around all day. You had a breakfast meeting at eight, followed by a conference with the world's terrorists at 10. At lunch time, you had frantic sex with your secretary and a goat, and in the afternoon you robbed a bank and two post offices. And you've been driving now for 121 minutes without a break. You're nicked.'

This is ridiculous. When I took my driving test, the Highway Code was full of sensible advice. It told me to indicate before turning and not to cross level crossings when a train was coming. It lived in the real world, but now it can't even get the braking distances right.

Next thing you know, it'll be telling us not to drive if we want a pee, and not to even so much as think of getting in the car if we haven't had sex for a while. I'm not kidding.

New research in Australia has shown that 42 per cent of drivers over there have 'dangerous sexual fantasies while behind the wheel'.

It goes on. 'We need to break the belief many drivers hold that they can automatically be safely in charge of a vehicle, irrespective of . . . what they are thinking about.'

I see, so now I face winding up in court charged with driving while under the influence of Kristin Scott Thomas.

'Your honour, I couldn't help it. I was just thinking about *The English Patient* and she popped into my head. I'm sorry. I'll be sure to have some bromide before I drive next time. Only not too much, in case I need a pee.'

Having been born white, I've no idea what it's like to be victimized for no good reason. But as a motorist in Blair's Britain, I suspect I'm about to find out.

Time to change Gear

To a great many people, *Top Gear* presenters have very possibly the best job in the world. Free cars, club class travel, no repercussions when you crash and large dollops of fame, fortune and foie gras.

So I'm sure a few readers may be a little perplexed to hear that I have resigned. Here's why. Now that I've gone, I don't need to drive a razor around my face every single morning. I don't need to buy new shoes every time the old pair start to look scruffy and, best of all, I have no need, ever, to set foot again in the armpit that masquerades as Britain's second city. Much as I liked Pebble Mill, I really did grow to hate, with unbridled passion, the city that surrounds it. Until you have driven through King's Heath on a wet Wednesday in February you have not experienced true horror.

You may have seen footage of the Colombian towns devastated by the recent earthquake. Well, King's Heath is like that, only worse. In seven days, God created heaven and earth and then, just to keep his oppo amused, he let Beelzebub do Birmingham. I pity James May, the man being touted as my replacement. He has been lured by the promise of untold riches, of motor industry obsequiousness on a biblical scale and of bathing in an intoxicating mix of public adulation and Dom Perignon. But he has not considered that his drive from England to Pebble Mill will mean getting through King's Heath.

There are, of course, other reasons why I needed to go. There was, for instance, surprise when I described the Corolla as dull, yes, even shock when I was seen to fall asleep while driving it. And again, there was surprise when I savaged the Vectra, refusing for seven minutes of televisual time to say anything good about it. By the time I got round to the Cadillac Seville STS, the Clarkson attacks were only mildly noteworthy. You had grown to expect them. The shock tactics had become predictable, and so weren't shocking any more.

And it was the same with the metaphors. The first time you heard me liken some car to the best bits of Cameron Diaz, you probably sniggered about it at school all the next day. But now, it's tedious. I never tired of trying to think up new ways to describe a car, and could regularly be found at four in the morning scribbling new lines on a piece of paper by the side of the bed. I thought of one only last night. 'It's like being left outside the pub as a child with a crisp drink and a bag of coke.' Great, but now I've nowhere to put it.

I will, of course, carry on writing for this magazine, and there's always Mr Murdoch to stand bravely between my front door and the wolf, but already I'm starting to miss *Top Gear*. I miss the banter with Quentin and Tiff, as we sniggered about Steve Berry and what he'd crashed that week. I miss Vicky's eyes and her ability to bring sex into absolutely everything. I miss climbing into a new car and thinking, 'Right. What have we got here then?' You may think that the best bit was the endless succession of new cars. But it wasn't. The best bit was sitting down at the computer with an expectant, winking cursor and then, four days later, handing over seven minutes of video tape to the producer.

The actual driving was always a drag. You sat in some Godforsaken hedge on a blind bend, waiting for the walkie-talkie to say the road was clear. And then you set off, only to find it wasn't or that the cameraman had lost focus and that you'd have to do it all over again, and again and again. I promise you this. It really isn't much fun driving a Ferrari when you are accompanied by a cameraman, a ton of equipment and a bloody great blinding light on the bonnet. I'm often asked what qualifications you need to work on *Top Gear,* and I've always given the same advice. Like cars by all means, but love writing. Love it so much that you do it to relax. See the new Alfa or whatever as nothing more than a tool on which your prose can be based. Don't worry about how quickly it gets from 0 to 60. Worry only about how you will explain this meaningless figure to your viewers or readers.

So what am I going to do to fill the void left by *Top Gear*? Simple. I'm going to write and write and write until the smiles come back.

Even soya implants can't make a great car

What a week. Gordon Brown decided to be all generous, only increasing the tax burden on home owners who have a car. Which means you and me, and everyone we've ever met. And all the extra revenue will be used by local councils to build stiles and footpaths in your garden so that ramblers can roam around your lawn and trample all over your geraniums.

Then we had the news that soya bean breast implants may leak and that no one really knows what this may do to the body. Oh, really? I've known for years.

Ingesting soya beans makes you grow a beard, don red socks and wander about in other people's gardens in a Day-Glo cagoule. My advice is simple: if you want boobs like Yorkshire puddings, put a couple of steaks in there.

There have been other shocks too. Eddie Irvine arrived in Australia with a car that was 1.5 seconds a lap slower than the McLarens and a clause in his contract that forbids him to win. But he came home first.

And here's a good one. I learnt that pretty soon Alfa Romeo is to launch a car with — and you should really sit down for this — a diesel engine. The girl I spoke to at Alfa's import operation said I'd like it and I didn't want to sound patronizing at the time, but I promise you this, dear, I won't. I will hate it. Putting a diesel engine in an Alfa Romeo is like putting chocolate biscuits in your breasts. It is mint sauce with beef, horseradish with pork. It's all wrong.

People who buy Alfa Romeos are enthusiasts. They've looked at the humdrum alternatives and decided that what they want, above all else, is design flair and engineering panache. They want steering that fizzes and a bark from the exhaust, and they're prepared to ignore the possibility that the car will take a mechanical siesta every afternoon. And I'm sorry, but anyone who is enthusiastic about their driving does not want to be dragged along by a diesel engine. Diesel is automotive soya, and Alfa Romeos are born to run on Aberdeen Angus. Alfa Romeos don't want to roam the countryside pointing out rare birds. They want to tear about, pulling them.

I can see why Alfa sells a diesel version on the Continent where the fuel is so much cheaper, but here there's no reason for buying derv unless you want to give your children something special for Christmas – cancer.

So what about an Alfa Romeo with an automatic gear-box, then? Surely this is also anathema: a catwalk model's double chin, an actor's stutter? Such a thing could never possibly exist. But it does, and it's sitting outside right now.

With the clear thinking we've come to expect from the Italians, there are two types, neither of which is conventional. Buy a 156 with a 2.0 litre engine and you get what's called the selespeed – a five-speed, push-button manual with auto override. Or if you go, as I did, for the 2.5 litre V6, you get the Q system – a four-speed automatic with a conventional gear lever for those moments when you want some manual control. Now I may not like diesels under any circumstances, but I'm not quite so rabid when it comes to auto boxes. In town, they are essential, and on the motorway it doesn't matter.

That leaves the 42 miles you do each year on open-mountain B-roads when a manual would be nice. So I have no problem with the idea of an automatic Alfa . . . in theory. But in practice the Q, I think, stands for question-able.

For a kick-off, the performance isn't simply affected. It's decimated. In the manual car you can get from 0 to 60 in 7.5 seconds, but in auto mode the Q car takes a second more. That's a light year. Even if you press the sports button, there's still precious little get-up-and-go, making overtaking time-consuming and precarious. Eventually I gave up and shifted the lever over to its manual setting, thinking that maybe there I'd get some zing. But I didn't. There are only four gears, and that's not enough. When-ever you change up, you drop off the wave of torque and power to find yourself in a slow-motion, soya bean sea of calm. And you? You're sitting shouting 'Come on, come on' as the Nissan Serena people carrier in front pulls steadily away.

So Alfa has tried to marry the ease of an auto with the tactile pleasure of a manual and, like everyone else, has failed. But don't worry. It takes more than a wonky gear lever to spoil the car to which it's fitted. I've raved before about the Alfa 156, but I see no harm in raving again. No car made costs so little, looks so good, handles so nicely and still finds space in the back for three baby seats. The 156 made my week bearable, and now I must devote some time to thinking of an excuse for borrowing it again. Even with a gammy leg, the 156 has to be the choice of the genuine enthusiast. I love it.

Lock up your Jags, the Germans are coming

Without wishing to sound too *Newsnighty*, I'd like to take you back to 5 February 1999. We're in Munich at a supervisory board meeting of BMW. And there's a bit of a row going on. The boss, Burnt Fish Trousers, is hugely supportive of Rover and wants to keep the Long-bridge factory open. But his No. 2, engineering director Dr Wolfgang Reitzle, wants to pull the plug and shut it down. For taking this stance, Reitzle is not popular among British trade unions, but then he doesn't seem to be popular anywhere. One BMW insider I spoke to described him as a 'complete bastard'. We first saw him in the BBC2 series *When Rover Met BMW*, marching around the Rover factory, ignoring the sandwiches so lovingly prepared by the canteen women. They'd even got some German pickles to make him feel at home, but he decided that instead of feeling at home he'd rather be there, and left. I met him some months later at the launch of the new 3-series in Spain, and frankly we got along rather badly. It seems he still hasn't forgiven me for saying in this column that he looks 'a bit like Hitler', and he was very peeved when I plonked myself down next to him at dinner in a seat reserved for his girlfriend. Still, we had a full and frank exchange of views, which ended with him banging the table and shouting something about Germany being made to pay for 100 years for what it did in the war.

Anyway, I digress. Back at the meeting in Munich, Fish Trousers has resigned and Reitzle has been asked to take over as head honcho. But when he doesn't get the support of the workers' council, he walks out too. The Long-bridge trade unions were delighted. And they weren't alone. I was so happy I broke into a case of Château Margaux and had a few friends round to celebrate.

But now, just six weeks later, he's back. And it really couldn't be worse, because he's been appointed by Ford to run Lincoln, Volvo and, most important, Jaguar and Aston Martin. A collection that Ford is now calling its Premier Automotives Group. This is a big worry. I have been driving a new Jaguar S-type this week, and to be perfectly honest it doesn't really feel as a Jaguar should. But then you can't make a chocolate mousse when the only ingredients you have are two sardines and some HP sauce.

Inside, the radio and air-conditioning readouts are bright green LEDs, such as you would find in any

American car, and despite all the wood and leather the switches feel as if they've come from a Fiesta, which isn't odd at all. Because they have.

Jaguar is trying like crazy to say it's all its own work, but I'm sorry: even the V6 engine has been lifted from a Ford Mondeo. And now they've been told they're part of the Premier Automotive Group, which means that in future the Jaguar distinctiveness will be eroded still further. I must say, though, that viewed simply as a car the S-type isn't bad at all. Mine came with Jag's own Welsh-made four-litre V8, which meant I could get from 0 to 60 in 6.6 seconds. And that, for a heavyweight four-door saloon with a (Ford) automatic gearbox, is pretty damned fast. I liked the view from the driver's seat too. The bonnet rises and falls like a kid's drawing of distant hills, and this somehow conveys a Volvo-ey feeling of strength and solidity to those inside.

However, there's nothing even remotely Volvo-ey about the handling. Although the steering is too light, you don't get a constant pitter-patter from the tyres that enables you to judge what's going on at the sharp end. And you have rear-wheel drive, which makes the handling balance just so.

But all things considered it's no match for the sharper, more handsome, more spacious and even better-handling BMW 5-series, a car that was designed by . . .

Why, step forward, Dr Reitzle. He may have the most stupid moustache in the entire world, but there can be no doubt that he is probably the cleverest automotive engineer working today. The trouble is that he's now being asked to put four completely different companies under one umbrella, and that's like saying to Brunel: 'Yes,

Izzy, it's a great ship, brilliant for floating around on the ocean and everything, but how about giving it some wings so it can fly as well?' A ship is a ship. And a Jag is not a Volvo.

Now we know from his dealings with Rover what Reitzle thinks of tradition and the plight of the British worker, Johnny. So when he realizes that he need make only one car with four badges, how long will it take him to open a factory somewhere cheap, like Namibia?

At best, I suppose, he'll simply close the Jag plant down, but if anyone dares to resist, who knows what might happen? Certainly, if I lived in Coventry I'd listen carefully at night for the sound of approaching Stukas.

Well carved up by the kindergarten coupé

Pretty well everyone on *Who Wants To Be A Millionaire?* would guess that the Ford Fiesta is Britain's best-selling car. But it isn't, you know; not by a long way. The accolade rests with a car that has one door, one seat, no windscreen and a steering wheel that isn't actually connected to the wheels. You know it well. It's that red and yellow shin-destroyer, the Little Tyke's Cosy Coupé, which in the past 20 years has annihilated the skirting boards in a staggering five million homes. Including ours.

I don't know how this happened, but our boy child has become something of a petrolhead. He can even tell the difference between a Mazda Demio and a Suzuki Baleno, which is useful. Because I'm buggered if I can. Each night he goes to bed with a copy of *Autocar*, and yesterday, on his third birthday, he was given a car that crashes into the wall, flicks on to its back and sets off in another direction. Well, that's the theory. In reality it bashes into your ankle, flicks over and then bashes into your other ankle, something it will do, nonstop, for up to 16 hours. Happily I was safe, because I spent the best part of his birthday in a faraway attic trying to assemble a radio-controlled car that quite clearly said on the box 'Ready to Go'. Foolishly I assumed that 'Ready to Go' meant we could get it out of the box, put it on the drive and spend a happy hour or so running over Mummy's new shrubs.

But it's only 'Ready to Go' once you've spent two hours hammering the batteries into the controller and a further three hours charging up the power pack for the car itself. Then you have to thread the aerial through a straw that goes through a hole in the bodywork and into a socket on the chassis.

Oh, no it bloody well doesn't. Well, not until you've shaved it with a razor blade, which will slip and take most of your left index finger off. Oh, how the kids all laughed! We'd got them a bouncy castle and a magician, but neither could hope to compete with the birthday boy's Daddy, who was running around in the garden with a toy bashing into his ankles and a big red fountain coming out of his finger.

Eventually, though, the radio-controlled car burst into life and for eight glorious minutes made the boy child squeal with delight. Then he remembered the Eddie Stobart truck he'd left in the dog bowl and was gone. This is normal. He is allowed to play with toy cars because he's three. I'll only worry if he's still doing it when he's 47.

We need to think very carefully about grown men

who buy toy cars for themselves. They may say they're impressed with the detailing on the engine, but it isn't an engine at all, you know. It doesn't work. They may say, too, that in years to come it will be a valuable heirloom, but come on. Show me a man who has a perfectly preserved James Bond DB5, with the little blue man still *in situ*, and I'll show you someone who, quite rightly, was bullied at school. We've all seen men on the Jerry Springer show who like to be dressed up in nappies. Well, that's what we're dealing with here: weirdos and oddballs. And, really, they shouldn't be allowed to roam the streets.

The name and address of any grown man who buys a toy car for himself should be fed into a police computer so that if there's an outbreak of child molestation in the town PC Plod knows who to visit. Don't get me wrong here. Spending £14.99 on a mass-produced Vietnamese toy car is insane, but I see nothing wrong at all with the man who actually gets out there with the Bostick and makes one. I once met a chap who has spent 20 years building a model Ferrari that was an exact, though smaller, replica of the real thing. He'd made not only every single part himself but also the mould for every single part. It even had a working 100-cc flat-12 engine, so that if he'd been nine inches tall he could have driven it. It was exquisite, and my respect for the man was boundless.

I admire, too, the man I once saw on television who had made a model submarine that would fire three-inch torpedoes. Obviously I wouldn't go round to his house in case he made me listen to James Last and told me 'dirty jokes', but remember: model-makers have gingivitis and a couple of nasty skin disorders, so instead of going out with girls they choose to spend all day in a shed gluing things together.

And let's be honest, this doesn't really affect the rest of us, does it? Then there are people who make model aeroplanes. Sometimes, as I sit in my garden on a summer day listening to their creations buzz and whine around the sky, I'm even tempted to take up modelling myself. In fact, I've already begun work on the plans for my first project. It's going to be a Little Tyke's Cosy Surface-to-Air Missile Battery.

Fruit or poison?

Last month, the road-test team on this magazine produced an advert-free supplement that listed the best and worst cars you can buy. Now the guys who wrote the supplement may spend all day talking about motorcycles, but they do drive every single new car that comes on to the market. They take them home at night. They take them away for cosy weekends. They take them to test tracks. In other words, these guys know what they're talking about.

Strange, then, that I read the supplement with a purple face and little bits of spittle at the corners of my mouth – a mouth that was gaping in disbelief.

It wasn't so bad to start with. They said Peugeot's 206 is the best small car, which is fair enough. Second slot was given to the Clio, which shows that even motorcyclists have some common sense. The Clio may not be as much fun to drive as a Fiesta, but it is cheap.

I had no real argument with the family car section

either, where they gave awards to the Focus and the Passat. And sure, I can see why the Jaguar XJ8 had to play second fiddle to the BMW 5-series in the executive car roundup.

But then we came to the off-roaders section of the supplement and everything went completely banana-shaped. The Mercedes M-class is built in America by Americans. It is too cramped, far too expensive, a bit ugly and apparently not even much cop off-road but, even so, the *Top Gear* magazine road testers put it on the top step of their podium.

This does appear odd, because the Toyota Landcruiser is the best off-roader in the world unless you live in Britain, in which case snobbery makes the Range Rover a better bet. And not the 4.6 HSE recommended by our team but the smoother and more economical 4-litre version.

Fuming, I turned the page to see that in the people carrier section, the Chrysler Voyager was praised for the power of its diesel engine. Hello, hello. Have you actually driven one? It is absolutely diabolical. And ... oh, my God, there's more. According to our boys, the Fiat Coupé is better than the Alfa GTV, which is just plain wrong, and the Mercedes-Benz CLK is better than the Nissan 200SX. Sure, in the same way that treading on a rusty nail is better than having sex with the entire sixth form of a girls' school.

But they saved their most magnificent piece of wrong-ness for the supercar section. What on earth is the Aston Martin Vantage doing in eleventh place, when the Lamborghini Diablo came in third? Given a choice, these guys would rather take a drug dealer's car than the Starship Blenheim Palace. Obviously.

Now I'm sure you read the supplement, too, and I'm sure you had a hernia from the stress it caused. Plus, I'm equally sure, you've read my views in this column and now you have full-on post-traumatic shock. This is what makes the automotive world go round. One man's poison really is another man's fruit of the forest. We may tell you that the Focus is by far the best family car that money can buy, but you may think it looks like the dinner of a dog. So you'll buy a Bravo instead. Or an Almera. And that's fine. Well, sort of.

So now we arrive at the new Rover 75. I understand that following our less-than-enthusiastic road-test report last month, the suits in the Longbridge division of Munich Central are apoplectic with rage. Because we didn't like it, the Midlands will have to be closed down. Three hundred million people will be thrown out of work and, as the money runs out, local businesses will close too.

Children will be forced to spend their formative years up inside chimneys, and their parents will wander aimlessly over rubbish tips searching for bread and guano.

But look. If I don't agree with our road testers on their choice of cars of the year, and you don't agree with either them or me, why should anyone agree with either of us on the 75? I might tell you that *Butch Cassidy and the Sundance Kid* is the best film ever made and you may say *Betty Blue* is better. And no matter how much we argue about it, we'll never, ever agree.

And so it goes with the 75. We looked at the overall package and decided that, while it offered submarine quietness and ocean-going luxury on the motorway, it fared less well as a driver's tool. And while we said it was good value, we didn't like the noisy wipers or the dash.

But what if you spend all day on the motorway, and you're on a shoestring? You're going to scoff at our findings and buy one.

There are only three objective reasons for not buying a particular car: it is unsafe; it is absurdly expensive; it is a Vauxhall Vectra. Bearing this in mind, there's no reason at all why you shouldn't rush out to buy the Rover 75.

And I hope you do.

Left speechless by the car that cuddled me

'We'll send a car to pick you up.' Whenever anyone says that, I get a little tingle down the back of my shirt, a little jingle of the ego glands. They're going to send a car for me. A car! And that means they're not going to send a Hyundai Stellar.

A Hyundai Stellar is a minicab. When they say a car it means they're going to send a Mercedes-Benz, and if you're really, really lucky it'll be an S-class.

Now for those of you who don't know, the difference between an ordinary car and the Mercedes flagship is as great as the difference between cattle class on an American airline and first on a BA777. An ordinary car will bash into your elbows and not stop boning its seatbelt warning light until you wake up. In a big Merc you snuggle down with a mug of cocoa while a man in grey flannel trousers gets you home. In an ordinary car, you motor. In an S-class, you travel.

No car on the road, not even a Rolls-Royce, has such ... what's the word? Such presence. If you want to poke the paparazzi with the metaphorical cattle prod, swish up to the kerb in the big Benz. But now there is a new S-class, and when that rolled up our drive this week the dog strutted outside and barked at it. For sure, it's much more handsome than the old model, but that's like saying Ralph Fiennes is much better looking than Lennox Lewis. He is, but I know who would make the better minder.

Basically, the new S-class looks like everything else on the road; the fist-in-the-face presence has gone, and get this: it doesn't even have double glazing any more. Pity. In 1998 Mercedes launched a £14,000 hatchback and merged with Chrysler, best known for the Talbot Horizon. It's dumbing down, and now even the S-class has gone all Big Breakfast.

But then I took it for a drive, and now I am speechless. Without a doubt it's the best, most complete car I've ever encountered. Whether you're in the back, slithering around W1, or in the front, doing a ton on the A66, it is utterly magnificent. Take the seats, for example. Naturally they move about electrically, and of course they're heated. But they also have little fans buried deep in the upholstery which cool your buttocks on a hot day. And they pulsate. As you drive along, little pockets of air move about in the fabric, kneading your weary back. This means you can get all the way from London to Bassetlaw without having to stop off in Northampton for a bath. So it's pretty comfortable, and that's before we get to the air suspension. You don't drive this car; you float around in it.

Which brings me on to the handling. You're probably expecting to hear that it's a bit of a liner, but it attacks corners with the agility of a small speedboat. If the *Titanic* had been built like this she would have missed the iceberg.

I'd love to tell you what happens *in extremis*, but way before the passengers are treated to anything so dramatic as tyre squeal, all sorts of electronic whiz-kiddery intervenes to slow you down. Good thing, too, because I was still playing with my seat, and the cruise control, and the satellite navigation, and the television, and the phone, and the trip computer, and the Tiptronic gearbox and all the other features you find listed in the three-inch-thick handbook. They even fit voice activation for various controls, and to make sure the computer is not baffled by accents Mercedes tested it on 180 people from every region of Britain. I'm told it even understands Geordie.

And then there's the keyless ignition. You simply keep what looks like a credit card in your wallet, and as you approach the car the door-locks silently slide upwards. Then, to start the engine, you press a button on the gear lever.

But what sort of engine should you choose? The 280 and 320 have six cylinders, a bit mean in a car of this size, and while the 5 litre V8 and 6 litre V12 may be sublime, they are also ridiculous. I'd go for the 4.3 litre V8. It comes with three valves per cylinder, offers 280bhp, gets you in complete silence to 140mph, and in my hands returned a remarkable 22 miles to the gallon.

And here's the clincher. There are 145 motors in the S-class – only one of which is the engine – and you are left in no doubt that for year after year none will break down. Not until the car has been sold 16 times and is finishing its days cruising the Melton Road in Leicester will there be any form of malfunction, and even then it'll probably amount to nothing more than loose stitching on the upholstery.

I do wish it still had the presence of the old model, but these are leaner, cleaner times, and anyway the weight loss is translated into a price loss. At £57,000, the new S430 is £3000 less than the old, which means you get unbeatable value from what is quite simply an unbeatable car.

One car the god of design wants to forget

Sometimes I send this column in to the newspaper knowing full well that it's not very good. I set out to make something as smooth as the Queen's lawn, but somehow I end up with northern Cornwall, all craggy and inaccessible. I go over it again and again, but all this does is create half a dozen meaningless oxbow lakes and a millstone grit outcrop. And before I have a chance to straighten things out, the deadline passes and I have to send it in anyway.

But let's be honest: everyone can look back over his work and know which bits are best forgotten. Even God, I suppose. With the south of France, he can say: 'I did OK there. I like the way you can ski down to the beaches and all the women have no tops on.' But we must never let him forget Australia, a vast and useless desert full of spiders that'll kill you and men in shorts. Or Florida.

Happily, though, life moves on and mistakes are buried in the mists of time. For God, earth is a distant memory as he busies himself with the planet Zarg. And me? Well, I'm writing this, and that rubbish I did last month about electric gates is at the bottom of your hamster cage. Even people who create something lasting are safe. The architects who did those tower blocks in the 1960s don't have to live in them. And an artist doesn't hang his most idiotic work above the fireplace.

However, when you're a car designer there's nowhere to hide. When you make a mistake with a car, it's going to come back and haunt you. Every single day it'll lunge out of a side turning and you'll be forced to say to your passenger: 'I did that.' So let's spare a thought for Giorgetto Giugiaro, whose company, ItalDesign, is celebrating 30 years as the car industry's most prolific design house. Remember the Maserati Bora? Well, that was one of his, and so was its six-cylinder sister, the Merak. Then there was the Lotus Esprit and the BMW M1. It may have had a German engine, and its plastic body may have been made by Lamborghini, but the styling: that came from Giugiaro. As did the DeLorean and more recently the Maserati 3200GT.

However, don't think his talent lies solely in the high-horsepower world of the supercar. He also did the 1970s Alfa Romeo GTV, the Subaru SVX, the Lexus GS300 and the Saab 9000.

I've just finished a book that lists his creations and it's incredible: the original Golf, the Scirocco, the Isuzu Piazza, the Renault 21, the Daewoo Matiz and, best of all, the Alfasud. All his. And the Ford Escort Cabrio. And the Lancia Delta. And the Fiat Panda. Been on a bus while you were in Italy? If it was an Iveco there's a strong chance Ital styled it. He does vans, trucks, tractors – even pasta.

I met him once and decided, quite quickly, that I'd like

to punch him in the face. He was punctual, polite, and though it was over 100 degrees up there on the roof of Fiat's Lingotto building he never broke into a sweat. His clothes were immaculate, and he was ridiculously handsome, despite some magnificently daft graduated sunglasses. We talked about our sons, how mine has a habit of mincing round the house with a pink handbag and how his has just designed a 12-cylinder roadster for Volkswagen.

He's funny, too. When Triumph launched its TR7 at the Geneva Motor Show, Giugiaro stared for some time at the profile, walked round the car, and said: 'Oh, no. They've done the same thing on this side as well.'

I just knew that I was dealing with a man who'd slept with more women than me, but despite everything I felt sorry for him. You see, his path to righteousness does contain one particularly large and virulent mistake. Flick through the book that celebrates his work and it's there: a small picture tucked away on page 46 – a verruca on the foot of greatness. I'm talking, of course, about the 1974 Hyundai Pony, which is almost certainly the ugliest car of all time.

Quite how this happened I have no idea. Maybe the design was inadvertently torn up by an overzealous cleaning lady and she glued it back together all wrong. Or perhaps the clay model was damaged en route to Seoul, and the people over there were too full of spaniel to notice. Either way, Giugiaro has to get up every morning and have breakfast knowing that on his way to work he might pull up at the lights alongside the result of his darkest hour. And as he peers inside, the occupants will peer back, their faces saying it all. 'You bastard. Why?'

Then, when he dies and gets to the pearly gates, there's a chance that all the receptionist angels have Ponys as company cars. And it doesn't matter how much he stands there saying he did the Bora and the Esprit, they're still going to put him, for all eternity, in a room next to the lift shaft.

Can a people carrier be a real car? Can it hell

This morning, pretty well everything went wrong. The electric gates broke again, trapping the postman in our garden − a garden that was being systematically eaten by some cows that had escaped from the paddock. The baby was screaming, the three-year-old had put an entire loo roll in the lavatory, the four-year-old was refusing to eat her cereal and the nanny was in Canada skiing.

Me? Well, I was lying in bed thinking that, all things considered, I was pretty damned glad to be a man. I suppose I wouldn't mind being a single girl, because I could tour the country, sleeping with all my friends. And there's more. Your stomach is flat and your teeth are shiny. But all this has to stop when you've calved a couple of times. I don't care what it says in *Cosmopolitan*; you can't be expected to have a job, clear up sick and, when the kids are in bed, come downstairs looking like Caprice.

I ran into an old girlfriend the other day after 20 years, and though she was still pretty enough to turn heads in the airport terminal she had the harried look of a woman who'd been up since six herding cows. It was as though someone had stencilled 'mother' on her forehead.

And this, I think, is a fitting metaphor for that automotive Alice band, the people carrier. No matter how many times motor manufacturers tell us that their new breeder wagon has 'car-like dynamics', we know they're talking rubbish. A people carrier may be built on the same platform as a car, but it is still desperately and unswervingly mumsy.

But that said, Peugeot is different. Tell Peugeot to design a small hatchback and they'll give you a sports car. Tell them you want a sensible family saloon and they'll give you a sports car. Explain that you're fat and that you want a slushmatic machine for getting to the golf club and they'll give you a sports car. Someone deep in the bowels of Peugeot's chassis department understands what the enthusiastic driver wants: razor-sharp turn-in lift-off oversteer, seat-of-the-pants message delivery, and a ride-handling balance that's just so.

Peugeot engines are nothing much to write home about, and they cannot compete with Toyota's on the important question of reliability. But when you stick your Peugeot into a corner and feel that passive rear-wheel steering kicking in, you'll forgive it anything. So, truth be told, I was expecting big things from the Peugeot 806 people carrier. I was expecting a bit of a Yasmin Le Bon, a car that manages to be mumsy and phwoar all at the same time.

To behold, the 806 may be identical to the Fiat Ulysse and the Citroën Synergie, but I knew that with a wave of his magic wand Peugeot's brown-coated Mr Suspension would have turned Wendy Craig into Mimi MacPherson. So I ignored the curious – some might say ugly – styling and climbed aboard. And then I ignored the cheapness of the trim, telling myself that this was the ordinary £18,000 2.0 litre CLX, not some motoring journalist special.

I had no sunroof, no air-conditioning, no leather trim and no CD player, but I wasn't bothered because you don't expect this from Peugeot. It may be a van, I thought, but it'll go like Van Halen.

And it didn't. It went like Van Morrison. I tried, really I tried, to push it hard, but driving the 806 quickly felt all wrong, and now I know we'll never have a sporty people carrier. If Peugeot can't do it, nobody can.

So what's it like, then, as a device for moving large families around? Well, it's got the usual array of flexible seating, the usual small boot, the usual oddment tray under the passenger seat and the usual woeful performance: 0 to 60 takes 13.7 secondzzzz. It's so inoffensive that given half the chance it would drive down the middle of the road. And this way you could test the usual airbags.

As you may have gathered, the 806 failed to light my fire, but, again, this is nothing unusual. People carriers just don't cover themselves in margarine and rumble around in my underpants. Making me choose the best is like making me choose which limb I'd most like to have amputated.

I can, however, tell you which ones to avoid. The diesel-powered Nissan Serena is a no-no because with a 0 to 60 time of 28 seconds it is officially the slowest car on sale in Britain today. Then there's the Chrysler Voyager, which is ghastly, and the Ford Galaxy, which is unreliable.

The trouble is, though, that you still have a list of possibles that stretches from here to the seventh seat way over there in the offside corner. I'm tempted to be obtuse and suggest you have a look at the Mercedes V-class because it's the biggest, but then the Seat Alhambra is just about the cheapest and comes as standard with air-conditioning. Or better still, avoid the need for such a car in the first place.

Might I suggest the rhythm method?

Hell is the overtaking lane in a 1 litre

Have you ever driven down the motorway at the speed limit? No? Well, don't, because it's not big, it's not clever and nor, surprisingly, is it desperately safe.

You may have seen me trundling down the M40 at 69mph with a bus fastened to my rear bumper and a face the colour of parchment. And I'm sure you wondered what on earth I was doing. Well, I'll tell you.

Since Gordon Brown decided to knock £55 off a tax disc if you buy a 1-litre car, I thought it might be a good idea to try one out, to see if an engine this small can actually be used to propel a car. I would expect to find a 1-litre engine in a cappuccino machine. I believe my hedge clippers have a 1-litre engine, and that seems about right. For pulling the leaves off a bush, 1 litre is sufficient, but for moving around I'd always assumed you needed 4 litres, preferably with some kind of forced induction.

Needless to say, there aren't that many 1-litre cars on the market. If you discount the ridiculous selection from our dog-eating friends in Korea and the stupid Wendy houses from Japan, there are, in fact, six. And the best is Toyota's Yaris. This does 50mpg and comes with a 3-year mechanical warranty, a 12-year guarantee against rust and whopping 20,000-mile service intervals. Prices start at £7500, but if you go for an £11,000 CDX you get air-conditioning, two airbags, a sunroof, a CD player and, if you want, satellite navigation and a clutchless gear change.

It's a handsome little car, too, which causes girls to go oooh and aaah as though you'd just driven past in a baby seal. Blokes like it, too, because it has alloy wheels and a badge saying VVTi. Which sounds aggressive. But it isn't. Sure, the engine, which is Welsh, comes with variable valve timing, but there's no getting round the fact it displaces just 1 litre.

Now, the quoted top speed is 96mph, so theoretically it could keep up with the traffic in the outside lane of a motorway in the same way that Stephen Hawking, theoretically, could sing *La Traviata*. But at outside-lane cruising speeds the Yaris is loud. You can forget about conversation and the fancy stereo, because all you can hear is a wall of white noise. Couple this to a digital dashboard that acts like a strobe and you have a mobile torture chamber.

After a mile I was ready to admit that I'm useless in bed and that Jeffrey Archer ghostwrites this column every week. After two miles you'd have learnt that I fancy Esther Rantzen. And after three miles, begging for mercy, I slowed down to 69 and sought sanctuary on the inside lane. I'd never been there before, and frankly I never want to go there again. You end up sandwiched between two trucks, and in a Yaris, with its miserable engine, you don't really have the power to build up enough speed for an overtaking manoeuvre.

I tried it once, lunging into the middle lane, and immediately my entire rear-view mirror was filled with the front of a massive, snorting coach. And what are you supposed to do then? You can't get back on the inside lane because you're overtaking a lorry. You can't slow down or the bus will come through your rear window, and because you only have a 1-litre engine you can't go any faster either. Still, you'll be saving a pound a week on your road tax, so I guess that makes it all worthwhile.

At this point I don't doubt that people who live in London are running around the room waving their arms and telling everyone who'll listen that the Yaris sounds just perfect for inner-city life. To which I say: Pah. If you only want to move around London, why have wheels at all? Ten grand puts a taxi outside your door 24 hours a day. The whole point of having a car is that it can get you away at weekends, and the Yaris can't. It's terrifying on the motorway, and on normal fast A-roads it's even worse.

You come up behind a tanker and a quick glance shows that the road ahead is clear for 2000 miles. So you drop down to third, bury your foot in the carpet and pull out to overtake. One hour later you're alongside the tanker's rear axle and there's a queue of cars behind, their drivers wondering why you're on the wrong side of the road, making no attempt to overtake. But you are. You've even gone down to second, but with the engine revving its head off and blood spurting out of your ears, you're still not making any progress. And now there's a car coming the other way, so with much apologizing to those behind you give up, back off and get behind the lorry again.

In my week with the Yaris I arrived everywhere 20 minutes late, bathed in sweat. It could be a really good car this, brilliant even, but it desperately needs a bigger engine.

And a better name. Yaris sounds like Paula Yates's dog.

Forty motors and buttock fans

Last weekend, Andy Wilman, that human carpet you sometimes see on *Top Gear*, asked if he could borrow the keys to a Mercedes S-class I had on loan.

No surprises there. People who come to stay are always asking if they can try out whatever cars are parked in the drive. And the S-class is big news. Some say it's the best car in the world. Some say it's even better than that, so Andy wanted to get out there to see if the reality lives up to the legend.

Strange, then, that after just a few minutes he was back inside the house having not driven the car at all.

'Why?' I asked him.

'Because there's no need,' he said.

And he might be right. When you're presented with a new Mercedes S-class, you sort of know it's going to be utterly silent and effortlessly fast. You can be assured there will be no twist in the tale or, thanks to the traction control, the tail. So, really, what's the point of actually driving the damned thing?

What you want, frankly, is to be amazed by the toys – and, believe me, the S-class amazes, and then some.

I mean, the seats come with a grand total of 40 motors apiece and small fans that cool or heat your buttocks as you move along. And it gets better, because as you adjust the temperature a small bank of blue and red lights illuminate. That's great. You don't have to sit there thinking: 'Is my arse hot or is it cold?' A simple glance will tell you.

And then your attention is drawn to the television, telephone, stereo and satellite navigation system, all of which are fitted into a six-by-six box which lives on the centre console.

Now, to those of us who are over 35 years old, this is deeply impressive – when we were growing up, your amp was the size of a washing machine, your TV was black and white, there were no satellites and your phone number was Darrowby 35.

Obviously, having been brought up in a pre-calculator age, I am completely baffled by computers. But that didn't stop me stabbing away at the various buttons, responding with an excitable shriek when the readout on the TV screen changed. Simply getting the radio to come on, and play music, gives hope to the world's old people that maybe one day they could buy an Internet and make it mow the lawn.

For all I know, the air-conditioning system in an S-class could mow the lawn and a whole lot more besides: bikini-wax your wife, make a pizza? Who knows? I certainly don't, because the controls made no sense to me at all.

In American cars, the function performed by a knob is written in English on the knob itself. The button to open the sunroof actually says 'sunroof'. Now in the rest of the world, people recognize that there's such a thing as a language barrier, and, as a result, they use symbols instead.

Again, this worked fine. Find a button with a drawing of a sunroof on it and, unless you're in an Alfa, it'll open the roof when pressed. But what happens when a car offers a new function you've never heard of before? The symbol on the switch will be meaningless.

There's one button on the S-class dashboard which appears to have a corn circle drawn on it. So you press it and – guess what? A small red light comes on. There's no whirring noise, no soft whoosh such as you'd get when the doors open on the USS *Enterprise*, just that little red light. And next to it is another button with what looks like a Breville snack and sandwich toaster stencilled on it. Again, when you press this, absolutely nothing happens. I would say that, of all the buttons in the S-class, and there are hundreds, 80 per cent appear to have no function whatsoever.

Obviously, the solution can be found in the handbook, but, look, it's the size of the Bible and makes even less sense. By the time you'd got to the chapter marked 'How to Walk On Water', your car would have rusted away. And anyway, I sort of know what all those buttons do. They change the driving characteristics slightly, making the car perhaps a little more lively in the bends or a little more prone to rear-end breakaway. And honestly, this is silly because you can't induce power oversteer when you're still at home, with all your friends in the back saying: 'Hey, what does that one do?'

Certainly, you should attempt to drive an S-class by yourself. What with Maureen lunging at you from every side road, and schoolchildren surfing on your back bumper, you have enough to worry about without having to translate ancient Egyptian hieroglyphics every time you want to turn the radio up a bit.

Of course, no one who buys an S-class ever actually does the driving. You have a driver, but from now on you're going to need two: one to drive the car and beat up pedestrians who want your autograph and another who must be computer-literate, skilled in satellite guidance and fully conversant with road-going avionics. So that's Andy Wilman and me out, then.

Audi's finest motor just can't make up its mind

When a new play opens for business, the reviewers give it one chance. They do not go back again and again just because the make-up lady's changed, or the auditorium's been vacuumed.

The same goes for food. A.A. Gill does not re-review a restaurant because one of the waitresses has been to the hairdresser. 'Yes, I know we're still drizzling your halibut with synthetic Norwegian truffle oil, but what do you think of my new bob?'

So it is with a sense of shame that I find myself writing this morning about the new Audi A8. I know I've written about it before and I know I finished that review by saying: 'Don't bother driving it. You won't like it.' But, truth be told, I've always had this thing about Audi's flag-ship. I don't much care that it's made from aluminium or that it has four-wheel drive. Nor am I bothered that the Audi badge is rather Fulham compared with Premier League names like Mercedes or Jaguar.

I like the A8 because it's so damned handsome. I used to see a black one kicking around Regent Street. It had blacked-out windows and polished chrome wheels, and the want-one factor was way up there in the red zone.

I used to think of it as the only real rival for Jaguar's XJR. But then I drove one and the dream fell apart, along with all my bones. The ordinary version was too soggy, and in the sport models the ride comfort was abysmal. A cat's-eye could remove your teeth, a pothole could sever your spine and a humpback bridge could bounce your passengers clean through the roof. It had very obviously been developed in Germany, where road surface irregularities are taken outside by men in leather shorts and shot in the back of the head. But here in Britain, where councils deliberately build bumps in the road, it didn't work at all.

However, Audi has just changed the suspension, and against all the rules of this reviewing game I was prepared to give the German underdog a second chance. The new model is still manly and handsome enough to hang a question mark over your sexuality, but now there are some snazzy wheels and a different radiator grille. The 4.2-litre V8 engine is also different. They've fitted five valves per cylinder so that it develops 310bhp – enough to get you past 60 in 6.9 seconds and on to a top speed of 155.

Me? I went rather more slowly than that because they've fitted a television in the dashboard which shuts off once you're going more that 5mph. Indeed, I'd like to apologize to everyone on the M40 for my glacial progress, but I was watching Countdown, trying to make a seven-letter word from 'telephone'. Sadly, the game ended when I reached London, because at this point the Audi's electronics went mad. In a traffic jam in Knightsbridge, the parking sensors began to beep, suggesting I was close to other cars – you don't say – and then the sat nav chirped in, saying: 'If possible, make a U-turn.'

At least, that's what I think it said, but it's hard to be sure because half way through the message the radio turned itself up to the sort of volume that can deafen dogs and told me of faulty lights in Hackney. Then the phone rang.

Now in an F-15 fighter-bomber your helmet is constantly filled with warnings about excess g, and an imminent stall, and missile lock, but this is a warplane. In a car, surely to God it is possible to engineer a system whereby messages come one at a time.

But what about the suspension? Well, for sure it's better than it was, but round town the A8 still crashes into potholes that Jag-man wouldn't notice. At speed things improve, but at speed another problem rears its ugly head. Turn the steering wheel and there appears to be a slight delay before you change direction. Hit the brake pedal and there's a pause before you start to slow down. The Tiptronic gearbox appears to be working in a different space-time continuum. And this means that despite the Sport Quattro badging, the big £56,000 A8 is not the driver's car it should be.

Now I know it's hard to blend comfort with sportiness, but Audi's boffins got both elements wrong first time. Now, amazingly, they've got them wrong again. If you want a sports saloon, you're better off with the blistering Jaguar XJR, which is much faster, more comfortable and £6000 less expensive. And if you want the ultimate big car, and to hell with driving dynamics, the A8 is soundly thrashed by the new 4.3-litre Mercedes S-class.

So much of the A8 is right. You will never find better seats, and should all eight of the airbags inflate you will find yourself rolling down the road in a bouncy castle. It's good looking, quiet, dignified and it appears to be beautifully made. But that suspension wrecks everything. Think of it as a cake. Perfect in every way except for the giant cowpat.

Keep the sports car, drive the price tag

All over the world there are human rights atrocities about which America does nothing. The Russians, for instance, went bonkers in Chechnya and all the while Uncle Sam got on with his beefburger. But then, out of nowhere, Mr Clinton decides to pick on Yugoslavia. Hurriedly, his generals consulted an atlas to find out where it was, and in the last four weeks of sustained action they've managed to hit a house in Bulgaria, a hospital, some Chinese, all the make-up ladies at Serbia's television station, a refugee column and Slobodan Milosevic's bedroom. At a time when he wasn't in it.

And in the process they've lost four big, fast, expensive planes, three soldiers and two Apache helicopters. But then this isn't surprising, because the Americans have a proud and noble tradition of being utterly hopeless at warfare. They lost in Vietnam, they lost in Somalia, they lost in the Bay of Pigs, and though they won the Gulf war they managed to kill more British soldiers than the Iraqis.

But then think about it. The Americans are the largest consumers of that most strangely outdated car, the Mercedes SL. And this says a lot. It should have said to Mr Blair that perhaps their weaponry was dodgy, too, and they'd hit the wrong country. But of course it's hard to hear messages when you're six feet up Mr Clinton's bottom. And he wasn't listening this week, either, because he's been touring the Balkan refugee camps with a brace of rather sweaty armpits and a wife who appeared to have put her mouth on inside out.

Ah, good, I thought. He's there to apologize for killing all those civilians. Or maybe he's there to hand himself over to the war crimes people. But no. He was there to offer some of these unfortunate refugees homes in Britain.

Now, Tony, have you thought about this? Have you asked the Italians what these Albanians are like? When they took in a few thousand after the last Balkan pugilism, even the Mafia was scared. Really, it never ceases to amaze me that people in positions of responsibility can be guilty of such muddled thinking. But then you don't need a war to bring the issue into sharp focus. You need only to see someone drive by in the aforementioned Mercedes SL.

Back in 1990, when this car was new, I used one to woo my wife. We tootled out of London on a sunny day, had lunch in Oxford, and on the way home called in at Henley to watch the regatta. It was all just too agreeable, and I remember the small crowd that gathered to watch as I raised the roof. You simply pressed a button on the dash and 11 motors did the rest. Nobody in Henley had seen anything quite like it. Nor had anyone seen such a heavy car go quite so quickly. The 3.2-litre straight six was nippy. The 5-litre V8 was a blast, and the 6-litre V12 could remove all your make-up.

I put my hand up and declared myself a fan. But then Princess Diana bought one and everything started to go wrong. In the same way that Bobby Ewing spoilt the previous SL by using one in *Dallas*, Diana brought the current SL out of petrolhead heaven and into the pages of *Hello!* As a result, it quickly found favour with the sort of woman who takes a photograph of a B-52 bomber to the barber's and says: 'I want my hair to be bigger that that.' I'm talking, of course, about the Cheshire wife. In a world where Stuart Hall is God and they model furniture on his blazers, the swanky, posh SL became as much a part of the Cheshire uniform as the gold shoe or the baggy knicker curtain. This meant that elsewhere normal people were getting out of their SLs fast. And these were then being snapped up by people in Southall, who were fitting Grateful Dead stereos and wheels from earth-moving equipment.

All this was bad enough, but then along came the much prettier SLK, which, at a stroke, made its bigger brother look heavy and unnecessary. Anyone with half an eye on the style mags would choose to use the baby, but up there in Wilmslow the only thing that mattered was 'how much?'.

At just £30,000, the SLK was far too cheap. Up there you could sell a dog turd if you priced it high enough. And the SL is currently going for anything up to £100,000. Now go on, enjoy a Biro-sucking moment. Think. If you had £100,000 to spend on a soft-top car, would you choose a 10-year-old design favoured by Americans and women with false breasts? Or a Ferrari 355 Spider? Or a Jaguar XKR? Or a Porsche 911? Anyone who chooses the Mercedes is saying to the world: 'Look, I may be rich, but even my smart bombs can't hit the right country.'

Mr Blair should have one. It may not signal to other road users that he's a psychotic war criminal with a BO problem, but you'd know something was wrong. And you'd remember, come the next election, to give him a wide berth.

Out of the snake pit, a car with real venom

It's the middle of the night and I can't sleep. For the past few days I've been driving a car so good, so exciting and, most of all, such incredible value for money that I simply have to write about it now. It's called the Holden HSV, and I was determined to hate it. First of all it's Australian, and Australia begins with the letter A. All the best countries begin with the letter I – Italy, Iceland, Ireland, India, Ingland and so on – while all the worst begin with an A: America, Austria and, of course, the godforsaken spider-strewn snake pit.

Why do you think God put it so far away? And why do you suppose he is now trying to remove its protective ozone layer? It's because God is British, and he's tired of being called a whingeing, dirty homosexual.

Ask Australians what makes their giant and useless continent so good and you'll always get the same reply: great climate and juicy steaks. Which is fine, but they should remember that we were brought up on a diet of drizzle and fish fingers and we produced the biggest empire the world has seen.

I dislike Qantas, Sydney, big prawns and the notion that if I go outside without a hat I will catch head cancer. So I was determined therefore to hate the Holden. And another thing: it's made by General Motors, which, in Britain, conjures up visions of the Vauxhall Vectra. GM may be the largest car company in the world, but so what? Richard Kiel is the largest actor, but he's a long way from being the best.

Then the car in question arrived. It looked like a cross

between the ancient Omega and the enormous Chevrolet Caprice. This was bad enough but, to make it worse, the HSV appeared to have been decorated by a 14-year-old boy. Maybe the Australians like silver side-skirts and red badging. Maybe this explains why they all have ovens but choose instead to burn their food in the garden. And maybe, because they spend so much time outside, they're not worried about the interior. Perhaps this is why it's grey and there's no ashtray. Good, I thought, this is a car I can savage.

And yet here I am at 4 a.m. dribbling the dribble of a man who's smitten. Just a few weeks ago I said the new S-class Mercedes-Benz is the best car in the world, but there have been times this week when I've doubted it. I suppose the key to this pant-wetting appeal is the 5.7-litre V8 engine. It may produce only 295bhp, but you get a colossal 350 feet/pound of torque so, at 70mph, in sixth gear, it is doing just 1500rpm. And don't worry if someone comes alongside to laugh at your silver side-skirts: with all that torque, a nudge on the throttle will put you two countries away in three seconds.

I was told, before the test drive, that the Holden is perfect for the TVR driver whose fruitful loins have forced him into a four-door saloon, and there's some truth in that. But in terms of character, it's more grown-up than a TVR. Indeed, it has an identical twin: the Aston Martin Vantage. They have the same gearbox. They make exactly the same noise. They're both big, and they both feel bigger still. Sure, the Holden doesn't look like a Castle Ashby, but then it isn't priced like a stately home either. A Vantage nowadays is £200,000, while the HSV is yours for less than £40,000.

Yes, the Aston is ultimately quicker in a straight line but, because the Holden's undersides have been tweaked by Tom Walkinshaw, the Bondi Beach bodybuilder can leave its aristocratic twin standing.

The grip is outrageous but, even if you go bonkers and decide to break it, even the most ham-fisted driver can get everything back in shape. On a twisting country road, you'd have to think mid-engined and Italian before you'd come up with a car that could pull away.

But do you know what sold it to me most of all? When you're tired of turning heads with the exhaust bellow, and bored with turning corners at Mach 2, you can settle back into the supremely comfortable seat, put it in sixth and waft home with the silence broken only by whatever you've put on the stereo. This car is a 162mph Meat Loaf or a 22mpg Brahms, depending on your mood. And because it's too big, there's space in the back for three children and room in the boot for their toys. And if you have a dog, that's no problem either, because the HSV is also available as an estate.

And ordinarily that would be that. I'd sign off now and you'd go and mow the lawn, saddened that such a great car is out there but on the wrong side of the world. However, because they do at least drive on the right side of the road in Australia, it is now being imported to Britain, where, I'm told, it can be serviced by any Vauxhall dealer. If you want one, and I assure you you do, forget the grass and call the importer on 01908 262623.

The Swiss army motor with blunted blades

Obviously, you would not dream of setting out at night in a pair of Rohan trousers, the stride of choice for those who value practicality over style – scoutmasters, mostly. This is a Swiss army trouser, tough and dependable, with a plethora of pockets and handy zips. It's good on Everest but, in all honesty, poor for the rather more common pursuit of social mountaineering. I suspect that if you turned up at Wentworth in a pair of Rohans the clerk of the course would ask you to leave. Unless they were egg-yellow of course. Which they aren't.

I bring this up because all week I've been driving a car designed specifically for Rohan man. It may be called the Zafira, which conjures up visions of Liz Hurley's new frock, but don't be deceived. This is brown Vauxhall, and you just can't get further from Versace than that.

To Rohan man, the brown Vauxhall is automotive ambrosia. It hides the dirt, like the patterned carpets in his sitting room. It was made in Luton, like his children, and the dipstick is clearly visible. This sort of thing matters to someone who goes out at night to talk tents. Rohan man loves all Vauxhalls, but the Zafira is going to get him more animated than a new piece of self-assembly furniture. This is because the Zafira *is* a piece of self-assembly furniture, with wheels so you can take it round to a friend's house and show off.

It comes in a box just an inch longer than a Vauxhall Astra estate. But inside it can be either a two-seater van, a five-seater estate or a seven-seater MPV. You can even carry a surfboard, so, ladies and gentlemen, I give you: the Swiss army car.

We've had people carriers before, of course, but they are unwieldy and unpleasant. They are vans with electric windows. They are for the terminally mumsy. But then along came the Renault Megane Scenic, a quite brilliant piece of design that put five fully adjustable seats in a car the size of a Ford Escort. It has been the sales success of the decade, putting the Megane at the very top of the sales charts in both Scotland and Northern Ireland.

And now Vauxhall has gone one step further with the Zafira. If the Scenic was brilliant, this is the work of a genius: seven seats in a car the size of an Escort. And when they say it takes 15 seconds to make the transition from van to car, they're wrong. I did it one-handed in just 12.

Of course, with the rear seats in place you have no boot; not that this matters on the school run. What does matter, if someone runs into your back end, is the proximity of your child's head to the rear window, but as this is a problem in all people carriers, it's hardly fair to single out the Zafira for a firing squad at dawn.

No, with a bench seat that slides forwards and backwards, legroom for adults all round, tons of headroom and a surprisingly handsome, compact body, I will admit that the Zafira is very possibly the cleverest family car out there. It's not bad to drive either. The 1.8-litre engine is a little gem, revving happily to its red line ... and staying there. For some extraordinary reason, Vauxhall has fitted sprint gearing, which means that, on a motorway, it's doing a noisy and wasteful 4000rpm.

However, as a result of this odd gearing, it's a zestful performer. It rides well, corners with precision and can – get this – even be quite good fun. So, well done, Vauxhall. Finally, you've come up with a half-decent chassis fitted in a people carrier – a car that will rarely exceed 40.

Really and truly, I should buy one of these. It can handle three children and a dog, and it won't even be flummoxed should I decide to take up surfing. Sure, with the range starting at £14,500, it's more costly than a Scenic, but it's cleverer and has more seats.

However, I'm not going to because it's a Vauxhall.

The very name is just so dreamy, and it's not going to get any better if they paint the press demonstration fleet brown. Brown is so much more than a colour; it's a way of life. Brown and Rohan: there are singers out there who could make them rhyme.

And the Zafira must surely be the last car made where the radio is not integrated into the fascia. It's like they couldn't be bothered – so what if it gets nicked? That's not our problem.

Time and time again, customers say that Vauxhall after-sales care is second to everyone, that they get hit around the head with a dead fish every time they walk through the showroom door. I'd like to take Vauxhall's senior management on a two-day fact-finding tour of London. I'd show them the fabulous new restaurants where the walls are blue and the service is instant and invisible. I'd take them to bars where the people have bright eyes. I'd show them the riverside developments, and then I'd say: 'You've got a good car. Now drop the brown, boys. Just drop the brown.'

Perfection is no match for Brian and his shed

Every year you can go along to the Motor Show and, on the final day, watch the cleaners sweep a sack full of dreams into the corporate wheelie bin.

Here's what happens. Someone called Brian, with leather elbow patches and a shed, decides to build a new car. So he borrows some money from a chap called Vince, who keeps pit bulls, and sets to work. The finished product has a plastic body, a Vauxhall engine, no antilock brakes and an ill-fitting hood, but Brian is so proud he actually calls it the Brian. Of course he does. Anyone who looks at the thousand or so cars on the market today and thinks 'I can do better' has a big, big ego.

Brian does market research by having his neighbours round for drinks. And when they make polite noises, he decides to borrow more money from Vince and take a stand at the Motor Show where, in 14 days, he sells absolutely none. And 10 days later, after Vince has repossessed the car and crashed it, Brian is found in an Essex wood. Well, bits of him are, anyway. And that's the end of the story until the following year, when someone called Colin turns up at the NEC with a car called the Edna, after his wife.

What these people need, more than anything, is a name. Ford, for instance, has just paid £6 trillion for Volvo, which breaks down like this: £1 for the factory, £1 for the staff and £5.9998 trillion for the badge and all it means to a million Gloucestershire antique dealers.

Then there's Lexus. That's a fabulous name. It sounds like a cure for cancer that NASA found on Mars. Call a book *The Lexus* and you'd have a Christmas No. 1. I'd buy a Lexus just to say I had one.

Some say Lexus has never won Le Mans and that there's no history, but I say, Pah! Lexus has an incredible history of never making a car that goes wrong. I'm not talking about the reliability you might get from Mercedes, where one car in a thousand breaks down. I'm talking total perfection.

If the US air force had let Lexus design their smart bombs, Nato wouldn't be in such a mess and they'd all be sitting around in the Chinese embassy today, eating snakes.

Jaguar has won Le Mans lots of times but, with its pace-maker build quality, Lexus hit them hard. Now they've decided to hit BMW, too, with a £20,000 car called the IS200. The figures suggest they haven't a hope in hell. Last year the BMW 3-series was the best-selling car in its class by an incredible margin. All on its own, the 3-series outsells the combined total for the Honda Accord, Audi A4, Volvo 540 and Alfa 156.

But here's a fact: anyone who sits in the new Lexus will want to buy one. Everything, from wacky chrono-meter dials to drilled pedals, is magnificent. The sat nav slides from the dash like something out of *Star Trek*, the gear lever is polished chrome and the leather is suede. Then the salesmen will click in and talk you through the 20,000-mile service intervals, the 3-year warranty, the low insurance and the amazing new 2-litre engine. It's called the VVT-I, and that's all you need to know.

Oh, well, all right: it develops 154bhp, which is enough to get you from 0 to 60 in 9 seconds and onwards to 134mph. It sounds sporty, too, which fits in well with the six-speed gearbox and the rear-wheel drive. In the show-room, and on paper, the IS200 looks good. On the road it looks even better. People will look at you go by and think, 'My, what a handsome car. Sporty, yet somehow restrained and tasteful. And I particularly like the way those big alloy wheels fill in the arches so nicely.'

And you? Well, sadly, you'll be fast asleep because, while the garnish is pretty and the price is nice, this is one of the most uninvolving cars I've ever driven. You'd get more driver satisfaction if you were beamed from A to B. You arrive at a corner, turn the wheel, with its natty silver-look handgrips, and a thousand high-tech, Japanese gizmos get you through to the other side. It's like frozen halibut. The ingredients are all there and the packaging's great, but, as a taste sensation, it's right up there with wood.

And do you know why? Because Lexus is a division of Toyota, and Toyota is a giant corporation where if one thing matters, it's the bottom line.

This is not rear-wheel drive because some driving enthusiast said such a move would make for a better balance. It's rear-wheel drive because the marketing department thought it would look good in the brochure. The IS200 is a cynical facsimile of the real thing. It's Virgin Cola and I absolutely hate it.

If you want a car that's good to drive, a car built by enthusiasts for enthusiasts, might I recommend that, when the Motor Show comes round, you make for Hall 73. Once there, you should talk to a man called Colin because, you never know, his surname might just be Chapman.

Waging war with the motoring rule book

Ever since the men from Austin went to help Datsun set up a factory after the war, the Japanese motor industry has slavishly followed where Europe and America have led the way.

I want you to think of one single Japanese motoring invention. Come on, I'm waiting. No, you're going to have to give up because everything from disc brakes to the windscreen wiper was developed in the West.

In a race to find the least inventive people on earth, Japan would line up with Australia and Burma in first place.

The trouble is, of course, that before the British boffin had a chance to show the Patents Office his new invention, some Japanese chap had copied it. And while British management prevaricated over who'd fund such a thing, thousands of perfect imitations were rolling off a production line in Yokohama. The Honda NSX was a shameless facsimile of the Ferrari 308. The Mazda MX-5 was a modern-day MG. The Datsun 240Z was a Capri, and the Toyota Supra an oriental Corvette.

But then came the Nissan Skyline, a car that didn't follow round-eyed rules. By using the sort of electronic whiz-kiddery we now expect from Japanese VCR designers, the world was treated to a car that pulled down its trousers and mooned at the laws of physics.

And this fire-breathing Datsun seems to have acted as a sort of cattle prod for the rest of Japan's car industry. Look at that Subaru Impreza 22B. There's no way that such a thing could ever have been styled in Italy, and if it were German it would weigh eight tons.

Then there's the latest generation of Honda VTEC engines, which sounds, looks and feels Japanese. And what about the spoiler on the back of an Evo VI? Was that designed in Longbridge? Yeah, and cod use breath-fresheners. Now all this, I think, is a very good thing. Five years ago there were maybe a couple of Japanese cars that I'd have actually wanted to own, but now there are several dozen. And topping that growing list is Mitsubishi's Galant VR4.

First of all, I quite like the idea of driving a Galant. I feel it would help little old ladies with their shopping. And, second, while I've never actually driven an ordinary Galant, I find myself drawn to what is simply terrific styling. It's like one of those women who, when you first meet them, don't appear attractive at all. But after a few hours, you're at her feet, slobbering.

And the VR4 is even better because it's had collagen lip implants. My five-door estate test car had a huge spoiler on the back, a deep front air dam at the front, fat wheels and sexy tyres. And if you don't believe a tyre can be sexy, you've clearly never studied the tread pattern on a Bridgestone S-02.

Basically, you look at this car and know it's Japanese. Which means you know that it won't break down. And then you go for a drive. Now we know that Chevrolet was first out of the blocks with a turbo and that Jensen was first with four-wheel drive. We're also aware that Audi was first to bring these technologies together in the Quattro. So you might argue, therefore, that the Galant is simply aping its four-ringed forefather. So what about the Mitsubishi's active yaw control, then? The car's rear end is fitted with a torque transfer differential system with an electronically controlled clutch that senses the condition of the road and the driver's style, then adjusts the yaw force accordingly. And to be honest, I don't remember seeing that listed in the spec of the new Rover 75.

And I haven't finished yet. The Galant's gearbox has the capability to learn what a driver is like, and then stores his shift patterns in its memory.

Without delving into the mysteries of electronic fuel injection, we know that what we're dealing with here is a motorized Canon Ixus. It's a bunch of super high technology, designed to wage war with the motoring rule book.

And it makes the Galant VR4 an enthralling companion. They say it develops only 280bhp, but that's a bit too neat, seeing as 280 is the limit under Japanese law. I mean, come on, chaps. It has got a 2.5-litre twin turbo V6; it does 0 to 60 in 5.9 seconds; it'll hit 150. Two-eighty brake horsepower, my arse.

It's let down only by a wretched interior. And why is it wretched? Well, in a bid to copy the European style, they've glued wood to the centre console and half of the steering wheel to create a symphony in DFS. It's World of Leather in there, too, and it's truly awful.

Mitsubishi has had the courage to make the car look and feel Japanese. And that's fine. I'll supply the passion every time I go round a bend fully 10mph faster than I could in the Jag. They really need to think of their own interior style. Seats on the floor? Foldout fans? I don't care – just make it Japanese, and not a Japanese interpretation of the Long Room at Lord's.

When this is done, European car makers will be in trouble, because the days when you bought Japanese for reliability and European for flair will be over. The Japanese will give you both. And all for less than thirty grand.

Evo's a vulgar girl, but I love her little sister

In the sort of circles where young men have earwigs on their top lips and wear their hats back to front, a new drum and bass superhero has cruised into town. It may be a small, 2.0-litre, four-door saloon, but the Mitsubishi Lancer Evo VI costs £31,000 and ranks even higher up the cool-o-meter than Gail Porter's pierced nipple. It has picked up the nation's youth where the Escort Cosworth left off and is taking them on an adrenaline ride to the planet Bad. In English, it's a road-going version of Mitsubishi's rally car, and it will accelerate from 0 to 60 and back again in 6.7 seconds. This is staggeringly, stupendously, gut-wrenchingly fast.

And thanks to four-wheel-drive and something called active yaw control, it will go round corners at the sort of speed they recommend only in the Exit handbook. Never, not once in 15 years of road-testing cars, have I found anything so intoxicatingly rapid.

Put your foot down hard and, with not a hint of lag, it lunges forwards with such violence you will grunt. And I'm not talking about some kind of self-satisfied post-orgasmic sigh. This is the 'Oh, my God' grunt of a man being scared half to death.

With the ride comfort of a skateboard on Chesil Beach, it just darts this way and that, ripping the heart out of any corner you throw at it, then exploding down the straights as though fuelled with Semtex. It's like going to the shops on a roller coaster. Then, when you get to town, the effect is even more incredible. Agri-yobs will stop vandalizing the phone box and put down their cider to salute you. You want respect from the nation's youth, but you don't want a stud in your stomach? Get an Evo VI.

However, in a normal motorway traffic jam, the reaction is rather different. People note its giant Tinsley Viaduct rear spoiler, its blue mudflaps and the front end, which is one giant radiator grille. And put simply, they laugh at it.

So while I appreciate the Evo's ability to generate one g while cornering on ice, I'd rather not have people point and say: 'Ooh, look, there's that bloke who used to be on the television. And he's driving the stupidest car in the whole world.'

The motoring magazines have called it a 'Porsche eater' and a 'road rocket', but in the real world, where people wear brogues, it is, I'm afraid, irretrievably vulgar. You may have one only if you are a drug dealer or a footballist. Pity.

But don't despair: the Evo has a sister, a car that combines the four-wheel-drive grip, the active yaw control and the turbocharged wallop in a body that's rather less Geri Halliwell and rather more Kristin Scott Thomas. It's called the Galant, which conjures up visions of Sir Walter Raleigh, a man who would happily lay down his freshly laundered cape so the Queen could cross some puddles. But a man with big pants.

And so it goes with the Evo's sister. Yes, it has spoilers and flared wheel arches, but they're small and discreet.

And inside there's a wood-look dashboard, automatic transmission and a cool, detached air. You almost expect to find Edward Fox in the glove box. You can pop into town and the agri-yobs will continue to destroy the bus shelter without looking up. You can sit in one of Mr Prescott's traffic jams and people will carry on picking their noses. It's just another car.

But peer into its trousers and you'll find a 2.5-litre V6 engine which, for that little extra something, has not one but two turbochargers. So it'll get from 0 to 60 in 5.5 seconds and on to a top end maximum of 150mph. It's not as fast as the Evo – what is? – but, amazingly for a car that's bigger, more comfortable and better equipped, it's cheaper. The more expensive five-door estate is £29,995. This is remarkable value for money.

Really, I want to conclude with a soaring eulogy, a rising crescendo of enthusiasm and affection, but I need to be a little bit careful. I told you only a month or so ago that the new Mercedes S-class is the best car in the world. And then, more recently, I typed a report on the Holden HSV in pure dribble. I know I should hold the few remaining superlatives back for the upcoming BMW M5 but I'm sorry, the Beemer can cocoa. With the Galant, I'm going for broke. It is utterly magnificent, an all-too-rare blend of Japanese high-tech with genuine good looks.

There's something here for everyone. Gail Porter's fan club will appreciate the active yaw control, in particular the green LED that comes on to tell you it's working. You will like it because it has a three-year warranty. And I like it because it's a 150mph estate.

You couldn't possibly buy an Evo VI, but the Galant

gives you a taste of high octane, high living without the heartburn. It is, in short, superb.

At last, a car even I can't put in a ditch

Learner drivers are always taught that, when a car starts to skid, you take your feet off all the pedals and steer in the direction of the slide. Yeah, right. And when you're learning to ski, you just nail a couple of planks to your shoes and whiz down the mountain.

Believe me on this one. When a car starts to slide, you can do whatever you damn well want. You can tear the steering wheel from its mountings or stamp on the pedals like you're playing the organ and it won't make the slightest bit of difference. You may as well eat your own nose because, in just a few seconds' time, you're going to be upside down in a ditch.

I should know. I've been test-driving cars for 15 years and that means finding out what happens when traction is lost. It means going to the outside of the envelope and then taking one more tiny shuffle into the great never-never land where you're ripped from the driving seat and replaced by the laws of physics.

I know the theory of handling as well as anyone. I know about weight transfer and tread shuffle. I know precisely what you should do in any skid, on any surface, but when I go out there and actually do it I always end up in a ditch, on my head.

I know, for instance, that it is theoretically possible to steer a car using the throttle. You open it up a touch and the back starts to slide. You keep it there to maintain the slide and then, when the road straightens, you ease off the power to bring everything back in line. For 15 years I've talked in the pub about cars that can be steered on the throttle and cars that can't. And it's all been nonsense because I'm not Michael Schumacher. Using the throttle in an attempt to steer a car simply determines how fast I go into the ditch.

But this week it all became clear. I was struck by a blinding flash of light and now there is no limit to my power. I am super-driver, a man who's no longer in harmony with his wheels. I am their master.

I was at Kemble airfield in Gloucestershire with the new BMW M5 and a photographer who was keen to capture some sideways action for a magazine feature. No problem, even for me. Anyone can get a car to go sideways. It's what happens when you've gone past the camera that's hard and uncomfortable.

But not this time. I got the car sideways, and it just stayed there. And then, when I'd had enough, I eased off the power and it settled back in line. Happy? No, no, no. I was wearing a grin so wide it shattered both the side windows.

There's more, too, because the M5 comes with a little button on the dash that sharpens both the steering and the throttle response. Press this and even Thora Hird could become Mistress Power Slide, a warrior princess in petrolhead heaven.

For hour after delicious hour I hurled that car round the perimeter road, its back end never quite in line with the front, and . . . blimey! What's that funny noise? Oh, no, the rear tyres have fallen apart.

No, really, in the space of a morning I managed to wreck two 18-inch Dunlops that cost £387 each. Oh, and there's no spare. BMW gives you a can of sealant and a pump, but this is of limited use when you're down to the canvas.

So let's think a little bit about the implications. Without any doubt, the M5 is the most flattering car a man can drive. It turns a ham fist into a sirloin of pork and handles, quite simply, like a dream. But if you peel away the handling prowess, like the handling prowess peels away the tyre tread, what are you left with?

You've got Sebastian Faulks with no writing skills. You've got George Clooney with a face like a horse's arse. You've got a £60,000 BMW 5-series made at a factory in Dachau, on the site of a former concentration camp. Sure, an incredible leviathan lives under the bonnet – a V8 that develops 400bhp – but you need to keep the nanny-state traction control on to preserve those tyres. You've got a fabulous interior, but you can have something identical in a 528i for half the price.

Certainly, your neighbours won't be terribly impressed by the M5, partly because the satin-finish wheels are vulgar and partly because it makes a deep, Brian Blessed booming noise that kills dogs and breaks all your finest Czechoslovakian glass.

It is demonstrably better than the 540, but the only place where you could possibly demonstrate its 'better-ness' is on a track or an airfield. And when your friends are all driving home afterwards, you'll be on the phone trying to find a tyre that no one stocks.

So obviously the M5 is a stupid car? Well, no, not really. Because you're never going to take it to an airfield, and you're never going to slide it round the corners, and who gives a damn if your neighbour's dog explodes? It pains me to say this, given my history of baiting BMW, but the M5 is magbleedingnificent.

Trendy cars? They're not really my bag

So there I was in the back of a cab with Rodney Bickerstaffe, general secretary of some union or other, on the way to Bond Street to buy my wife a handbag. Pretty surreal, huh? And then the nightmare began. There are two criteria that must be met by a handbag, each of which is mutually exclusive: it must be fashionable and it must be practical.

While I was driving along with my wife the other day, she asked me to find her sunglasses. This meant diving into her bag, where I discovered she had a normal pair of spectacles, a spare pair, a normal pair of shades and a pair she got on prescription. And none of them were right, so the search continued, down past the make-up bag, the mobile phone, enough keys to baffle a warder at Brixton jail, two wallets and a foldout photograph frame. Further down, below the plasma, there was a medicine chest, another mobile phone, more keys and then the sanitary area. And at this point I gave up and said: 'Look, darling, I really can't find them. Can't you just squint?'

This is a woman who doesn't need a handbag so much as a binliner. But that would never do, because in *Vogue* I see the modern woman sports something no bigger than a tea bag. Well, she does today, but I know enough about fashion to know that, by the time I'd got to the till with a microscopic blue Versace, it would have become as up-to-date as an ox cart. And by the time I got back to the shelf to change it for something in grey, there'd have been a punk revival and I could get a binliner after all.

I began to understand how Prince Philip might have felt when he discovered that two homosexual priests had been playing tonsil hockey at one of his garden parties. Bewildered, in a what's-the-world-coming-to sort of way.

I finally settled on something with studs and then stood back in horror as I watched my wife transferring the contents of her old bag into the new one. It was like watching the Queen Mother move all her furniture into a two-bedroomed terraced house.

And within a month it'll be out of date. But what the hell? Staying fashionable, on the handbag front, is not that expensive. The big problem is staying fashionable in cars. Cars never used to strike much of a chord with the terminally trendy because everyone had a Ford Cortina and you'd have to wait maybe six years for a new model to come along. That and the sheer expense of a car meant there was never a sense of here today, gone tomorrow. But now car makers have reinvented the idea of fitting a wide variety of different bodies to the same basic plat-form. So a whole raft of supposedly new models can be made, economically, in small numbers. It's called niche marketing.

When Volkswagen launched the new Golf-based Beetle back in January 1998, people with black clothes and tiny handbags ordered one straight away. It was the car of the moment, but American demand was so massive that British deliveries are only just beginning now. And I'm afraid the moment is past. To review the Beetle as a car would be as pointless as reviewing the latest La Perla knickers on the quality of the stitching. It was designed solely to be a fashion statement, to be a bandwagon on to which the 1960s revivalists could jump.

Thanks to the Abba thing, we're now in a 1970s time warp that should have been good news for Rover. Only they got muddled and thought it was the 1470s. Really, I'm surprised their new 75 isn't offered with a thatched roof.

Today, you're far better off with the Mercedes Smart, which is positively *de rigueur* in St-Tropez. But when the next registration prefix comes along, you should be thinking very carefully about Fiat's six-seater Multipla.

It's a long, long way from being the best-looking car in the world, but handsomeness rarely has anything to do with making a fashion statement. I saw a girl in the Style section of this paper last week wearing what appeared to be a wastepaper basket.

Even in the countryside, where the Hermès headscarf has been the crowned head of state for 2000 years, people are starting to get the idea that cars can be fashionable. So it's off to Cheshire with the Range Rover and into the courtyard with Toyota's Landcruiser.

Flares may be out in the Voodoo Lounge but, believe me, flared wheel arches are in on the range. And don't worry about the Made in Japan sticker because so is sushi.

This is complicated, I know, but things could be worse. With the trendification process now enveloping everything from music and cars to clothes, food and entire postal districts, it can't be long before houses are swept up in the style tidal wave.

'Oh, my dear, Georgian is so last year. You've got to pull it down and build something in yellow.'

Why life on the open road is a real stinker

Have you any idea what life is like for a travelling sales-man? The traffic jams. The endless parking tickets that eat into your commission. And all the while sweat is pouring down your back because the fleet manager was too mean to put air-conditioning in your hateful, diesel-powered Vauxhall Vectra. On Wednesday you drove all the way to Carlisle only to find the chief buyer had gone out for lunch with his secretary and – snigger, snigger – they weren't expecting him back.

I know a bloke who used to sell franking machines for Pitney Bowes. And to win over secretaries, he'd ask to see their tongues, pull a face and say they'd been licking too many stamps. Can you imagine that? Can you imagine having to chat up a fat temp just so you can flog her boss a crummy franking machine?

We like to dream the American dream of life on the road, sailing across Montana with the warm wind in our hair, but the reality is somewhat different. Because life on the road in Britain means you're a rep, and you have a Vauxhall and you're stuck in a jam watching prime ministers flash past in the bus lane.

You bought a Lion bar when you last filled up with petrol and bits of it have landed on your shirt; you failed to close that last deal and now, at 6 p.m., you're hunting for somewhere to spend the night. You're on a budget of £50, it's pouring with rain, and you're in Cardiff.

I know exactly what it's like because I've done it. You crawl round the one-way system, wipers smearing the neon, in a desperate search for that elusive grail, a two-star hotel where the sheets are made from natural fibres.

That's my abiding memory of life on the road. Nylon sheets, waking up every morning with my hair on end, and spending all day pumping 4000 volts into anyone with whom I shook hands. That, and being laughed at in Indian restaurants for eating on my own. 'Look at that bloke with the funny hair. He's got no mates.'

Other reps used to eat in the hotel, with waitresses called Stacey tottering about in micro-skirts. But I'm allergic to patterned carpets and, when you've heard 'Stairway to Heaven' being murdered for the thousandth time by the Mike Twat Singers on the piped Muzak system, you have to get out.

Then, when it's two in the morning and you can't sleep because of the sodium streaming through the curtainless window, you're awoken by a drunken bloke called Dave who's been given a key to your room by mistake.

I'd lay there all night, listening to the Doors on my Walkman and doing my best Kurtz impression. The horror. The horror.

However, this week I found myself staying at a Travelodge hotel and, oh, how times have changed. A plaque outside said it was 'the 100th Travelodge to be opened at Hickstead' and that the ribbon had been cut by Judith Chalmers. All Travelodges are opened by a top celeb. They got Annabel Croft in Hemel Hempstead and

Torvill and Dean for Nottingham. But the hotels, so far as I can tell, are all the same. For a flat rate you get a room that, I'm told, can sleep up to six people. Well, that kept me busy for a while. I found beds for only four and, after an hour, gave up looking for the others and turned on the television, hoping to find some German pornography. But there was none, and I was rapidly running out of distractions.

Bored out of my mind, I played hunt the telephone, but there wasn't one, so I went in search of a bar, but there wasn't one of those either; just a cold drinks machine in the unmanned reception area. Yup, you pay when you get there and, by ten, the hotel is bereft of staff.

There wasn't even a restaurant but, according to the booklet in my room, I was welcome to traipse across the car park and eat in the Little Chef. Eat. In a Little Chef. Interesting concept.

The following morning I watched my fellow inmates trooping out of their cells, climbing into their super-heated Vectras and setting off for another day of sweat and disappointment. They looked relaxed, but then they would. I mean, there was nothing in their rooms to have distracted them from the business of sleep, that's for sure.

I don't doubt that the Travelodge idea is a good one. By not providing the guest with anything at all, staff costs are kept down, and this can be passed to the customer: £49.95 is good value for a clean room with cotton sheets.

But what if you had to spend every night on the road, in the same featureless room? Eventually you'd find those other two beds, but then what?

And all you have to look forward to is another day on the pleblon upholstery in your Vectra. I stink, therefore I am a rep.

Cotswold villages and baby seals

We have the builders in at the moment, so time is tight. Just when you think you have spare minutes for some coffee and a cigarette, the headman wanders over to say that the walls are all out of kilter. Or that the water pressure isn't good enough for a new bathroom and that we must dig what amounts to the Suez Canal. 'It can be rushed through in about 11 years and it'll cost £4000 million.'

However, while I don't particularly like having a boiler that runs on peat and blows up every time there's a cold snap, I would much rather live somewhere with a bit of history than somewhere new, somewhere faultless, somewhere Barrattish. And that, I suppose, sums up all that's wrong with the new BMW M5, a car that you can read about elsewhere in this issue. It's just too perfect, too well sorted, too damned smug for its own good. Had we been at school together, it would have played in the first XI and been excellent at physics. And I would have stolen its milk at playtime.

I will happily admit that it beats the XJR on pretty well every front in the same way that, dynamically speaking, a brand-new house beats an old one. Perhaps this is why new estates are littered with BMWs just 10 minutes after the last of the JCBs trundles home. And all the old cottages with leaky taps and ancient wisteria have Jags outside.

That said, I have noticed a growing trend in petroldom which is to be welcomed. Car makers are building new cars that have the 'Oh, I must have one' appeal of Daisy Cottage.

I'm thinking primarily about the Smart. This little car is riddled with the sort of faults that simply would not be acceptable to Mr and Mrs G-Plan. In a World of Leather it's DFS – Downright Frigging Stupid. You can't take it out of town because a passing bread van will blow it straight into the hedge. The six-speed semiautomatic gearbox takes an age to shift, and the car corners like Bambi.

There are upsides, though, like it's just about the best inner-city car I've ever found. You can park it nose-on to the pavement, prices start at only £6000, the panels are indestructible and it does 60mpg. Great stuff, but immaterial.

What matters is that it's just so damned cute. If you crossed a Cotswold village with a baby seal, you'd be only halfway there. You'd need to garnish the mix with a teddy bear and a primary-school ballet class to match it for aaah-ness.

Then you've got the Fiat Multiplex Cinema. I have it on good authority that even Fiat boss Gianni Agnelli thinks this clever six-seater is 'absolutely hideous', and he's right – it is. But then so was the Elephant Man, and that didn't stop us crying when he croaked.

There's more too. I have never been able to watch Michael Elphick in anything – *Boon*, chat shows, whatever – since he broke the Elephant Man's matchstick cathedral, and I will not be able to speak to anyone who criticizes the Fiat. It may be new and modern, but I want to own one even more than I want to own Blenheim Palace.

At car industry press conferences five years ago, all we would ever ask was whether the four-door saloon that we'd just been shown would be available with a diesel engine, or four-wheel drive, or as an estate. And that was all we wanted to know. But now we could say: 'Will it be painted pink and have the engine mounted in one of the wheels?' And they'd probably say yes.

I suppose that Renault must take the credit for having started this trend by introducing the Twingo, but now there are characterful cars on virtually every street corner. The Mercedes-Benz A-class. The Rover 75. The Honda S2000. The Jaguar S-type. The Evo sisters from Mitsubishi. And, yes, even the Daihatsu Move. Alfa Romeo is back in business, and Ford – the most conservative of all car manufacturers – has recently caught the bug, too, so now, instead of giving its designers pencils and tracing paper, they all break at five for a cup of tea and an E. And have you seen that new Vauxhall sports car? Well, like, wow.

But this wave seems somehow to have bypassed Munich, where the Teutons are still struggling with their track rod ends and their camshaft technology. BMW seems to be struck on the idea that a car is just a device for moving people around. Look at those television ads that pick on one tiny engineering detail and hammer it home. Great. We know that BMW pays great attention to the tiniest nut and bolt, and that the cooling system in the M5 is to be admired.

But loved? I rather think not.

It's just like the *Financial Times*. What great insight. Not only do the people who produce this paper understand the City but they can write about it all in a clear and concise fashion. They do the job entrusted to them quite brilliantly . . . but that's all they do.

So come coffee time, which do you settle down to flick through? The *FT*? Or the *Sun*? And which car would I rather take home tonight – the M5 or the Jag? Absolutely no contest.

Shopping for a car? Just ask Rod Stewart

Time. For centuries, mankind's greatest minds have tried and failed to explain its secrets but, so far, only Rod Stewart has come close. 'Like a fistful of sand, it slips right through your hand,' he once sang.

Pink Floyd reckoned that each time the sun comes up you're shorter of breath and one day closer to death. But for those unschooled in the ways of 1970s rock, it remains a mystery with no tangible beginning and no foreseeable end. All we know is that, despite its abundance, there's never enough around when you want some.

If you live to be 70, you have just 600,000 hours to play with, and that isn't enough, I'm afraid, to mess about on mainland Europe trying to buy a cheap new car. I'm sorry, but the only way you can save money on a car is by wasting time. And time is the most precious commodity you'll ever have.

The experts say that 'all' you have to do is phone round

a few dealers in, say, Holland, getting quotes. Then you 'simply' wire them some money and, six weeks later, fly over to pick up your shiny new right-hand-drive car. Never mind that it will have had its stereo stolen; you bring it back to Britain, contact the VAT man, fill in some forms, pay Customs whatever they want, go to the post office, write to Swansea, fill out some more forms, buy some registration plates and – hey presto! – job done.

Well, most of us haven't got time to clean our teeth properly, leave alone take two weeks off work to save £4.50 on a poxy Rover 216.

I'm forever watching programmes on the television where men with improbable hair busy themselves with money-saving tips. Don't call in a plumber for that burst pipe; 'pop' over to your local DIY store and buy a welding torch. Then 'simply' dig up your drive using a pile-driver, repair the pipe, encase your handiwork in concrete and you're back in business.

I read a report in the newspaper last month which said that men spend only 15 minutes a day playing with their kids. Well, of course we do. It's because we're all in the shed building DIY microwave ovens to save money.

Now, look. The clock is ticking. You've only got 300,000 hours left and you'll spend 100,000 of those asleep, and another 25,000 watching period dramas on TV. Then there are the queues caused by Mr Prescott's bus lane and, whoopsadaisy, your arterial route map has just exploded. Tilt. Game over.

I wouldn't mind, but I'm not certain you can save money by buying abroad. This is because, when we want a new car, we usually decide how much to spend, rather than what model we want. Who cares that you can buy a

Rover for £10,000 in Copenhagen? You don't want a Rover, and you don't want to spend even a tiny fraction of your short life in Denmark.

Say you have a £12,000 budget and, wisely, you decide to buy a Ford Focus. But then you hear that such a car is available in The Hague for £8000. So are you going to buy the Ford and save four grand? Or are you going to go over there and get something a little more tasty? An Alfa 156, perhaps, or a swanky, posh Lexus IS200.

Or – and this is what 92 per cent of new car buyers actually do – are you going to read about the great deals over there, then buy your car in Britain because you've got better things to do with your day than haggle with a dope-smoking pornographer in a bad jacket?

Well, don't worry, because help is at hand from Kia, a Korean car company that has just announced that it will slash its prices to bring them more in line with the EU norm.

Of course, buying a Kia in Britain has its drawbacks – you end up with a Kia for a kickoff, but, hey, there are worse things in life. Certainly, you should avoid the Pride, which will endow you with none, and I'm not sure the four-wheel-drive Sportage is terribly good either. The new four-door Shuma I can see would work well on the cab rank at Nairobi Airport, but here in Britain it belongs on *Call My Bluff* with Alan Coren trying to convince his opponents that a Kia Shuma is a sort of kebab.

Kia has never managed to get its names right – a trend that goes back to its first vehicle, the Bongo. But the prices are low, so I've had a good look through the range to find the least nasty and have chanced upon something called the Clarus 1.8LX. It's a four-door saloon that comes with air-conditioning, seats and a big, ugly radiator grille.

Now sure, when you park it on the driveway your neighbours will come round for a good laugh, but you'll be able to wipe the smile off their faces when you explain that it cost just £10,995 and that you bought it in England. Then you'll be able to go inside, slam the door and spend some quality time arguing with your children about Doris Troy.

Gruesome revenge of the beast I tried to kill

The Chrysler Voyager is like that creature from Hallowe'en. You can stab it, shoot it, throw it out of the bedroom window and six months later it'll be back in the sequel – *School Gates II.*

You were told in 1997 that Mr Blair and the wide-mouthed frog were using one to ferry their children to and from school. And even though you know the man is a war criminal, you still went ahead and bought one.

In 1998 you were told that it came stone-dead last in a survey to find Britain's most environmentally friendly cars. And in early 1999, in this very column, I announced that the Vauxhall Vectra had lost its crown. 'The worst car you can buy now,' I wrote, 'is the Chrysler Voyager.' I hated it.

But still the queue to buy one stretched out of the dealership, round the corner and halfway up the inner ring road past Asda.

Last week, though, I really thought its time had come.

In official crash tests, part-sponsored by the government, it was described as 'appalling'. In a frontal impact at 40mph, the steering column was forced up into a driver's head and the footwell split open. It scored zero.

Surely, I thought, this would be it, the end of the road. No more rearing up out of the bath to stab Michael Douglas. It would be dead. Finished. Roll credits; not that it has ever earned any.

But no. An owner told me yesterday that he wouldn't be getting rid of it because it was only the driver who'd be killed, not the kids in the back.

I don't think I've ever encountered such collective lunacy. You know it will kill you, your family and the planet. You know that it's uncomfortable and ungainly. You know it's the choice of Mr Blair, and still it is the second best-selling people carrier.

Of course, those with four children will say there is no alternative, that no people carrier is desperately safe, except the Toyota Picnic, and there's no way you will drive around in something called a Picnic. Quite right, too.

Well, I've just bought an eight-seater that could run into a house at 40mph and everyone inside would look up quizzically and say: 'Did anyone feel a bump just then?' It is a Toyota Landcruiser, and it is so vast that, if our dog were to die in the back, it would be three months before the smell reached our noses in the front. I saw some footage the other day of what happens when a big off-roader hits a normal car, and it was incredible. It just rides up the ordinary car's bonnet, ripping the roof clean off and severing the heads of anyone inside.

The commentator was trying to tell us we should all buy Golfs as a result. But me? I was jumping up and down

on the sofa shouting: 'I have got to get me one of those.'
I wanted the biggest off-roader that money could buy.
You can keep your Land Rover Uzis and your Shogun
AK47s. It's a war zone out there and I wanted the Toyota
Howitzer.

New, a Landcruiser costs £44,000, but I got round
that by buying a P-registered model with 30,000 miles on
the clock. It cost £22,000 and, no, I didn't spend extra
on a warranty. As it was designed to go from Adelaide
to Darwin, it should be able to manage the school run
without exploding in a maelstrom of cogs and wire.

So, short of buying a tank we now have the safest, most
reliable car in which to move our children around. Which
makes me all warm and gooey and new-mannish. I may
even bake a cake this afternoon and do the hoovering.

But there are drawbacks, chief among which is the
sheer cost of keeping it going. The school is 18 miles
away, which equates to 72 miles a day, and that adds up to
£150 a week in petrol. I worked out yesterday that if
I drive one mile into town for papers, it costs 50p for
the papers, 60p for the fuel and £400 for the remedial
dental work.

I'm sure you need hard suspension for trips across the
Nullarbor Plain, but it doesn't half get wearing on the
A44. The Landcruiser's like a suit that's been lined with
sandpaper. With its air-conditioning and leather seats, it's
outwardly smooth, but the slightest bump and you'll be
needing a Band Aid.

It is so uncomfortable and thirsty, in fact, that I tend to
avoid driving it, and, to make matters worse, so does my
wife, who says it's a big, ugly grey box and refuses to go
anywhere near it. Last week she hammered the point

home by taking our four-year-old to school in her BMW Z1 that offers all the protection of a Kleenex. It doesn't even have proper doors, for crying out loud. And yesterday, when a near neighbour called round and offered to do the school run on our behalf, we agreed with the sort of vigour that dogs display when you offer them a quick snack.

And guess what she was driving? Yup, a Chrysler Voyager.

Out of control on the political motorway

At the British Grand Prix last Sunday, 150 helicopters ferried 11,000 photocopier salesmen into the circuit, making Silverstone, officially and for one day only, the busiest airport in the world. But after the race finished, there was a lengthy and unexpected pause in the incessant takeoffs and landings. I sat in our Twin Squirrel listening over the headset intercom to our pilot endlessly telling the tower that 'Brown Pants One' was loaded with passengers and ready to go.

But the clearance just wouldn't come because away to our left, in the takeoff zone, a chopper was straining at its leash, swivelling this way and that as its pilot turned to the wind for help. The engines were screaming and the blades seemed set to break free from their mountings, but it was obviously too heavy to get out of everyone's way. Because there in the back was Jabba the Hutt himself. Man of the People 1999: his fatness, John 'Chopper' Prescott.

Yes, having messed up the roads with his traffic-calming flowerpots and bus lanes, Taffy Two-Jags was now doing his level best to screw up the airways as well.

Of course, he may well have been supported in this quest by the vast hordes of ordinary grand prix fans who faced a four-hour wait to get out of the car parks. But as he'd gone on television before the race to say he was secretly hoping the event would be won by someone called Damien, I rather doubt it.

Most people at Silverstone, I suspect, would quite like to see Chopper replaced by the sinister Mr Spock, who this week outlined the Conservative Party's proposals for getting Britain moving again.

They may have taken the greyness out of the leader, but now they want to put it back – on the M4. And that's good enough for me. I don't care that William Hague seems to be growing cress on his head; I'd vote for anyone who promised to tear up that stupid pinko bus lane.

They also say they will freeze petrol prices, abandon plans for a workplace parking tax and ban local authorities from building scaled-up models of K2 on every suburban back street.

It's all good news, but there are no proposals to paint speed cameras yellow and relocate them outside schools, where speeding matters, rather than hiding them in bushes in the middle of nowhere. And this motorway business is absurd. They want a minimum speed limit of 50mph, which is unenforceable because everyone will claim the traffic was heavy, and a maximum of 80, which means everyone will do 95. And that's way too fast. Quite apart from the speed differential between caravans, which will be exempt from the minimum restriction, and the faster traffic, there will be no time to react when a pensioner drives the wrong way down your carriageway.

There would be carnage, and we'd face the horrifying spectre of Chopper boarding his Huey and riding back to power to the strains of Jim Morrison singing 'The End'. Which it would be.

He'd want to replace cat's-eyes with tubs of luminous flowers, put petrol up to £7 a gallon and use Apache gun-ships to police the bus lanes on Silverstone's Hangar Straight. And where would that leave the new BMW Z8?

I know I say that everything is the best-looking car in the world, but this new two-seater convertible really is. Honestly, I mean it this time.

It's part Austin Healey, part 507, and all man, with its chromed door mirrors, its aggressive air intakes in the front wings and its cat-like haunches at the back. When you see Pierce Brosnan use one in the next 007 film, *The World Is Not As Big As My Hair*, you will want this car so badly that it aches, and you'll keep on wanting one right up to the moment when you hear it's likely to cost between eighty and one hundred thousand of your pounds. That puts it in direct competition with Aston's new DB7 Vantage and Ferrari's 360 – two cars I'll be comparing next week, incidentally.

Now I know the BMW badge cuts ice at the lodge and makes golfers churn up the tee, or whatever it's called when you miss. But in the real world, BMW is in Division Two, an easyJet in a squadron of F-15s.

You might but struggle therefore to think of the Z8 as a rival for the 360, but under the aluminium body is the same 400-brake V8 you get in an M5. And the same electronic whizbangs to keep the driven rear wheels in check.

So, it will be just as fast as the Ferrari. And yet it looks even better. Which means it's hard to find a reason for saying no. Yes, the steering wheel's on the wrong side, but that's less of an issue than you might think. And yes, it's German, but so what if it just sits there in the Fast Show, not laughing.

I can think of only one reason for steering clear. The only way you could enjoy it is by voting Conservative.

Old sex machine still beats young fatboy

Let's be honest: the Ferrari 355 is not a comfortable car. The headroom is tight and the driving position awkward. And you can't have lightning without some thunder, so it shrieks and creaks and bellows and rumbles. Certainly, you can't hear the radio.

I've had a 355 for three years and in all that time I've done just 5000 miles. Which is a problem for Ferrari, because when it's sitting under a dustsheet for 355 days a year, it won't go wrong, it rarely needs servicing and it's unlikely to crash.

To make more money out of its customers, Ferrari needed to make its cars more usable. Dealers can't survive when they see the cars they sell coming back for a paltry service only once a year. They want things to go wrong. They want us to go out and crash into a wall. They want to sell us spare parts.

So the 355 has gone, and in its place stands the bigger, softer, more user-friendly 360 Modena, which even comes with space behind the front seats for golf clubs. It's a car for commuting, for trips to the shops. It's dipping its toes in the real world . . . and that's unfortunate, because they will be bitten off by Aston Martin's new DB7 Vantage.

The DB7 has always been a looker, but its 3.2-litre Jaguar engine was never quite good enough. Well, the Vantage comes with a monstrous 6-litre V12, which churns out 420bhp. In a straight drag race with the 360 there's almost nothing in it, and flat out both will exceed 180mph. Only when you turn the nanny state traction control systems off and get to a corner will the Aston pull ahead. Sure, it's a big soft old Hector but, when you reach the limits of adhesion, it's so damned easy to control that even someone with the anatomical properties of Kali could manage.

The Ferrari has more grip, for sure, along with less roll and a more precise steering setup, but when it reaches its higher limits all hell breaks loose. It is almost impossible to handle and, as you fight the wheel, you will knock the paddle-operated gear shifters, making matters worse.

People go round corners, not at the speed the car will go, but at a speed at which they feel comfortable. And the Aston feels comfortable at a higher speed. It's as simple as that.

Then there's the question of price, and again it's a victory for the Brit. The Vantage costs £92,500, while a 360 is £101,000. And you have to add another £6000 if you want the stupid, jerky, F1 semiautomatic gear change.

So what about comfort? Well, amazingly the Ferrari is more spacious, but it's bloody noisy and, no matter what you do with the adaptive suspension, it's always more vicious over the bumps. For a trip to Bulgaria, I'd prise myself in with a shoehorn and take the Aston.

The Vantage is a spectacularly good grand tourer. You ride around on a wave of torque waiting for the road to open up, and then you drop a cog on the six-speed gearbox and let the awesome power strut its stuff. It is Dr Jekyll with Mr Hyde seats. It's like drinking ultra-hot Bloody Marys in a gentleman's club. It's just great.

So what of the 360? As an everyday car it is birched, to within an inch of its life, not only by the Aston but also the Porsche 911 and the six-speed Alfa GTV. For the drive to work, you'd be better off with a Nissan Micra.

However, despite Ferrari's silly attempts to take it into the comfort zone, the 360 would still be my choice for a two-hour blast on a sunny Sunday morning. Its new 3.6-litre V8 isn't heavy metal, but it's not soft rock either. In fact, as the revs soar towards the stratospheric red line, you're left wondering how on earth something this loud can possibly be legal in softly softly Euroland.

As you pull on the left-hand paddle to change down, the engine management system double-declutches on your behalf and the exhausts bark like amplified dogs. Then you turn the wheel and find Ferrari's party piece. The steering may be light, but the diamond-sharp precision enables you to put the car exactly where you want it on the road.

As a toy for high days and holidays, the 360 is beaten only by one other car: its predecessor, the 355. The older car is what a mid-engined Ferrari V8 should be: harder, with a more aggressive bite. It's less tail-happy, easier to control *in extremis* and, to my mind, a thousand times better looking. Sure, the 360 has 400bhp to the 355's 380, but when you look at the power-to-weight ratio it's a dead heat. The new boy, I'm afraid, is fat.

So the conclusion is simple. If you want just one car,

buy a Vantage. But if you have a normal car already and want something for the weekend, the 355 is still 'the best car in the world'.

Whatever happened to the lame ducks?

Unless something is done, and soon, the motoring journalist will line up in the history books alongside the thatcher and the dry-stone wallist. We'll become national tourist attractions; put on display in farm parks and expected to entertain tourists with our red faces and our witty anecdotes about the brake drums on an MG.

In the good olden days, it was the motoring journalist's job to smoke a pipe. And then afterwards he'd don a patched tweed jacket and head for the hills in a new car, determined to ascertain whether it was good or bad. Back then, even the axle on every car was different. There were swing axles, beam axles, seesaw axles and slide axles. It was a veritable children's playground under there, with cart springs trying, and usually failing, to isolate occupants from the bumps.

Not even the doors were uniform. We had the Rover 90 with hinges in the wrong place, and the Renault 14 with a door at the back! So much to talk about. So much fun to be had.

And now we've got this platform-sharing business. Why try to whip up enthusiasm for the VW Lupo when the readers all know it's nothing more than a mad-looking version of the Arosa?

No car maker will take a chance with a radical piece of

design because the financial risk of failure is too great. So the only advances we get these days are new engine management systemzzzzzzz.

As the millennium draws to a close, I'm running out of bad cars to savage. No, really, in the last month I've driven a Ford Mondeo ST200 which was brilliant, a Smart car which was brilliant, an M5 which was brilliant, an Evo VI which was brilliant, and a Lexus IS200 which was dull (but brilliant). Desperate to find something to maul, I booked a Vauxhall Zafira, which turned up at Telly Towers sporting a brown paint job. 'Aha!' I thought. 'Perfect – a brown Vauxhall people carrier. I'm going to rip it apart, bit by bit.' And I would have done, but it turned out to be brilliant.

No, honestly. Even if you forget all about the patented Flex-7 seating system, which is inspired, you're still left with a car that's just really nice to drive. Smooth, fast, economical and with a deftness of touch that belies its family-man aspirations.

Seething, I turned my attention to Volvo, who did everything in their power to make me hate their new C70 convertible. They flew me in a no-smoking aeroplane to Rome, where we loaded into a no-smoking bus that took us on an hour-long trip to Italy's only no-smoking restaurant, where I had to sit outside to enjoy a Marlboro. Just to pay my hosts back, I was pretty much determined to feed the C70 to a literary shredder, but I can't do it. If you want a soft, vaguely luxurious four-seater convertible at an affordable price, it is truly brilliant.

I suppose if I gathered together the equivalent BMW convertible and the soft-top Mercedes CLK, I might be able to dig up some microscopic differences, causing me to declare one of them a resounding winner, but, honestly, it would be like comparing fish fingers with baked beans. They both taste nice and they both fill you up, so eat whatever you damned well want. Unless you're having drinks with the Queen afterwards, in which case best stick to the fish.

Yesterday I drove a car that actually looked like a baked bean. It is made in Korea, where they eat dogs, by a company called Hyundai, and it was called the Accent GLSi. Obviously, I would rather pull my own ears off than buy such a thing, but after a yard or two it became apparent that it was indeed the wheeled equivalent of Ronseal's five-year wood guarantee. It did exactly what it said on the tin. Farted, mostly.

But for the man who needs his food to be mashed before it goes anywhere near his mouth, it was absolutely perfect; well, it was until I went banger-racing in it, but there you go.

As a result of all this, I have a remarkably simple request. Because of platform sharing, car makers are now able to make new cars, economically, in small numbers. It's called niche marketing and is best explained by the Golf, which has so far given rise to all sorts of diverse elements, from the Audi TT to the newly updated Beetle.

Well, look. If VW can make a new Beetle just to keep a handful of trendy fortysomething urbanites happy, how about deliberately making a really bad car to keep the journalists who work for car magazines happy?

Ford did it with the '92 Escort, and Vauxhall followed up with the Vectra, but since then there's been nothing. So here's an idea. BMW obviously hasn't a clue what to do with Rover, so why not use it to produce a selection of overpriced joke cars, simply for the press? Something with a thatched roof, perhaps, and dry-stone doors. You could then line them all up on the north bank of the Thames and play amusing tunes with their horns, and we'd love you for it.

Bikers are going right round the bend – slowly

My elderly neighbour popped round recently to say that gangs of motor-bicyclists are going past our houses much too quickly and that something must be done.

Gulp. It was one of those moments when you pray for a meteor shower or an earthquake or any damned thing that would provoke a change of subject. But nothing was forthcoming so, with a sweaty neck, I was forced to agree.

Here's the problem. Urban fortysomething motor-bicyclists have discovered that the roads in north Oxford-shire are wide and largely unencumbered by functioning speed cameras. So on sunny Sundays they come along with their sports exhaust systems and go bonkers.

It's horrid, but as I have spent the past 15 years challenging the notion that speed kills, it seemed unwise to start a campaign for curbs in my back yard.

To stay away from the petition, we've spent the past few weeks indulging in some 'neighbour avoidance'. This largely involves mooching around the house with the shutters closed. I'd have been happy if she'd stretched cheese wire across the road, or mined the cat's-eyes, or anything really, just so long as the bikers went away and I wasn't publicly involved.

But instead she wrote to the council, which arrived to paint red marks across the road. The result was both dramatic and immediate. Last weekend I noticed the average speed was down from 180mph to somewhere in the region of 165.

And now it's my turn to do something. So here goes. First, I should explain that being heterosexual with no fondness for rubber clothes, I don't like bikes very much. And I don't want my children to like them either. Some parents say drugs are the biggest threat to youngsters today, but I disagree. Every weekend, everyone under the age of 25 takes crack, smack and E and very few are harmed as a result. Bikes are far more dangerous. So, to put my kids off, I take them into the garden on peaceful Sundays and we watch them hurtle by. 'See that man, kids?' I say. 'He likes looking at pictures of other men's bottoms. And by tonight he'll be all dead.'

Am I getting through yet? Well, let me try another tack. Bikes are not that fast. Only last month I went to the Thruxton racetrack in Hampshire with a Porsche 911, which is by no means the speediest car in the world, and a Yamaha R1, which can knock on the door of 200mph. But over a lap the car was faster by 0.75 of a second.

As the flag dropped, the motor-bicycle tore off at a truly breathtaking pace but, as the first corner approached, it came over all weedy and pathetic. To get it round a corner at more than a brisk walking pace, the rider had to lean right over so that his knee was actually touching the road. They call the scuff pads on the end of the footpegs 'hero blobs', but I have no idea why because halfway through the bend the car, despite the burden of four-wheel drive, just sailed past.

Yes, the motorized bicycle was faster on the straight bits, but where's the fun in that? If you want to go fast in a straight line, try easyJet, which, for as little as £29, will take you pretty damned close to the speed of sound.

I'm regularly overtaken by bikers, and I'm sure they feel manly as they sweep by, but come the next corner I like nothing more than getting right up behind them and flashing my lights. Get it right, and without wishing to be too lavatorial, you can almost smell their fear.

For sure, it must be pretty damned scary to ride around on something that will skid into a hedge if you sneeze. But look, guys, if it's a cheap thrill you want, why not rent *Armageddon* from the video shop? That way, there's a zero chance of running over my dog.

Seriously. Hit a dog in a car and they'll make you a freeman of Seoul. Hit a dog on a bike and you get a one-way ticket to the pearly gates.

Bikers fight back by saying the motorized bicycle is faster in traffic, but it isn't. When I want to go somewhere in London, I pick up the car keys and set off. You have to go upstairs and spend 45 minutes dressing up like Freddie Mercury. By which time I'm there.

And I've got the girl, who assumes, as you struggle in, dressed in chains and rubberwear, that you'd be more interested in her brother.

So look, instead of hacking out to the Cotswolds this afternoon, why not mow the lawn or play tag with your mates?

If not, then would you mind awfully fitting an exhaust

system that masks the sound of the engine in some way. Because if you don't, I will set up a stereo in your garden tonight. And play Barclay James Harvest at full volume until dawn.

Freedom is the right to live fast and die young

It's strange that new Labour should choose to take its holidays in Tuscany. Italy, for heaven's sake: it's the very antithesis of grind-your-own potpourri, nanny-knows-best Islington.

A friend who teaches sociology at Rome University once said that it is not difficult to govern the Italians, just unnecessary. 'You can have as many laws as you like,' he explained, 'just so long as they are not enforced.' And you have only to look at Italian history to see why. One minute they had Caesar nailing orders to the village notice board, then along came Hannibal with a different set of instructions: 'Kill a Florentine: Win an Elephant!'

And it's still going on: governments never last more than a year, so new rules that come along one day are overturned the next. Best, really, to just get on with living and ignore the edicts. That way you have anarchy, which before a change in the 1929 OED, meant 'a perfect state where no government is needed'.

So I wonder what Mr Blair and the wide-mouthed frog thought when they found people openly smoking in Pisa's no-smoking airport. What went through their minds when they found that, contrary to Euro-rule 277/4b, the lavatories in the restaurants do not have two doors separating them from the kitchen? And how on earth did they cope on the roads?

They have bus lanes in Italy. But cars go in them, too, because, it's said, the police are more interested in crime. Rubbish. They are only interested in their uniforms, which were designed by Fendi. Except the bags: they're Gucci. Maybe if British police had Reebok shoes and Paul Smith suits they might worry less about your dirty rear numberplate and whether you've had a wine gum. It's a thought.

Contrary to popular belief, though, you can get stopped by the Italian police. A friend was hauled over for jumping a red light but, when she explained that this was permitted in Britain after 11 p.m., the officer said 'Good idea' and let her go. It *is* a good idea. My professor mate says it is an act of 'monumental stupidity' to sit at a red light when nothing is coming. And he can't fathom the new Italian law that says you must wear a seatbelt. 'It's like betting against yourself,' he says. 'Putting one on is like saying, "I am, at best, a mediocre driver. I may crash."'

The Italians do crash, a lot. Every bend on every road is garnished with floral tributes to poor old Gianni, who, at a crucial moment, ran out of talent. But in Italy to die in a 100mph fireball is to live.

Certainly, I pity poor old Mr and Mrs Blair if they chose to drive around at the speed limit, because other road users will have thought them mad. Indeed, in Sicily once, while trying to discover just how fast the Sierra Cosworth would go, I chanced upon a police van trundling down the inside lane. Naturally, I braked hard, but it was no good, and out of the van's window came an arm, followed by a head and then a whole torso. He was waving at me frantically, demanding that I put my foot down again. Sure, he was a policeman, but he was, first and foremost, an Italian bloke and therefore he, too, wanted to know if the 150mph rumours were true.

And it isn't just the police who encourage you to go faster. I was once struggling down the outside lane of the autostrada in some godawful diesel Fiat with a large Alfa ramming me gently, but repeatedly, up the rear. And in the back of it were two nuns.

Go to an Italian motorway service station in the holiday season and the car park will be littered with tourists, usually American, crying their eyes out: 'We just daren't go back on the road.'

To stay out of trouble, you need to go fast. Really fast. People will dive into a hedgerow when confronted with an onrushing Ferrari. There's none of the mealy-mouthed Terry and June pettiness you get in Britain.

Of course, they have drink-driving rules, but, so far as I can tell, there's no specific limit, so nobody gets prosecuted. Then there's parking: to find a space, you use your bumpers to move another car out of the way. And hands up anyone who's seen a double yellow line in Italy. Or a speed hump. Or a Gatso.

But here's the thing: it works. And while I'm sure Mr Blair was initially horrified by the freedom bestowed on the subjects of a fellow Euro-state, I do hope that, as he lay awake at night, digesting the fagiano alla caruso and listening to the crickets, he realized that ungoverned traffic is like an ocean: it will find a natural level that swells like the tide at rush hour and then, all by itself, ebbs away.

Unfortunately, Canute Prescott is still in charge of the roads, and I think he takes his holidays in Mablethorpe.

A shooting star that takes you to heaven

A senior Ford executive told me the other day that modern customers will make up their minds about a car in just five seconds. Five seconds? What planet is he on? Five seconds is far too long. *The Big Breakfast* can give you an entire news bulletin in five seconds. And I certainly didn't need five seconds to make up my mind about the new Nissan Almera GTi. I saw a picture and knew immediately it was horrid.

Oh, I'm sure its fuse box wasn't nailed in place by an Indian, and I'm sure it's jolly fast. But telling someone at a drinks party that you have a Nissan Almera is like telling them you have ebola. And that you're about to sneeze.

Nissan says in its press advertisements that the Almera is better than the Golf GTi, but I don't need five seconds to find out that it isn't. It's a bloody Nissan, for crying out loud. In the 20-year history of hot hatchbacks, only one car maker has ever been able to take the fight to Volkswagen. Peugeot.

Peugeot, of course, is French, but I see that as a good thing. While we wail and wring our hands at American trade shenanigans, they just up the price of Coca-Cola to £50 a can. Marvellous.

Then we must consider Peugeot's latest advertising campaign, where we see Gatso cameras with long lenses and cars taking off from humpback bridges. These are designed specifically to annoy the 'speed kills' lobby and that, too, is a good thing.

And, finally, their new pocket rocket, the 206GTi, is made in Britain. But spend just five seconds in the cockpit and you'll be overcome by a need to get out again. It's vile. The dashboard is made of deeply veined plastic so it looks like an elephant's arse. And you cannot find a comfortable driving position. Unless you are an ape.

Then you'll note there is no sunroof and, despite a dashboard-mounted readout that says GPS, no satellite navigation either. After five seconds you'll not only be out of the car but out of the showroom, on your way to see that nice man at VW with his solid-as-a-rock Golf.

Mistake. Big mistake. What you need to do is turn the key, ignore the engine scream and set off. Eventually the overactive choke shuts down and you are in what can best be described as an asteroid. All the other cars on the road become big lumbering planets. But the 206? It's doing 2000mph on the back roads, where neither Mr Prescott nor Isaac Newton can get at it.

You know that ball that Will Smith found in the *Men in Black* laboratory, the one that Tommy Lee Jones said had caused the New York blackout? Well, that's what the 206GTi is like. You don't so much drive it as hang on and hope for the best.

For those who haven't seen *Men in Black*, it's like taking a terrier for a walk.

It's like body-surfing down an alpine ravine. It's an extreme sport with wheels.

It's fitted with frantic, sprint gearing so that, at motor-way speeds, the engine is doing 4000rpm. This means it's loud, but it also means you're right in the power zone and even so much as a breath of wind on the throttle pedal will give you a whole new hairdo.

It's not that the 2.0-litre engine is particularly powerful, but Peugeot has used every trick in the book to eke as much as possible from it. And it's the same story with the rear suspension. Front-wheel-drive cars understeer, right? Turn into a corner too quickly and the nose will want to plough straight on. Not in a Peugeot. If you lift off the power, the back swings round and you get armfuls of oversteer, which is exactly what the enthusiastic driver wants.

Seven years ago it seemed the hot hatchback was about to breathe its last. The performance meant they became a must-have accessory on the Blackbird Leys housing estate. And they were getting bigger and fatter, which meant that their engines, strangled by catalytic converters, were not capable of justifying the insurance premiums.

However, engineers overcame the catalytic menace, and clever alarms brought the insurance premiums down again. And I'm delighted, because I reckon the hot hatch is man's cleverest automotive achievement yet.

The Peugeot costs just £14,000 and for that you get a car that provides as many smiles per mile as a £140,000 Ferrari. It carries as many people as a Jaguar, but fold the rear seats down and you can get a filing cabinet in. It's small, nippy, easy to park and classy enough to cut ice with a liveried doorman. It's the little black cocktail dress of cars and, in my opinion, the best label right now is Peugeot.

I've just been out for a 100-mile spin in the 206Gti, and I'm still beaming. I'm terribly, terribly fond of it.

Congratulations to the Cliff Richard of cars

At school, a very simple punishment was meted out to anyone suspected of liking Cliff Richard. They were pushed into a plunge pool eight feet deep and eight degrees below zero. And they weren't allowed out again until they admitted that 'Devil Woman' was drivel.

I recall the night when a close friend inadvertently put 'Carrie' on the pub jukebox. He was taken outside and stoned. Cliff, we explained, was not only a fully paid-up member of the St George's Hill godsquad, which made him about as cool as an Indian jet fighter on reheat, but he was musical sediment.

Given one bullet and dispensation from Jim Callaghan to kill one person with no fear of retribution, I'd have gone for Colin Welland. But given two bullets, Cliff would have had a whole new design on the front of his zany blazer.

Yet today Cliff is the only artist to have had a hit record in five separate decades. He's been knighted for not smoking, and when he plays at the Albert Hall, Kensington just stops. He can enliven a dreary day at Wimbledon, and I'd be lying if I said I'd never walked down the street humming 'Living Doll'.

It's 40 years since Cliff's first No.1 but, as everyone says, he doesn't look a day over 25. Well, no, on television he doesn't, but he forbids any camera to be positioned below the height of his nose. If you saw his neck, you'd think he was an iguana.

And so it goes with the Mini, which is celebrating its fortieth anniversary this weekend. There's a big party at Silverstone today, and upwards of 70,000 people are expected to turn up. So let's just get that straight. A party . . . for a car. A car that was designed on the back of a napkin, a car that was in production for just six days before strike action shut down the lines.

They say it was the first car to offer a transverse engine, but so what? The Wright brothers' aeroplane was the first to become airborne, but that doesn't make it better than an F-15.

At school, anyone found in possession of a Mini was pushed into the plunge pool and made to sing Cliff Richard songs until they drowned. The Mini was a BL thing and, by the time I was old enough to care, even the cute charm had been replaced by an Austin Maxi radiator grille.

What we wanted as the 1970s drew to a close was one of the new Golf GTis, not some relic that Twiggy used to drive. This was the Farrah Fawcett-Majors generation.

The only reason the Mini survived is that BL kept making it. And the only reason they kept making it is that they were too stupid to stop. Even when the superior Metro came along in 1980, they didn't dare pull the plug. Today the Metro has gone but, incredibly, the Mini lives. And, strangely, I'm glad, in a cosy, nostalgic, changing-the-guard sort of way.

There's a new ad campaign for this odd little throw-back, but the nation's vicars won't let us see it. It shows a bunch of naked men being judged by a panel of women in a game show. They work out which one has the Ferrari and which one a Porsche, then say in unison, as they encounter one particular crotch: 'Aah. This one drives a Mini.'

And that's the point. The Mini started out in life as a damned clever response to the Suez crisis. Then it was a silly joke, and now, just by hanging around, it's become a cheeky chappie: a wheeled Terry Wogan, as important to UK plc as the Queen. Today, Clint Eastwood has one, and half are sold in Japan, where they sit on the shelf alongside tins of Harrods shortcake and posters of Michael Caine. It's a lava lamp, with a chequered flag on its roof and pop socks on its funny little wheels.

Of course, if they were to launch it now, with a price tag of £9300, we'd laugh and want to know where the hatchback was. But we don't. We turn a deaf ear to the transmission whine, and we even manage to ignore the relationship between the pedals and the steering wheel that forces drivers to adopt a vaguely lavatorial driving position.

Next year, all this will be swept away when the new Mini breezes on to the market wearing what some are saying will be a price tag of £14,000.

Great. But will it be better than the original? Well, lots of cars have tried to beat it and, dynamically, all have succeeded. But if all my motoring were limited to urban roads, where comfort and silence and performance don't matter, I'd buy a Mini and 'boing' down the road with men going 'aah' and women thinking I had a big one.

I hope it doesn't rain today for what is this little car's last birthday. But if it does, I have no doubt that the 5000 Minis on show will rear up on their hind wheels and belt out a few tunes to keep everyone happy.

David Beckham? More like Dave from Peckham

Cruise the neo-Georgian executive housing estates of suburban Britain and you'll find a BMW 3-series on every driveway. While Cheryl is inside, tending to her baggy knicker curtains, Dave is outside with his carriage lamps and his stripy lawn burnishing his 318i.

BMW is so successful that while every other European car plant shut in August, it kept right on going. Three shifts a day, quenching the thirst of photocopier salesmen everywhere.

BMW is Manchester United, a single cohesive force, a perfectly synchronized robotic being that moves around the world destroying everything in its path. And me? Well, some say I'm biased when it comes to BMW, that I'll support Leeds or Chelsea or anyone who wants to take them on, and it's all true. But even I'm able to recognize that they do field some truly great players.

The 529i, for instance, is sublime. It embodies the whole mission statement of BMW, combining Teutonic quality with a zestfulness you just don't expect in this class of car. It is a world-beater, a David Beckham with wheels. It even has a skirt.

Then you have the M5, which is perfect, and the M coupé, which looks like a bread van but goes like a pepperoni pizza. And that's it. BMW makes three great cars. And the rest?

This week I've been driving around in the new BMW 3-series coupé. But it isn't a coupé. Like its predecessor, it's more of a two-door saloon that costs more than the four-door. And if you're not paying a premium for style, then what, pray, is the point of paying a premium at all?

It looks like a BMW, which, in The Close, is obviously a good thing, but compared with the Peugeot 406 coupé and the knuckle-bitingly gorgeous Alfa Romeo GTV, it looks half-hearted. You do get pillarless doors, though, and therein we find the cause of a gigantic bruise that has now enveloped the entire right side of my face.

Here's why. You open the door and, with an end-of-the-day sigh, drop into the carefully moulded driver's seat. But halfway through the drop, the heavy door starts to swing shut and you don't see it coming because there's no frame round the window. Scream? I sounded like Jamie Lee Curtis in *Hallowe'en*.

BMW must be aware of this problem because, by pressing the remote buttons on the key twice, the window is lowered out of harm's way. That's called retroactive engineering, which means responding to a problem that shouldn't be there in the first place.

This is a 3-series that is supposed to be sporty. Not only that, it's a 3-series coupé, which is supposed to be sportier still. And on top of that, it's called a 323, which harks back to the car that, in 1984, was the holy grail for every estate agent in SW6. After the summer of love and the winter of discontent, we burst into the 1980s with a spring in our step. We were playing hopscotch on the heads of the homeless, and the badge of belonging was a 323i.

The new car, actually, has a 2.5-litre engine, so it should be called a 325, but hey. With property prices again heading for the heavens, BMW's marketing department wants to milk the mood of the moment, and who cares if the end result makes no sense.

No, really, it doesn't. The old 323i was light, fast and

agile, whereas the new one is a suet pudding. Sure, the engine is a deal more powerful than the old one, but compare the crucial power-to-weight ratio of both cars and the problem is plain to see. The 1990s 323 coupé offers up just 120 horsepower per ton, whereas the 1980s equivalent delivered 140.

This shortfall is devastatingly obvious on the road, where the new car has about the same get-up-and-go as a boulder. It is not even remotely fast. And there's no point turning to the 2.8-litre 328i either.

So what about the handling? The 50:50 weight balance? The rear-wheel drive? The years of racing pedigree honed and tamed for the road? Well, I'm sure it's all there, but I couldn't find it. I looked hard for an hour or two, and all I found was an airbag in every single nook and cranny. In many ways, this 323 reminded me of an old Volvo. Except, of course, no Volvo ever tried to slash my face off.

The 323 coupé is more than just a disappointment. It is genuinely a poor car, a bad effort from a company that teases us with greatness but, as often as not nowadays, delivers a plateful of offal.

What this car should have to suit the needs of its fan base is bull's-eye glass and a couple of hanging baskets dangling from the door mirrors. And what you should have if you want a lemon-sharp coupé is the Alfa GTV.

A prancing horse with a double chin

The Road Test Editor of this magazine and I have been friends for more than 20 years. In that time we've spent every single New Year's Eve in one another's company. We've been on holiday together. We're great friends. But this week I told him to f★★k off and slammed the phone down.

So what's brought this about, then? Has he been sleeping with my nine-month-old daughter? Or have I inadvertently urinated in the petrol tank of his motor-bicycle?

Sadly, not. I'm afraid the row is about a car. The Ferrari 360 Modena. Tom says it's motoring nirvana, and I say it isn't. Tom says the F1 paddle shi(f)t gearbox doesn't jerk, and I say it does. Tom says I can't drive, then. And that's when I put the phone down.

Tom thinks I have an agenda and that I'm being controversial simply to make a name for myself. But I'm not. I've read what the road-test department has to say, and, while I respect their opinions, mine differ. And now I'm going to explain why.

First of all, the 360's new 3.6-litre V8 develops 400 brake horsepower and that, I'll admit, is an achievement. I doubt very much, for instance, whether Cilla Black could design an engine that churns out 111bhp per litre. I know I couldn't. But if you look at the power-to-weight ratio, you'll find it's no better than the 355. So, we can deduce from this that the new boy's a bit of a fatty.

Certainly, it has a double chin and a dumpiness around the arse which weren't there in the 355. Dare I suggest that in Ferrari's endless quest to keep the signs of age at bay, they've gone a face-lift too far and created Zsa Zsa Gabor in aluminium? And what's that smile all about? Ferraris are supposed to snout down the road like angry bloodhounds, not cruise into town looking like Jack Nicholson's Joker.

They say they've raised the front end so owners don't graunch it quite so often on pavements and the like, but I don't want practicalities sticking their nose into the equation. I want my V8 supercars to hunker down and snort the white lines right off the road.

Same goes for the interior. Yes, it's a good deal more spacious than it was in the 355, and now Tiff has somewhere to put his woods. But I'm not bothered. Just so long as there's enough room for me, and Radio 2, I couldn't care less.

Then there's the question of comfort. Ferrari has tried to make the 360 as user-friendly as a Porsche 911. That way, owners will actually drive the car, rather than putting it in an armour-plated garage under a carefully laid ermine dustsheet. And that way, they'll need more services and more spare parts, which is good for Ferrari's bank balance. But look. If you use a car every day, it will cease to become special. In three years I've done only 5000 miles in the 355. Each one has been under a blazing sun, on the way to somewhere agreeable and nice.

You've read, I'm sure, that despite the new comfort, the 360 makes a bloodcurdling noise as the revs climb round towards the stratospheric red zone. And for sure it does, but the howl you get from a 355 has now gone, and that's a pity. So's the razor-edge sharpness. A 360 in its 'sport' setting feels exactly the same as the 355 in 'comfort'. And why, pray, does a 360 have traction control? I thought the whole point of a Ferrari was that you spent your time not only fighting the road but the machine itself. Wading into battle with a silicone nanny tips the balance too heavily in the driver's favour. It's like putting Nato up against the poor old Serbs. They never stood a chance.

Needless to say, I turned the traction control off within six seconds of starting the engine ... and unlocked the key to another problem. When you go past the limit in a 355, it is surprisingly easy to control. But the 360 isn't. You end up sawing away at the wheel, and it's only a matter of time before you nudge one of the gear shifters, making the problem even worse. I'll admit the F1 gearbox works well on down-changes, but it is ferociously jerky on the way up, and it's a nightmare when the rear lets go.

It's like a Psion Organiser. To input a vital piece of information takes well over a minute, whereas carrying out the same task with a pen and paper, you can have the whole thing jotted down and stuck to your wall in – what? Five seconds. So why not save yourself six grand and stick to a proper gearbox?

In fact, save yourself thirty grand and buy a used 355, because it is more aggressive than a 360, more lithe, more moreish. Sure, the 360 is user-friendly, but if you want an everyday car, buy a 911. Or an Alfa GTV V6.

As a once a month, high days and holidays funster, which is the whole point of a Ferrari, the 355 is still the best car in the world. And if you don't agree, I couldn't give a damn.

£54,000 for a Honda? That's out of this world

If a Martian walked through the door right now, wanting to know about Earthman's motoring habits, I'd be weak on, I think, two areas. I'm a bit shaky on why the M4 has a bus lane. And I'd need at least an hour to explain why the new Nissan Skyline, a two-door saloon, costs £54,000.

I mean, a Nissan is hardly a prized possession, like myrrh or a Myrrhcedes. Quite the reverse. It's down there with Primark and those tree-trunk house-name plaques.

So Mr Martian would ask if perhaps the Skyline was a comfortable touring car, able to swallow vast distances with the careless disdain of a similarly priced Jaguar. And I'd have to say: 'Gosh, no.'

Without any doubt, this is the most uncomfortable car I've driven. On urban roads, where the surface has been mutilated by cable TV companies, the Skyline can devastate your entire skeleton. You learn to weave round potholes, veering on to the wrong side of the road if necessary. And if you encounter a speed bump, why, you just turn round and go home.

The problem is threefold. First, it runs on tyres of such incredibly low profile that it appears someone has simply smeared a veneer of black paint on the wheel. They offer no give whatsoever. Then there's the chassis, which has all the flex of a steel girder. And finally there's the seat, which is padded like Kate Moss. Driving this car is like being dragged across Iceland behind a horse.

If the Martian had eyebrows, I'm sure they'd be raised at this point as he sought the reason for that £54,000 price tag. Is it fast, he'd ask. To which I'd have to reply: 'Er, not especially.'

Japanese law forbids any car maker from producing a car with more than 280 brake horsepower – I haven't a clue why – so that's all the Skyline can muster. It's fast, yes, but not spleen-shattering, not so fast that your ears start to move about.

Bald figures show it does 0 to 100mph in 10.5 seconds and will carry on accelerating all the way to 155mph. That sounds good, but in reality the 2.6-litre engine suffers from whopping turbo lag that the six-speed gearbox does little to mask. Put your foot down at 70mph and it behaves just like a Nissan: 15 seconds later you're still doing 70.

No matter what sane, level-headed question the alien asked, I'd be stumped. It costs at least £1000 a year to insure. It's gaudy to behold. It has a fuel tank the size of a yoghurt pot and, because it does only 18mpg, it needs filling up every 35 yards. There's no sunroof or sat nav.

I'd be forced to explain that the only reason it costs so much is that it goes round corners quickly. That's it. That's the Skyline's party piece. It's good at roundabouts.

I lent the keys this week to Colin McRae who, after five laps of Silverstone, climbed out and smiled. And called it 'pretty impressive'. Which is like a normal person saying it's 'unbebloodylievable'.

I remember being amazed about 10 years ago when Chevrolet announced their new Corvette could generate 1g in a corner. Well, the Skyline manages 1g without so much as a chirp from its tyres. I know this because in the middle of the dashboard is a TV screen that tells you how much g is being generated at any one time. Or, by pressing buttons, it will say how much power is being transmitted to the front wheels and how much to the back. It gives you readouts on the exhaust temperature, the intercooler temperature, the percentage of throttle travel you're using, even the state of your fuel injectors. And now we're starting to get to the nub of this incredible car.

You see, I looked round a London flat recently, which was on the market for £450,000. A lot when you consider it had no rooms. I mean it. There were no walls, floors, heating, lighting, anything, the idea being that buyers tailor it to suit their needs. That's what you get with a Skyline. Nissan sells you the bones and you, by plugging in a computer and changing the odd chip, add the flesh. I know a bloke who's taken the engine up to 800bhp.

You can play around with the thinking four-wheel-drive system, you can alter the aerodynamics – and that's probably why the Skyline has become the car of choice in Formula One. Ask Johnny Herbert what he drives and, like a good Ford boy, he'll say: 'A Ford.' But really he drives a Skyline. They all do.

And the new version? Well, it's like the old one, only a little bit stiffer, a little more aggressive, a little more fantastic. You can't compare it to any other car because there's nothing remotely like it.

So if the alien asks how a Nissan could possibly be worth £54,000, I'd just toss him the keys. And after a mile or two, he'd be trying to part-exchange his spaceship.

It's Mika Hakkinen in a Marks & Spencer suit

I met a food stylist the other day and wondered, How did that come about, then? How do you start out in life wanting to be an astronaut or a film star and end up with a Davy lamp on your head, using surgical tweezers to arrange sesame seeds on a bun?

And then I wondered some more. What a sham. It is this person who builds up my hopes in hamburger restaurants. I see a photograph of a bulging, steaming snack that bears no relationship whatsoever to the tired old cowpat I'm actually given. Apart, perhaps, from the steam.

And that brings me neatly to the Audi TT. When they first showed me a photograph of this Bauhaus barnstormer, I was positively moist with anticipation. But then I went for a drive and, within half an hour, found myself wearing that detached, middle-distance expression normally reserved for dinner parties when I find myself next to a man who services reservoirs.

The Audi TT looks like a sports car, but it isn't one. It's an automotive Ginger Spice, superficially lithe and speedy, but beneath the clothes all droopy and loose. Like a soggy walnut.

Interesting, then, that I've fallen madly in love with the new Audi S3, a car that shares the same turbocharged engine as the TT along with the same four-wheel-drive system and the same six-speed gearbox.

This is because the S3 doesn't try to look like a sports car. Apart from bigger wheels, wider arches and a more crouching stance, it looks like a normal A3, which is an unpretentious hatchback. And because I wasn't expecting it to garnish the road with Tabasco sauce, I didn't really mind that the gearbox was vague and that the brake pedal acted like a switch.

And so what if it doesn't have electric responses when you turn the wheel? Audi, bless them, have never been able to make a car that handles properly but, for the thousands of doctors and solicitors who buy such things, it doesn't really matter.

If you want a sharp suit, go to Subaru and buy the Armani Impreza. If you want Boss badging, buy a BMW, but if you just want something for work, there's always good old Audi & Spencer.

But then I pressed the accelerator pedal and thought: Whoa, hold on a minute. The S3 may not be up to much in the bends, but in a straight line it is positively explosive. Even in sixth gear at 70mph, it hurtles off towards the horizon like a rabbit.

I simply wasn't ready for such vivid performance from what is basically a 1.8-litre, three-door hatchback. And that's where the S3 really scores. By maintaining low expectations, you're constantly being delighted – by the epic night-time dashboard that glows like Los Angeles, by the blue-tinted headlamp beam and, most of all, by the Recaro seats. Not since I drove an old Renault Fuego have I ever been quite so comfortable. In a car, that is.

It's also been a while since I felt so comfortable with a car. While it doesn't actually turn heads, it has real-world good looks. What I'm trying to say is that it isn't Brad Pitt or David Beckham; it's just a handsome bloke on the other side of the bar.

And that four-ringed badge comes with no unpleasant

baggage. When I see an Audi coming up our drive, I'll rush to the door to see who it is. When I see a BMW, I close the shutters and pretend to be out.

You buy an Audi because you want a practical, well-made tool to convey you, and some passengers, sensibly and with the minimum of fuss from your agreeable house in the country to, let's say, Assaggi in Notting Hill. People with Bee Ems go to Quaglino's, so they can shout.

And finally we get to the price: £27,000. Which is a lot for what, as I said, is basically a hatchback. But it is not a lot for a car that does quite so much, quite so well. For the same money, you could have a Mitsubishi Evo VI, but you'd arrive everywhere looking like Gary Rhodes. Or you could have the BMW 323 coupé. But you'd arrive everywhere late.

For the past year or so I've been singing the praises of Alfa Romeo's GTV6, which is £28,000. In fact, I've come awfully close to telling strangers in traffic jams that they've bought the wrong car. 'Oi, you. Why are you driving around in that p.o.s. when you could have had an Alfa? You are a moron, and I hate you on a cellular level.'

Well, now there is an excusable alternative. If you really, really need back seats and you absolutely must have a boot that can take more than one prawn, you may buy an Audi S3. It's the second-best car in this class, which is like being the second-best racing driver after Michael Schumacher.

There you are. The S3 is Mika Hakkinen. Cool. Detached. Handsome. And much, much faster than you'd think.

Like classic literature, it's slow and dreary

I don't like patterned carpets, but I know why people buy them. They may be unrestful on the eye, but turquoise and gold squirls are to be found in smart restaurants like the local Harvester. So they're seen as posh. And practical too. The *Torrey Canyon* could crash into your sofa and only the most eagle-eyed visitor would be able to spot the stain.

I know why people buy Agas too. They can't cook food, choosing instead to heat the kitchen to a point where cutlery melts. And when they go wrong, you have to spend half an hour listening to some gormless customer-care woman who says that all her engineers have personal problems. But Agas bring a certain country goodness to your kitchen, and you get an owners' club magazine that features other Aga louts like Felicity Kendal.

I know why people live in Wilmslow. I know why people become burglars. I know why the M4 bus lane was built. But I never got to grips with the Vauxhall Vectra.

There's a new version now and, even though it looks much the same as before, Vauxhall says there have been 2500 alterations, prompting those wags at *Viz* to suggest the old one must have been a 'right pile of crap'.

They're right. It was. When I was asked to review this hateful car for *Top Gear*, I adopted a philosophy that took Ronan Keating all the way to No. 1: 'You say it best when you say nothing at all.'

It was shamefully dull, enlivened only by a tool to get the dust caps off the tyre valves. And guess what? The new Vectra is no better. Oh, I'm sure its chassis is more responsive and it'll break down less often, but this was never the problem. The problem was the shape, the dreariness, the sense that someone else had styled this car while the proper designer was at home waiting for the Aga engineer to call round.

So I don't care that the new model has one-piece head-lamps or a new grille. It's still boring. We want our cars to be like airport best sellers. We want the cover festooned with swastikas, guns and girls, but instead Vauxhall gives us Thomas Hardy. It's the Penguin classic of cars. I bet if you peeled away the bodywork you'd find an orange spine.

And I've been testing the lavishly equipped V6 GSi, which is supposed to cast a halo of sportiness over the rest of the range. Basically, it's a normal Vectra that appears to have been magnetized and driven round a motorists' discount shop. Hundreds of cheap bits have just sort of stuck to it, so you now have Thomas Hardy in a tracksuit.

The problem is money. The GSi costs £21,500, and Vauxhall knows full well that everyone with even a modicum of sanity would buy an Alfa Romeo 156 instead. Or a BMW. Or a wheelbarrow.

So, to make the Vectra more appealing, it is decked out with gizmos. Inside you get satellite navigation, traffic master, which is like 'ask the audience'. It even has a 'phone a friend' button. Press it and you're connected to an operator who can tell you whether the male seahorse carries its partner's eggs. And where the nearest break-down truck is. This is all very nice, but it means the entire glove box is filled with a machine.

Then there's the air-conditioning. Looks good in the brochure. Doesn't work properly. Rather than cooling the whole car, it delivers jets of ice-cold air in narrow corridors so that your nose is fried while the wax in your left ear is turned into an uncomfortable icicle.

And then there's the steering wheel, which is metal. So, on a hot day it's like driving with your hands in a toaster. This car has everything, but nothing works properly. Not even the engine. The 2.5-litre V6 develops 170bhp, which means the Vectra goes from 0 to 60 in 7.5 seconds and onwards to a top speed of 143mph. Theoretically. But in my test car, it felt like the fuel was being delivered not as a fine mist but in large lumps. Under hard acceleration, it felt like it was trying to drink minestrone through a straw.

And even if you leave this unsavoury element out of the equation, you're left with a car that, despite the spoilers and the gravelly exhaust note, is really not very fast at all. It's merely adequate, like the handling and steering and brakes and interior space and styling and fuel economy and ride comfort. There were only two points that could be classified as good: the Recaro seats and the shape of the door mirrors. And that isn't enough, not by a long way.

I'd like to say that, despite the 2500 alterations and the V6 power, the new Vectra is still the most horrible car you can buy. But the Chrysler Voyager diesel is nastier, and I suspect the new Kia Clarus is even worse.

So there we are. The new Vauxhall Vectra: not even any good at being bad.

Prescott's preposterous bus fixation

Earlier this month I wrote a column for *The Sunday Times* in which I might perhaps have said motorcyclists were a tiny bit gay. Certainly I claimed that they liked to look at photographs of other men's bottoms.

Well, there's been an awful lot of fuss and bother, with e-mails flying hither and thither, flicked V-signs in traffic jams and a piece in *Motorcycle News* which said I was being deliberately controversial. As opposed to what, I wondered? Accidentally controversial?

It also said that I only wrote the column to annoy the Road Test Editor of *Top Gear* magazine. 'A tad wasteful', they suggested, to devote an entire column in a national organ to one man.

Oh, really? Well, they devoted a whole column to me, and now I'm going to devote what's left of this one to John Prescott, who has a brilliant new wheeze. Basically, if Railtrack don't get the trains to run on time, they'll be fined £40 million. Which is more than you get for urinating in a public place. I find myself wondering what good this will possibly do. Certainly it'll ensure that money, which could be spent making the network better, goes to the government, where it will be spent on a few more focus groups. And big penalties like this will scare away investors. So, the trains will get worse. And then they'll get fined again.

I wouldn't mind, but it's not like the people at Railtrack sit around every morning thinking up new and exciting ways to bugger up the network. I'm sure they're doing their best, and the last thing they want is Jabba the

Hutt and his ginger-haired, rhubarb-shaped sidekick at the Rail Regulator acting like a brace of school bullies.

I should have thought it would have been more helpful if Taffy Two Jags had said: 'Look, if you can't do anything to make the trains better, we'll give you £40 million to spend on new signals or better coffee or something.' But, oh no. Chopper Prescott has decided to spend all his money on another lunch. And a diving holiday in the Maldives. And a helicopter to get to the Grand Prix, where he cheered wildly for someone called 'Damien'. And what little there is left over is being spent on turning the road network into a giant f******* bus lane.

Now, look. Trains are a good idea. They help alleviate the pressure on Britain's roads and work well in tandem with the car and truck. Buses don't. Buses are stupid.

With the power of hindsight, everyone can see that Beeching was wrong to rip up the railways in 1963. It may have seemed like a wise move at the time, what with the coming of the car, but now we can all see it was madness. And I will bet everything I own that in 30 years' time we'll all be sitting around saying: 'Prescott was an arse when he made all the roads buses-only.'

Actually, I'm saying it now. It's all very well claiming that each bus is full of 50 smiling motorists who've left their cars at home, but that's simply not the case. If you look at a bus after, say, 10 o'clock in the morning, it is almost always empty. And if there is someone on it, you can just tell they've never owned a car in their lives – not with that hairdo. And that coat.

Prescott doesn't seem to understand that no one will buy a car, tax it, insure it, pay to park it somewhere and then use the bus to go to work. But then we should remember that he failed his 11-plus and was described by his mother as 'not very bright'. But even he, surely, can see that a car is far more comfortable and far more convenient than a bus. A car goes where you want it to go and comes home when you're good and ready. A car offers you peace and Terry Wogan. A bus offers you nothing more exciting than the opportunity to sit on someone else's discarded chewing gum.

And buses are not fast. All the coach operators who use it say the new M4 bus lane has made no discernible difference to their journey times. And one operator, in Reading, even cut services after it was opened because there was 'insufficient demand'.

Only 50 buses an hour use the M4 between Heathrow and London – that's less than one a minute – and they now have a lane all to themselves. While the 16,000 cars that use the same stretch are hemmed into the remainder. It's idiotic. It's insane. It's the product of a damaged mind.

And it gets worse because a quick survey of the 50 buses using the new lane reveals a nasty surprise. Most are airline coaches ferrying flight crews into central London for a little light sex.

And then we have the 350 taxis. Well, that's really helping the road network and its overtaxed users. Sitting there watching American businessmen whiz past you into town at 50mph while you just sit there and sweat.

History, I assure you, will not be kind to Mr Prescott, and I suggest that history starts right now. So drop him a line, explaining exactly why next time round you'll be voting for . . . well, anyone, really, just so long as he goes back to serving gins and tonics on the *QE2*.

Take your filthy, dirty hands off that Alfa

Did you know that there's such a thing as a summer truffle, and that it's nowhere near as good or tasty as a winter truffle? No? Well, don't worry, because neither did I until I tried to order dinner the other night at a restaurant where this sort of thing matters.

I had to sit there, nodding sagely, while the waiter guided me through truffle technology. We'd gone through earth and moisture and pigs when, all of a sudden, he adopted the look of a man who's just been stabbed in the back of the neck with a screwdriver. His open-mouthed, wide-eyed expression showed he was in deep shock, but analysis showed there was bewilderment too. Maybe a hint of anger. And all because here, in Michelin central, a man on the next table was putting Tabasco sauce on his fish.

I understood this expression well, because I had worn something similar two days earlier, when Alfa Romeo delivered to my house a 156 with a diesel engine.

This is like teaming white socks with your new suit. It's like playing Mozart at 45rpm. And, yes, I imagine, it's like spending eight hours preparing the perfect fish only to have someone with an asbestos mouth pour nitro-glycerine all over it.

I'll tell you some things about Alfa Romeo which will outline the preposterousness of such a thing. We think of McLaren as a dominant force in Formula One today, but back in the 1950s Alfa was so far out in front it once pulled in its car on the penultimate lap and polished it. So it would look smart when it crossed the finishing line.

Enzo Ferrari began his career with Alfa, a company that has given the world some of the most exquisite cars ever made. Have you ever seen a 2900B? Have you heard one? Henry Ford did, and said later: 'When I see an eight-cylinder Alfa Romeo, I take off my hat.'

And it's still going on today. Oh, sure, people who want four wheels and a seat can buy a BMW or a Vectra, but those who know, those who care, those who want the steering to talk and the engine to howl: they buy an Alfa. So what in God's name were they thinking of when they fitted a diesel engine to their magnificent 156? A filthy, carcinogenic, rattly diesel! In a work of art!

Yes, I know that in Italy diesel fuel costs 3p a ton, and the savings make up for the catastrophic loss of self-worth, but why export it to Britain? Why? Here, diesel engines are for mealy-mouthed, penny-pinching, open-toed beardies in Rohan trousers. They're for people who absolutely don't care about cars or motoring, only the need to do it as cheaply as possible.

Diesel Man yearns to be a parish councillor. He fits yellow headlamp covers and a GB plate when driving in France. He studies road maps before he sets off rather than on the motorway, and he always fills up when the tank is still a quarter full.

You can always spot the son of Diesel Man in the playground at school. While all his mates are telling one another how fast their dads' cars go, he is warbling on in a nasal whine: 'Yes, but my dad's car does 50 miles to the gallon.' And then they steal his milk, and rightly so.

Because despite the wild claims of Diesel Man, diesel cars rarely average more than 35mpg. If he says he's getting

50 or 60, you can tell him from me that he is a liar. And then punch him in the face.

Alfa Romeo has done its level best to enliven the concept of diesel motoring, droning on and on about its new five-cylinder turbocharged 2.4-litre five-cylinder engine. But the simple fact is this: at 4000rpm, when a normal Alfa would be rolling up its sleeves for an all-out, spine-tingling assault on the upper reaches of its blood-curdling rev band, the diesel version is out of puff and begging for a gear change.

Yes, the diesel has torque, but where's the power? Where's the zing, zing, snap, snap, whoa-that-was-close excitement of a Twin Spark. Or the would-you-listen-to-that bellow of the V6. Where's the fun?

You sit there, on your Recaro seat, clasping a Momo leather steering wheel, gazing over a carbon-fibre dashboard, listening to an engine that belongs in a bloody tractor. They say it's eight decibels quieter than a normal diesel, but that's like saying Concorde is quieter than a Harrier. It's still noisy enough to give you a nosebleed.

And at £20,300, it's not cheap. The Twin Spark 2.0-litre version is £100 less and completes a double whammy by being about a million, billion, trillion times better.

PS. Oh, and before I go, A.A. Gill wants to buy an Alfa, so if you have one for sale drop him a line. Doesn't matter what model. He can't tell them apart.

Yes, you can cringe in comfort in a Rover 75

I've just finished reading this month's edition of *GQ*, which is a style magazine for men, and it seems 1970s kitsch is very much in vogue at the moment. Beanbags are back, and so are lava lamps. Then we find page after page of furniture that is made from black leather and brushed aluminium, such as you'd have found on an old Akai tape deck. Or wood, which is so dark and so heavily grained it actually looks like Fablon.

So, if the 1970s are in, then the new Rover ought to be the car of the moment. It's even called the 75, to remind us of a time when 10CC were not in love, and it is festooned with all sorts of natty throwback styling details. If this car could have its hair done, it would probably go for an Afro.

Seriously, it's actually very handsome and, though it's big, it's not at all tank-like. No more than a tank top anyway. But, strangely, this is an acutely embarrassing car to drive. Maybe it's me. I'm the first to admit that I don't like Ben Sherman shirts or those new shoes which look like punts. I buy into fashion only when I'm absolutely sure it isn't fashionable any more.

I can't abide the idea that I might be setting a trend because – who knows? – it might be a trend nobody else will follow, and I'll be left out there with a halibut on my head and big pink kneepads. Well, that's how I felt in the new Rover. Idiotic. Out of step. Not sure whether I was Dr Finlay or Dr Feelgood. Did I want milk or did I want alcohol?

The problem is simple. The 75 has been on sale for

months, and I have not yet seen one. The new Jags, which are a deal more expensive, are everywhere, but nobody is buying the 75. So people were looking at me, and that's unnerving.

I think I see why Rover has taken on Sophie Wessex to help get the nation 'on message'. According to Brian Sewell, the art critic who was used in commercials for the 75, she will get high-profile, trendsetting opinion-formers into the car, so the rest of us will breathe a sigh of relief and follow suit.

But I fear it won't work. Sewell cites A.A. Gill as a prime target for Sophie, but I know he'd rather pay for an Alfa than be given a Rover. And when you look at all those smiling faces at GQ's Man of the Year party, you can't help thinking: How many of you lot would buy a Rover. Jamie Oliver? Johnny Vaughan? TPT? Not a chance.

Above all, you see, it's Rover, and that is just about the least cool badge in the business. At best, it is associated with tweedy doctors in Harrogate; at worst, it conjures up visions of Red Robbo dancing like a Cossack in Lickey End. Rover, the name, is a dog.

But what of the car? Mine came with a 2.5-litre V6 that went with the automatic gearbox about as well as a marriage between Harold Pinter and Scary Spice. Do not think this is a fast car and you will still be disappointed. It is woefully lethargic, unwilling to kick down, and, even when it does, a lumbering barge.

Then there's the interior, which is even more wrong. I liked the piping on the seats. I liked the seats themselves, and I liked the creamy dials. But why have they put ultramodern LCD displays alongside ancient LEDs and set them all against a wood 'n' leather backdrop?

That said, my car had every conceivable toy, which caused me to guess its price at £35,000. In fact, you could buy such a thing for just £25,000, and that's good value. Good, but not amazing.

The handling, however, is neither good nor amazing. I suspect BMW ordered Rover's engineers to stay away from 3- and 5-series sportiness and, as a result, we've been given a wheeled suet pudding. But because of this the Rover does have one trump card. After a hard day at work, when your head is pounding and the traffic is awful, there is no better car in this class for getting you home. It is as comfortable as a Rolls-Royce, soaking up Mr Prescott's speed mountains like they're just not there. And it's eerily quiet, too, so that as you get on to the motorway and hit the cruise control, you simply cannot believe you're in a machine that goes head to head with a BMW 3-series, let alone a Ford Mondeo.

So if you're in the market for a car that drives like a candlelit bath, the Rover 75 should be your first choice. But, of course, if you're in the market for such a thing, you are almost certainly old. With Volvo out of the way, and Nissan now importing the Skyline GTR, Rover has a clear run at the Saga louts. 75? It should be the minimum age limit for buying one.

Don't you hate it when everything works?

I'm writing this on a new computer, which has decided that all 'I's shall be capitals and that occasionally it's fun to type the odd word in Greek. So I've spent most of the day on the phone to a man who explained, with a lot of sighing, that it's all very simple. And I suppose it is, if you've spent the past 14 years in an attic.

Even now, the Internet isn't working, there's a new machine on my desk which apparently does nothing, I can't send e-mail and, every time I ask the computer to print something out, it says I have performed an illegal operation and will shut down. What I should do, of course, is take the whole damned thing over to Seattle and shove it up Bill Gates's arse. But I don't have the time because I'm learning how to use my new mobile phone, which is the size of a pube.

It's funny, but all I want from a telephone is the ability to converse with people who are a long way away. But this mini-marvel can do so much more. I can have conference calls, receive the Internet and, best of all, there is voice-activated dialling. However, if you record someone's name and number in your house, it won't work if you try to recall them in the car or the street. So that's handy.

And tell me this. My last phone was an Ericsson, and this new one is an Ericsson. So why are the connection pods completely different? Why have I got to buy a new hands-free set for £40? Not that I will, because hands-free is one of the most dangerous inventions since the shark. No matter how carefully you lay out that wire on your passenger seat, and no matter how steadily you drive, I guarantee that, when the phone actually rings, the earpiece will be in the seatbelt clip, the microphone will be stuck under the handbrake and the wire itself will have tied itself in a double reef knot round the gear stick.

I'd like to send someone a strong fax about this, but I can't because all fax machines don't work. I am on my third this year, and even the latest version, which is the size of a helicopter gunship, makes origami animals out of every piece of paper that goes near it. Or if you stand there with a hammer in your hand and murderous intent in your eyes, it pulls the paper through neatly, but 84 sheets at a time.

The simple fact of the matter is this. No piece of modern technology works . . . except your car. No, really. Think how cross you'd be if your engine died as often as the signal to your mobile phone. Or if your heater broke down with the regularity of your printer. Nowadays, we just get in our cars and expect them to work. And they do.

Unless they're Toyotas. I bought a three-year-old Landcruiser in the summer, partly because I thought it would be safe for the children, partly because it is an eight-seater, partly because you don't need to slow down for Mr Prescott's speed mountains and partly because I knew it would be more reliable than a Discovery. Which only goes to show how much I know.

Now I know it could have been owned previously by a man with butter fingers and ham fists. I know it could have spent its entire life taking fat people up Ben Nevis. But it's a Toyota Landcruiser, for heaven's sake. It's built to head-butt the Kalahari and arm-wrestle the Outback. This car is designed so that it can sidle up to the Sahara Desert, call it a poof and escape with its differentials intact.

So why is it the most unreliable piece of junk I've ever had the misfortune to own? It judders, creaks, lurches every time it stops, the electric windows have broken, and last week I was faced with a £400 bill for new brake discs. If it really has only done 30,000 miles, that's pathetic.

Obviously, I'm angry but, conversely and rather perversely, I'm also delighted. You see, like all the other motoring writers, I've always been happy to peddle the story that Japanese cars are reliable because ... well, because they just are. But I now have first-hand experience, and it's rather nice to find that our inscrutable little friends make mistakes too.

Obviously, it has to go, but what should replace it? My first choice is a Range Rover – from Belgium, where they cost 40p – but some of me fancies an M-class Mercedes. American colleagues tell me that early models were made by mad people in blindfolds and that an endless catalogue of faults would give me acres of good material for this column. And you'd be surprised to find how important that is.

I mean, I've clocked up 20,000 miles in my Jaguar XJR and absolutely nothing has gone wrong; 15,000 parts continue to work in perfect harmony, which makes it great to own. But not so much fun to write about.

The kind of pressure we can do without

I love this time of year. As the temperature drops, Jack Frost does dot-to-dot drawings on our windowpanes and we're greeted every morning with a visible reminder that we've woken up breathing. Even the countryside manages to look interesting, with a Technicolor blaze in the treetops and a crispness that makes the air almost brittle.

Yes, I love the autumn, but what I love most is the torrent of advice we get from motoring organizations about the preparations we must make if our cars are to survive the winter. We're told that, before every journey, we should check our shock absorbers, our headlights, our wipers and that there is a thermos of hot coffee in the boot in case we break down on a moor. They even say we should clear all the frost from all the windows before setting off, but that's stupid. As soon as I have made a hole big enough to see through, I'm off. It's far too cold to stand around doing pre-flight checks when the most I'll be doing on the way to work is 4mph.

Only this morning, Goodyear sent me a missive saying that in wet and possibly freezing conditions the only contact your car has with the road is four small patches, each no bigger than a postcard. So . . . what exactly are we to do about that, then? Well, it seems we must check our tyre pressures regularly because, as the thermometer falls, so does the pressure in our tyres. This increases the rolling resistance, meaning fewer miles to the gallon and curious handling anomalies.

Well, now, look. I'm very sorry, but I'm a busy man, and I really don't have the time to check how much air is inside my tyres. If the steering goes all wobbly and the car starts to veer wildly, I'll be aware of a problem but, until then, leave me alone.

Going to a garage is one of the most unpleasant experiences a human being will ever encounter. It's so awful that, when my petrol gauge is down beyond the red, and I've just driven past a sign saying 'Services 1m and 27m', I will always, always, go for the gamble. And when I get there, I'll gamble again.

I have driven past garages with the fuel needle bent around its bump stop. I have felt the first cough of doom and still kept right on because the station forecourt looked a bit dirty. This drives my wife insane with rage. Indeed, I'm on my final warning. I've been told that, if she climbs into my car once more to find the tank is full of nothing but air, she will kill me. And, to put that in perspective, an affair will only get me broken knees.

I hate filling up. And there is nothing in life that annoys me more than a slovenly petrol pump. Or one that cuts off every two seconds. Or an attendant who won't reset the counter until you've been into the hut and called him a fat, gormless waste of the world's resources.

You can imagine, then, that after I've put £50 of petrol in the tank, quite the last thing I want to do is buy a token for the air machine and grovel around in a sea of diesel getting brake residue all over my fingers. And why, when you drop a dust cap on the floor, does it evaporate? They do. They just disappear.

I wouldn't mind, but it's all so pointless. I remember reading a report recently which said that all garage air pressure gauges are out by as much as 20 per cent, which means you stand absolutely no chance of keeping pace with the law, leave alone microscopic fluctuations in barometric pressure.

Not that there are any nowadays. Goodyear paints this picture of winter as a looming asteroid, an extinction-level event heading our way, and that there's diddly squat we can do to prevent impact some time in late November.

Well, look. I have not seen a single snowflake for four years and, even if we do get a light dusting, the radio will immediately fill with police messages warning us all to stay at home. Why do they do that? We pay £30 billion a year to Mr Prescott for our road network, and we expect him to provide, in return, a selection of gritting lorries and snowploughs. I mean, they can keep the roads open in Alaska and Lapland, so surely it isn't beyond the wit of a nation that gave the world Brunel to clear a motorway in Kent once every five years.

The fact is that cars have helped to make the world nice and snuggly warm these days, which in turn has made the roads much safer. But if, by some miracle, you encounter some black ice, or perhaps a little sleet, don't think, as your car slithers towards a ditch: 'Oh, no. I wish I'd checked my tyres.' Because it wouldn't have made the slightest difference. You're in a two-ton car, and a bit of air pressure here and there is no match for the laws of gravity and momentum.

Three points and prime time TV

Why don't you go catch a burglar? For 50 years or more it's been the automatic, kneejerk reaction of any errant motorist who's been pulled over by plod. It's even become a music-hall joke, a ritual in the tired old plots of 8 p.m. sitcoms. But it's true. Why don't they go and catch burglars? I mean, the only reason we drive too quickly is because we want to get home and catch one before he deposits a large turd on our Bukhara rug.

CCTV has driven teenagers out of the city centre, so now they queue outside remote farmhouses waiting for their turn to defecate on an heirloom. And where are the police?

Well, they know you're only after a crime number for insurance purposes, so they're about 40 miles away, trimming their moustaches so they'll look good on next week's exciting edition of *Police. Stop. Kill.* Today, the police spend all of their resources on JetRangers and sophisticated infrared cameras so they can get action-packed footage of car-azy motorists. And the Crown Prosecution Service? That's busy sorting out the video rights.

Small wonder, then, that people are beginning to take the law into their own hands. In France recently, a much-burgled home-owner left a radio on his kitchen table and a note that said: 'This is not a radio. It is a bomb.' He came home later to find a burglar spread evenly around his kitchen and was promptly arrested. And now it appears to have happened here with the news that a Norfolk farmer has been charged after two youths were shot in the garden of his house.

Friends say the poor man had been driven to despair by an endless stream of burglaries and that the police weren't interested. Well, they wouldn't be. It's hard to see a marketing opportunity in painstaking house-to-house enquiries. It's late-night BBC2 at best. Nah, let's go get another speeder.

Certainly I'd shoot anyone who broke into my house. Then I'd bury them in the garden and carry on with life as though nothing had happened.

No, really, a friend and I once staked out our street in Notting Hill, saying that if we caught the youth who'd been breaking into our cars we'd chain him up in a shed and invite other victims of car crime round to spend some time with him. And we meant it. Of course, this has to be against the law. You can't condone vigilantes in a civilized society. If you let home-owners shoot intruders willy-nilly, burglars will tool up to meet the threat. Then you'd arm the police, who'd shoot motorists for ratings, and it would all be like America. We might even end up with the wide-mouthed frog as president.

So what's to be done? The police no longer feature. I know nowadays it says they're 'Fighting Crime. Slashing Fear. Filming Disorder' on the side of their cars, but that's just a slogan to inspire Bruce Willis action from the men. The reality is that in the countryside, one policeman has to cover more than 200 square miles – impossible when his superiors demand 35 motorists a night, DVD footage and format rights.

We can't expect tougher sentences for the tiny minority that are caught, either, because the prisons are full of speeders and people who've shot burglars. And all the while, IslingTony is being lobbied by inner-city councillors who plead for leniency.

So the burglars who are daft enough make faces in front of CCTV cameras end up with 10 minutes of community service. Little deterrent for someone who's being driven out of his mind by an all-consuming need for heroin.

There's the nub of the problem. Eighty per cent of all crime is drug-related. No one breaks into your house because they need funds for music lessons. They break in because they need some smack 'n' crack.

And I'm sorry, but we've got to give it to them. Legalizing drugs will bring the price down, and cheaper drugs will mean less crime. It is as simple as that. And to argue that we'll all become junkies as a result is nonsense. You can buy drink, but we're not all alcoholics.

The police have lost the war on crime because they've

been diverted by the lure of fame and fortune on television. And we aren't allowed to blow the little sods to kingdom come, so let's see the root cause on sale in 24-hour filling stations. Alongside the cigarettes.

You could even tax them. This would surely generate enough to get the police out of their Volvos and into something a little more big-screen friendly. But until this happens, I'm afraid, there's a better-than-evens chance that you'll come home one night to find a burglar peeing into your grandfather clock. And all you're allowed to do is offer him some buns.

Every small boy needs to dream of hot stuff

Dream all you like about one day owning a helicopter and a Bentley, but the sad fact is this: nearly four in every ten new cars sold in Britain come from Ford or Vauxhall. Ford is the market leader, and they may think this has something to do with a combination of good cars, a dealer on every corner and a bit of natty television advertising. Well, they're wrong. Ford is ahead because of something that began 30 years ago ...

One fine morning in June 1969, I set off for school dimly aware that my father would pick me up that night in his new car, a nothing-special Saluki Bronze Ford Cortina 1600 Super. Had he done so, I might never have become interested in cars. I might have become an astronaut. Or a homosexual. But he had a last-minute change of heart and swished up to the school gates in a 1600E, which had Rostyle wheels, extra fog lamps and a bank of gauges set into the wooden dashboard. Well, I was smitten. From that day to this, I've been a Ford man.

This means my first car was a Ford, and we have a Mondeo now. But elsewhere in the English-speaking world, it can mean so much more. In Australia, for instance, during the annual motor race at Bathurst, Ford and GM supporters have the sort of full-scale battles that would make British soccer hooliganism of the early 1980s seem restrained. Only once have they reached a peace, when a Nissan won and they all joined forces to jeer at its driver.

Then we have America, where you can buy T-shirts saying 'I'd rather push a Ford than drive a Chevvy'. But then, I know a chap in Texas who says he would shoot anyone who came on to his premises in a vehicle that bore the blue oval. He hates Fords; says they're made by communists and long-hairs.

I suspect the reason this kind of thing doesn't happen here is that Vauxhall has been so catastrophically dreary for 30 years. Whereas Ford has always had a little something to tickle the ticklish bits of the nation's nine-year-olds.

After the 1600E went west, they were quick out of the blocks with the RS2000, a hot Escort that neither handled nor went quite as well as legend suggested. But it looked good and had wrapround seats and a steering wheel the size of a shirt button. This car softened the blow of failure. 'Oh, well, I'm a bank manager so I'll never have a Bell Jet Ranger. But who cares, because if I can just close this deal I could have an RS2000. And that's not so bad.'

In the 1980s we had the XR3, the RS1600i and latterly the RS Turbo. Even the Fiesta leapt on the bandwagon, sprouting a wholly unnecessary turbo to match the mood.

Vauxhall tried to keep up, with hot Astras and Novas, but their hopeless attempts were in the end blown away by, first of all, the Sierra Cosworth and then its four-wheel-drive Escort sister. Executive power for middle-management money made it a hit in the real world, while a spoiler the size of Devon made it a must-have for any nine-year-old's bedroom wall.

Believe me. Anyone who was under 12 when Ford and Cosworth started making babies is a Ford man and always will be.

But since the Escort Cossie was removed from the scene by EU noise regulations, Ford has been strangely quiet. Oh, I know there's a mildly tweaked Puma coming out this Christmas, but it's only a chicken korma and, anyway, only 1000 are being made.

I've looked into the future and I see no rip-snorting, muscle-bound brute. They came up with a modular 6.0-litre V12 engine that was nice, but that's been given to Aston Martin. They bought a Formula One team, but they've handed that over to Jaguar.

Where's the next GT40? Where's the turbo nutter bastard Focus? My three-year-old is entering the phase where his motoring future is about to be mapped out and there is nothing around to stamp a blue oval on his heart. If anything, he's erring towards Luton. Vauxhall has revealed plans for a two-seater sports car, and he likes the look of it very much, especially the orange paint. Me, I'm more interested to know it will be built by Lotus and will share many components with the spectacular Elise.

It'll be called the VX220 and is designed to be just as much fun as the Elise but with a little more comfort. At

£23,000 or so, it could be the first car in 10 years that even gets close to Mazda's MX5.

Ford would, of course, explain that they own Mazda, and that's true. But what's the point of buying up other car companies when the mother ship is left out in the cold. If it doesn't offer up a chicken chilli jalfrezi with extra-hot sauce fairly soon, my boy's going to end up in a Vectra. And that's a big worry.

Footless and fancy-free? Then buy a Fiat Punto

Should I ever be banned from living in England, I'd go to France, partly because the houses are cheap, but mainly because you're never more than four feet from an ashtray. Italy, sadly, is out of the running, partly because it has been colonized by Mr Blair and his cohorts and partly because the Italian notion of an emergency plumber is someone who can be there in less than seven weeks.

France works. Italy doesn't. Explain, for instance, to an Italian hotel receptionist that all the lights have fused in your bedroom and she'll say it's time you were in bed anyway. Arrange to pick up a hire car in Milan and, when you get there, the man will look at you as if you're mad. A car! From me? At the car-hire desk?

Then there's the noise. In rural France, you are woken every morning by a hundred schoolchildren swarming past your bedroom on scooters, and this is annoying. Vespa, remember, means wasp. But they're soon past. Whereas in rural Italy, the dawn chorus comes early and noisily, not from the birds but from the nation's dogs, all of whom think they're Pavarotti. Then, round about six, you get the two-stroke descant. In Italy you are never more than four feet from that bane of countryside living, a strimmer.

Italy is a nation of extremes: extreme beauty, mood and fashion, tarnished with extreme noise and disarray. That's why you should always think twice about buying an Italian car.

Of course, I've got an Italian car, but then I am mad, and anyway I'm not talking about that sort of Italian car. I'm talking about the urban buzz bomb, the wheeled mule, the metal donkey, the covered wagon with a dashboard. The small Fiat.

Last time I looked, Fiat had a 60 per cent share of the Italian market, which tells us that Italians love them. Which means they are ideally suited for people who train their dogs to sing; people who were born to strim. And that's not us. Take a look at the new Punto. It looks great, as chic as the changing room at a Milanese fashion house, and it handles with the sort of verve and aplomb Italian drivers demand.

When the plumber takes seven weeks to get to your burst pipe, the delay breaks down like this. Six weeks, six days, twenty-three hours and fifty-nine minutes sitting around drinking coffee. And one minute to drive his van the fourteen miles to your house. Italians like to drive quickly, and the Fiat responds to this challenge well.

It makes you lean a little further forward in your seat. You stretch the revs a little more between gear changes. You brake later, steer more vigorously. It makes your heart beat a little faster. So, yes, it has the gut-wrenching, come-back-for-more appeal of the country that makes it.

At £13,000 or thereabouts, the 1.8-litre HGT model

I drove is also good value, compared, that is, with similar cars in Britain. Obviously, it is available in Belgium for 25p.

So you may be tempted to go over there and buy one. Well, yes, and in the showroom it'll look good. You'll like the lights, the seats, the dash, the cheekiness. And you'll keep on liking it until you get to an oblique junction, where you find the rear pillars, so fine and fluted from the outside, mean you can't see if anything's coming. They should really provide a rabbit's foot on the dashboard, or a sprig of heather. Something you can rub before closing your eyes and lunging out into the traffic flow.

And then we have the pedals, designed for someone whose feet have been amputated. You have the same problem with an Alfa 166 in that, when you go for the brake, you will also press the throttle and clutch at the same time.

And what about the gearbox? Well, go for second in a hurry and it's like you've stuck a steel rod and a bag of gravel into your blender.

All things considered, then, I'd love to buy a Punto and drive around watching girls, but all the little day-to-day faults would leave me cross that I hadn't bought a Fiesta.

Let me put it this way. If the Punto is San Gimignano, the Fiesta is Bolton. It is a department store, not a boutique. It is a supermarket, not a delicatessen. The Fiesta just does everything well and, if you accept the car is not a two-weeks-a-year holiday home but a device for moving around, it's what you want. And if you don't, I suggest you turn to Peugeot, because even the 206 is a better everyday bet than the new Punto. It can be Calais when it's raining and you just want to go to work, but it can also be Paris chic and a hoot in the Alpes Maritimes. It has a huge ashtray, too, and – who knows? – it may even run on sewage.

To sum up, then. If you want a small car, buy a Fiesta. If you want a small car with style, buy a Peugeot and, if you have no feet, buy the Fiat.

Now my career has really started to slide

Audi are in all sorts of bother at the moment as they try desperately to fix a handling problem on the TT sports coupé. Two people are dead, and the newspapers are littered with stories from drivers whose £30,000 cars have head-butted beech trees.

Here's the problem. You're barrelling along when all of a sudden the road tightens unexpectedly. Worried, you take your foot off the throttle and brake, which causes the nose of the car to dip. This raises the rear, causing the back tyres to lose grip, and now you're going sideways. This is called oversteer, and, in Ladland, the petrolheads love it more than life itself.

In motoring magazines every car is photographed going sideways, its driver dialling in lots of opposite lock to counter the problem. This is deemed to be fun. When you apply for a job on a car magazine it doesn't matter if you can't spell, or even if you have personal hygiene problems, just as long as you can take a ton of metal and noise and make it dance. If you can, you're man. If you can't, you're gay.

This was a big worry in my early days because I simply couldn't do it properly. I never had big enough balls to drive into a corner faster than was prudent; often, when presented with my puny efforts on film, art directors would pull my hair and call me Mandelson.

Eventually I mastered the art of making a car go slightly sideways for a thousandth of a second, just long enough to get it on film, then I'd undo my seatbelt and get in the back, where I'd lie, whimpering, until I coasted to a halt somewhere. But this didn't work well on television. With a *Top Gear* camera pointed at me, I was expected to make the car slide and keep it sliding. I just couldn't do it.

I will admit now, for the first time, that I used to take demonstrator cars to an airfield at night and practise. But it didn't matter whether they were front-wheel drive, rear-wheel drive, four-wheel drive or even side-wheel drive: I'd always come home with nothing more than four bald tyres.

I read up on the theory, buried my nose in books about physics, and talked to racing drivers. But it was no good. The years rolled by in a cloud of wasted tyre smoke and pirouetting steel. By the time I left *Top Gear* I had summoned up enough courage to make the car slide, but then I'd run out of talent and it would always spin. Time and time again I needed a tow truck and a bandage for my ego.

But then, while filming the Aston Martin DB7 Vantage for my new video – called *Head to Head* and out now – it all came together. I turned into a corner doing about 110mph and lifted my foot off the throttle. As usual, the back started to swing wide and, as usual, I applied some opposite lock to the steering. But this time the car didn't go into a spin, and I had all the time in the world to put my foot back on the accelerator and hold the slide. Finally, I learnt what it was like to 'steer the car on the throttle'. And it was great.

At 39 I became a man, and to celebrate I got out of the Aston and showered it with big, sloppy kisses. I bought it flowers and am thinking of moving with it to a little house in Devon where we could rear geese.

Of course, the DB7 Vantage has all the right ingredients for stunt driving like this – fast steering and a colossal 6-litre V12 engine that drives the rear wheels – but since that glorious moment I have the confidence, and now I can make anything oversteer. Front-wheel-drive Golfs, no-wheel-drive Hyundai Accents. I bet I could even make my Aga go sideways. All you need is a 600-acre airfield.

Here's the thing, though. It's taken nearly 15 years and about 15,000 sets of tyres to reach the point where I could confidently handle an Audi TT. But the question is: could you? You know what it says in the Highway Code. You know you should steer into the skid. But could you be relied on to get it right, at 80mph, in a rainstorm, with a tractor coming the other way? Probably not.

Car makers need to think about this before they make those final tweaks to a new model's suspension. It's all very well providing oversteer for the road-test department of a car magazine, but when normal people break free of Mr Prescott's traffic jam and put their feet down, you really have to offer completely fail-safe handling.

Few cars have it. The Golf GTi is one and the Alfa GTV is another and, er . . . that's about it.

In the interests of balance, I should say that the Audi

TT is by no means the most tail-happy car you can buy.
You should try a Peugeot 306GTi. If you think you're
man enough.

The best £100,000 you'll ever waste

If I'd designed the new Mercedes S-class, I'd have packed
a lot of towels and headed for the beach. I'd have spent a
while jet-skiing and barging in queues, safe in the knowledge that it would take the rest of the car industry years
to catch up. But then I come from a country whose car
industry is now restricted to a wooden sports car from
Malvern and a plastic one from Blackpool. The Germans
think differently. When they finished work on the new
S-class they immediately handed it over to AMG, the in-
house tuning division, and ordered them to make it better.

Where do you start? I drove the standard car six months
ago and described it in this column as the best car in the
world. So being ordered to 'make it better' would be like
asking a plastic surgeon to make Kristin Scott Thomas
prettier. First of all you'd ask how. Then you'd ask why.

Unless you were German, in which case you'd take the
engine out and start work. First of all, it was increased
from 5.0 to 5.5 litres, then the crankshaft was redesigned.
They added forged aluminium pistons with oil injection
jets to keep them cool. Then a twin stream intake system
was fitted, along with a new intake manifold.

Every single thing under the bonnet was changed, and
the results speak for themselves. The new car churns out
360bhp, making it precisely no miles an hour faster than

the standard model. In a dash from 0 to 60, however, the gulf is obvious to anyone with a brace of laser beams and an atomic clock. The standard car does it in 6.5 seconds; the AMG derivative does it in 6 seconds dead. Oh, dear. All that work for half a second.

I'd have given up at this point and gone skiing, but, no, Hans took off his bad jacket, rolled his sleeves up even further and began on the styling. He developed all sorts of plastic add-ons, spoilers, aprons and skirts, but each one spoilt the purity of the original.

So the car he lent me this week came with none of it – just four big wheels that were just about acceptable and a brace of chromed exhaust pipes that looked ridiculous.

And what about the suspension? What wizardry has been woven here? What little tweaks? Well, they did a lot of head scratching and decided that the original couldn't really be beaten, so, er, well, they did nothing. Just fitted some bigger brakes that reduce the stopping distance by about an inch, probably.

So there you are. The new AMG S-class Mercedes-Benz: £25,000 more than the standard car, and for that you get absolutely nothing. Not even peace of mind that it'll still be there in the morning.

No, really. It comes with what's called keyless entry. You keep a credit card-sized transmitter in your wallet so the car can sense when you're getting near. It then opens the doors, allowing you to slide in and start the engine by touching the gear lever. But when you get out again and walk away, how do you know if the doors have locked? Every time I walked back to check, they were open, but that's because the transmitter was in my wallet. So I have no idea whether the system works or not.

I have no idea either how much this car would be worth after a year. Standard S-classes will always find a home with one of the endless upmarket minicab firms who see it as a short cut to all the plum jobs and big tippers. But I can't see Mr Patel going for a model with oil injection jets.

All in all, then, the new car is daft. You pay £100,000 now and, in a year, you'll get £1 back. If you're lucky.

That said, however, the world is a daft place. You could provide Michael Winner with hot and cold running flunkies and he'd still complain that the sun was in his eyes. And I know plenty of people who do a £30 million deal and think, 'Right. Where's the next 30 mill. coming from?'

For some people, enough is never enough. You could put them in the first-class compartment of a 747 and they'd spend the entire journey fidgeting because they weren't far enough forward. They wouldn't even be happy with a deckchair nailed to the nose cone unless it came with an extending arm that enabled them to fly along 40 feet in front of the plane itself.

These are the guys at whom the AMG S-class is aimed, and they will lap it up. It's no better than the standard car, but it's no worse. And it is a lot more expensive, which is what matters most of all.

Put an imaginary billion in the bank and you'd have a car like this. I would. And if next year they came out with a special Myrrh edition with panda bear ear upholstery, gold pedals and a Jacuzzi, I'd have one of those too.

In the real world, the normal S-class is still the best car in the world, but on planet Plutocrat the AMG is even better.

Styled by Morphy Richards

It's happened again. Just months after Mercedes was forced to recall all its A-class hatchbacks because they had a worrying habit of falling over, Audi has had to pull the TT. It seems that if you lift or brake while cornering at high speed, the back will snap into violent oversteer and you will slam into the crash barriers. Already, in Germany, two people have been killed, and Audi has had to act fast.

Remember, Audi was pretty well wiped out in America 10 years ago by rumours that their cars suffered from 'unintended acceleration'. Dim-witted Yanks said that even if they had their foot on the brake, the car kept on accelerating at full speed until it slammed into a child/pensioner/dog.

I must confess that I felt rather sorry for Audi on that one. It was, let's face it, the fault of America's education system and a proliferation of lawyers rather than an engineering problem over in Ingolstadt. A car simply cannot accelerate unless its driver hits the wrong pedal – easily done in a land where the smart bombs can't even be guaranteed to hit the right country.

And now I feel sorry for them all over again. Quite apart from the major redesign, they're having to pay a small fortune to fit the 40,000 examples they've already sold with different stabilizers, altered dampers, modified wishbones and a new rear spoiler. They've got two deaths on their hands, and they're looking down the barrel of a serious public relations disaster.

The trouble is, the TT never really knew what it was. If it had been billed as just a motorized jacket, a poseur's pouch with no delusions of racetrack glory, they could have fitted wooden suspension and all would've been well. But it was, after all, going to carry the Quattro badge, and it was going to have a 225bhp engine, so it needed to be sporty as well. And that meant it had to have some lift-off oversteer.

In motoring magazine land, we'll tolerate front- and four-wheel-drive cars only if they give us this handling quirk in spades. Lift-off oversteer is more vital than customers. If we can't get a car to go sideways while careering past a photographer, it is dismissed as a hopeless dud. Understeer is for wimps.

It's a saloon-car thing. It's pants. And the car makers know this, so they dial it into their sports cars to keep us happy. Oh, sure, they know full well that in the wrong hands, in the wrong weather and on the wrong road it can be fatal, but they want good reviews . . .

The thing is, though, that the Audi was by no means the worst offender. If you want real lift-off oversteer, try a Peugeot 306 for size. That thing behaves like a hungry puppy, wagging its tail at the slightest provocation. And while this is a huge hoot on an airfield, it can be down-right scary in the wet.

Just think. You're barrelling along, snicking through the gears, feeling the tyres scrabbling for grip when, all of a sudden, while going round a corner, you find a tractor coming the other way. So, in a panic, you lift off. And whoa, now you have to miss the tractor while controlling a lurid tail slide.

Only recently I was called old and fat for saying I'd rather have a Golf GTi, which always ploughs straight on, than a 306. But it was for this very reason. On a racetrack, the 306 kicks the Golf's arse, but in the real world, I'm telling you, it's the other way round.

I congratulate VW for ignoring the pleas of us motoring journalists. And Alfa Romeo too. Back in the summer, I went to an airfield with a GTV and tried everything in my limited repertoire to make it misbehave, but it wouldn't. So, if you're faced with an emergency, there's one less thing to worry about.

And what about the Focus? Car of the Year. Best-selling car in Britain. Darling of the motoring press corps. And why? Because when you lift off in a bend, the tail swings out.

Audi was only trying to get some of this glory with the TT, and that is probably why the company was angry with Tiff and me when we came back from the launch and said it was a dog. We said there wasn't enough feel and that the oversteer, when it came, was rather cynical; a bit of icing to disguise the fact that the cake itself was a bit stodgy. *Autocar*, of course, raved, saying the handling was in fact superb. Just like they did with the Mercedes A-class.

And Audi pointed this out to the two TG boys who wouldn't toe the line. Everyone else likes it, they said. We've had rave reviews in Germany, they said. And now they are admitting that the car's handling 'in certain circumstances' has been criticized. I can't tell you how good that makes me feel.

But I still feel sorry for them, and that's why I have spent the last few minutes working on a solution. Cars that oversteer need to be ugly, and that way people who want a car to pose around the harbour bar are not going to be caught out when the damned thing starts doing the waltz at 150mph. Enthusiasts are forever saying they don't care what a car looks like, so fine. Get someone from Morphy Richards to come up with some poseur-repelling styling and all will be well.

The terrifying thrill of driving with dinosaurs

This week I was going to tell you all about the new Rover 25, but after my less-than-flattering review of its bigger brother recently Rover says that all their press demonstrators are booked out until February. And no more reservations are being taken. Roughly translated, this means: 'Get lost, sunshine.' Undaunted, I asked if I could perhaps borrow one of the new Land Rover Defenders. Finished in original Atlantic green and equipped with snazzy alloy wheels, I'm seeing quite a few in south-west London and, to be honest, they look rather good. But guess what? Rover has only one demonstrator, and it's booked out for ever.

I've been down this road before. Toyota once banned me from driving its cars, and Vauxhall made life difficult after the Vectra episode, but there are other ways of getting test drives and, in time, I'll explore them. So be afraid, Rover. Be very afraid.

For now, though, let's talk about the Lamborghini Diablo, which *Autocar* says is 'the last genuine supercar on sale'. It's a thought-provoking argument that made me think about the definition of a supercar. I've always taken it to mean a car where practicality and cost worries are crushed in the quest for speed and style. And, on that basis, lots of cars fit the bill, but I sort of know what they mean about the Diablo. You can't drive around in this heavyweight brute with its rippling abs if you have a concave chest and spectacles. No, I mean it. If you have limbs like pipe cleaners, you will not have the strength to push the clutch pedal down and, even if you could, you'd think the gearbox had jammed every time you tried to move the lever.

This is bad enough, but it must also be pointed out that you can't drive the Diablo if you have a neck like a birthday cake and arms like ship's pistons. Oh, sure, you might be strong enough, but you'll be too big to fit in the cockpit.

I know only two people who have bought a Diablo, and one is Rod Stewart, so, really, it's pretty pointless talking about the new model. But since there's no Rover, I shall plough on regardless.

It's called the GT, and it is able to whisk small, strong people from 0 to 100mph in under nine seconds. Scientists call this level of acceleration 'bleeding scary'. To achieve this, Lamborghini's engineers – and they are the maddest bunch of people I've ever met – lifted a few quid from their new owners at Volkswagen and threw the old Diablo's body away. They replaced it with one that looks exactly the same but is made from ultra-light carbon fibre. This added about £100 million to the cost of making the car but saved 70 kilogrammes.

Then they set to work on the V12 engine, which was taken up from 5.7 litres to 6.0 litres and equipped with all sorts of titanium wizardry so that it now produces perfect silence and geraniums at town-centre speeds but a colossal, thunderous, ear-splitting, tree-felling 570bhp further up the rev range.

I could tell you about the top speed, but I have no idea what it is and no intention of finding out. I do know that the official fuel consumption figures say that, around town, the Diablo GT will return seven miles to the gallon.

It won't. Cars never match the official figures in real life. So actually, it will probably return no more than 4mpg on a busy, stop-start Friday night. And that is so bad it's quite funny.

But then this is the point of the Lambo. The whole car is so bad it's hysterical. The air-conditioning works with the punch of an asthmatic blowing at you through a straw. The rear visibility is almost completely nonexistent and, while I see the new model has a backwards-looking video camera and a screen on the dash, you'll still have to say a few Hail Marys before pulling out of an oblique junction. There are no gadgets and gizmos either. Just an angry snarl and a big right fist.

And yet. When you put your foot down hard and that engine girds its loins for a full-frontal assault on the horizon, there is an 'Oh, my God' moment that no other car can quite match.

I once drove a Diablo at 186mph, not because I wanted to, but because I lost the ability to move my feet. Ferraris have lost this raw terror factor in recent years, and Porsches never really had it. The only other car I know that can do this bowel-loosening, supersonic baritone thing is the Aston Vantage, and that's nearing the end of its life.

So *Autocar* may have a point. It seems the chill wind of environmentalism has created an ice age in which dinosaurs like the Diablo find it hard to thrive. Only six of the new GTs are being imported to Britain, and I suggest that, if you want a last-chance power drive, you give it a whirl.

Perfect camouflage for Birmingham by night

Eating out in Birmingham was always one of life's more disappointing experiences. First, you had to find Birmingham, which is located above a series of tunnels, and then you had to find some food. Usually, this meant parking in a multistorey and then being beaten up a lot. Eventually, you'd have your wallet nicked, so, bleeding and hungry, you'd go back to the multistorey to find your car had been stolen as well.

When the ambulance finally took you home, six weeks later, you'd had time to ruminate on your night out, and most people usually reckoned that going into Birmingham city centre after dark was a 'bad thing' and that they wouldn't be going again.

To entice them back, a number of restaurants are now opening. You must still zigzag from your car to the front door, making use of whatever cover you can find, and your car still won't be there when you come out, but at least you won't go home hungry.

Le Petit Blanc opened recently, and in the next few months there will be Bank and Fish. But my attention was drawn this week to the launch of a new 'independent' called the Directory that offers food which is described as eclectic modern British. At the bar this includes a club

sandwich, Cajun chicken or a chargrilled vegetable sandwich that could be modern British were it not for the
addition of pesto and crème fraîche.

Therein lies my point. Modern British is entirely
eclectic, a selection of ingredients that are only British
insofar as they were nailed together here before being
served. I even found out this week that the chip was
invented in Belgium and that fish in batter was introduced
to us by an Italian.

And so it goes with the eclectically British Nissan
Primera. To get round EU import restrictions, it was
assembled in Tyne and Wear, but the parts came from
Japan. Oh, sure, Nissan will argue that a huge percentage
of the car's total value was British, but this included the
lavatory paper in the gents and the flowers given to
customers on delivery. The gearbox and engine were as
Japanese as sushi.

I used to hate this notion: that you could employ half
a dozen former dockers to fit a car's windscreen wipers
and it would suddenly become all John Bullish and strut
around shouting 'two world wars and one world cup'
every time it saw a BMW. But I note that the new
Primera is not only being built in Britain. It was designed
here, too, and it isn't even being sold in Japan. Nissan has
woken up to the fact that Ford and Vauxhall are perceived
to be European because the cars they make here are
designed here and, generally speaking, are not sold in
America.

So the new Primera: what's it like? Well, I've just spent
a week with one and, to be honest, it's like a Japanese
saloon car. It looks like a Japanese saloon car. It drives like
a Japanese saloon car. And this is not a criticism. There's

nothing wrong with Japanese saloon cars providing they don't pretend to be Moroccan or Portuguese. This new car is European. It just doesn't feel it, and that's not the same thing at all.

Like the old model, the new one is almost wilfully boring to behold. They've fitted a new nose, but it looks like a tongue, and round at the back it looks like ... do you know, I can't remember. And I can't picture the side either – just that tongue at the front.

Inside, it's grey. It was probably black or brown, but I remember it as grey. As far as space is concerned, it is fine. I could sit behind myself, if you see what I mean, and there were lots of little nooks into which the sales rep could put his electric razor and gum.

But you know, sitting there in the showroom, the new Primera is like the old Primera. Just another car. Just another way of spending £15,000 on four doors and a seat. To see why this car is so good, you need to take it for a drive.

Now, I know I had the 2-litre super-sport ripsnorter, so of course it was good. But the chassis on this new car is little changed from the chassis on the old, so even the lowly models will have handling way above their station. This is perhaps the only repmobile out there that is genuinely good fun to drive.

I must say the five-speed gearbox was a bit vague, but you could always opt for the hyperdrive auto with six-speed sequential override. I have no idea what this is, but it sounds fab. As does the engine. At high revs it makes an angry, growly noise which urges you to explore that handling prowess.

Indeed, this is exactly the sort of car you should use if

you wish to eat in one of Birmingham's new restaurants. It's big enough to take all your friends, and good fun on the way. But best of all it looks so terribly dreary that nobody will nick it.

Another good reason to keep out of London

The first time I drove a Porsche Boxster, everything was just so. I was on my way to Scarborough to film it, with nine of its closest drop-head rivals, and I was crossing the Yorkshire wolds, which play host to some of the best driving roads in Britain. And, boy, was I having fun, slithering round the corners, enjoying the metallic rasp of that 2.5-litre engine as it passed 5000rpm and generally doing the sort of speeds that aren't allowed.

And then I noticed a pair of headlights in my rear-view mirror, a way off to start with but getting closer. Eventually, they were right on my tail and, obviously, I reckoned this was one of the others on its way to Scarborough – the BMW Z3, perhaps, or the fearsome TVR Chimera.

But no. When it finally overtook, it was a 1.3-litre Vauxhall Nova. And from that moment I've always rather hated Porsche's attempt at a mass-market sports car.

I suspect that, when the original idea came along to do a small, two-seater convertible, the Stuttgart marketing boys in their tartan jackets were well aware that such a car might pinch sales from the 911. So, to create a gulf, they insisted that the Boxster should be de-tuned to the point where its engine would struggle to mix cement.

And quite apart from the fact that it couldn't pull a greased stick out of a pig's bottom, it was far too expensive. Why pay more than £30,000 for a two-seater car when, for half that, you could have a Mazda MX5, a car that manages to have front and rear ends that are distinctly different? You could drive a Boxster backwards and nobody would be any the wiser.

Given the choice of any two-seater sports car, I've always put the Boxster in about ninth place, just ahead of the three-wheeler Morgan but behind pretty well everything else. Even the dreadful BMW Z3.

However, Porsche's engineers must have been aware that their baby was out there being minced by Novas, so they walked into the marketing department, taped everyone to their chairs and set about righting some wrongs. Thus, there's now a 2.7-litre engine in place of the 2.5 and an S-version that costs £42,000. And at that price it had better be unbloody-friggingbelievably good.

Let's start with space. There's plenty, if you're a suitcase. In fact, there's a choice, if you're a suitcase. You can go either in the back, behind the engine, or in the front with the spare wheel. If you're the driver, things are not so good. Obviously, there's plenty of headroom, if the roof is down, but if you are cursed with a brace of legs I'm afraid you've had it. They simply won't go under the wheel, which is like the London Eye, only bigger.

My first instinct, on climbing into the new Boxster S, was to climb right back out and use the Jag, but in the name of research I persevered. And now, a week later, I'm glad because, truly, the car has been transformed.

Sure, it still looks like something out of *Dr Dolittle*, and the engine sounds like it came out of a Hoover, but there is fun to be had here. It is what the old Boxster wasn't. A sports car that's capable of outrunning a Vauxhall hatchback. And this is important.

Whereas the original car was fine at dinner parties, where you could walk into the room brandishing a Porsche key-ring, this new one can cut it out in the real world. Up here in the Cotswolds, or down here if you're Scottish, there's a meatiness to the power delivery and an unusual crispness to requests from the helm. Yet none of the old car's rigidity or comfort over bad road surfaces appears to have been lost. This means that, on the motorway, it doesn't interfere with the job in hand: thinking up new nicknames for Mr Prescott, mostly. I guess since he's now in charge of second homes and building in the south-east, we'll have to call him Two Houses.

And before you know it you're within the M25 and ready for the cut and thrust of London, where it goes all wrong. Because it looks the same as the old car, nobody knows you've bought a serious flying machine and, as a result, everyone gobs at you.

Really. In a BMW, people won't let you out of side turnings, and rightly so, but in a Porsche, people deliberately get in your way and, if you ask them, politely, if you could squeeze by, most indulge in the most fabulous hawking before letting fly with a docker's oyster the size of a cabbage. If this is something you find undesirable, then don't buy a Porsche.

Or if you must, make it the S and stick to the countryside. Out here, you'll be going too quickly for anyone to realize what you were in. And you'll be having far too much fun to care, even if they do.

My favourite cars

With the possible exception of the Vauxhall Vectra, every single new car that comes along is better, faster, safer and more reliable than the model it replaces. So, on that basis, the best car ever made must be in production right now.

Obviously, it's something compact and fuel-efficient, like the Volkswagen Lupo. And yet somehow the Lupo misses the point completely. It's a tool, a device, a white good that happens to be blue or yellow. It is bought with the head, rather than the ill-gotten gains of some rash moment when you stood bolt upright and said: 'I have just got to have one of those.'

If cars were like Black and Decker workbenches, people wouldn't talk about them in the pub, drool over them at motor shows, yearn to own them so much that it actually starts to hurt. And that's why the Lupo, excellent though it may be in the Co-op car park, actually comes pretty close to the bottom whenever I'm asked to name the three best cars ever made.

Number one on that list is the Ferrari 355, and I really don't think I can be bothered to explain why. Not again. So let me put it this way. Until quite recently I didn't actually own a car. There seemed little point when, every Monday morning, a raft of new models would be delivered to my door, fully fuelled, insured and ready to go. Obviously, I enjoyed driving them, but not once did I ever think of actually buying one. Some were good, but none was ever that good.

Until one day I climbed into a 355 and, within an hour

or so, I knew my standard of living was about to fall dramatically. I bought one a month later and, really, that says it all. Actions, you know, speak so much louder than words.

Not that you can hear either in my next choice, the Aston Martin Vantage.

Now I know there's a new DB7 Vantage, and I know that, dynamically speaking, it eats the big old bruiser, bones and all, for breakfast. It's prettier, easier to handle, nicer to drive, more reliable, and all those other things that just don't matter.

I'd love to own a DB7, but I fear I'd spend my entire time beating the steering wheel in a silent rage, angry that I didn't buy the real thing. The most powerful car on the market. Its big brother. The old V8.

Aston likes to say the DB7 Vantage is made by hand, but in reality the 6.0-litre V12 engine comes in a box from Ford – well, Cosworth to be precise, but let's not split hairs. My point is that the 5.4-litre V8 that goes in the old car is beaten into shape on site by men in brown store coats.

And there's more. The old engine delivers 600bhp, which might sound like overkill, but remember the car into which it's fitted weighs more than two tons. That's really why I like it so much: it's all so excessive, bigger than it should be, heavier, faster, more brutal. You just know that, if it were a person, it would have gout.

Choosing a third car was hard. Every fibre in my body said it should be the new BMW M5, but I couldn't. I just couldn't do it. I went back to the reference books, kicked some twigs round the garden and racked my brains for something less . . . less German.

The car I've come up with is a Datsun. Well, actually, it's a Nissan, which is nearly as bad, but then it's also a Skyline GT-R, and that's not bad at all. What I love about this car is that it's so completely Japanese. It's as though the designers just gave up and said: 'Look, we've tried for 50 years now to copy European style, and we're hopeless.' So the roof is there simply to keep the rain out, the doors exist only to facilitate entry, and the bonnet is a device for providing access to the engine. This car has absolutely no style at all. And it's even worse on the inside, where you're treated to acres of grey interspersed with lots of shiny black. Horrid is too small a word.

However, what the Japanese can do is technology. So the Skyline wades into battle sporting every gizmo known to man. It has four-wheel steering, variable four-wheel drive, a g-meter, ceramic turbos, the lot.

And it all works. I spoke last week to a chap whose last Skyline did 160,000 miles before it blew up. And that was only because he drove all the way from London to Val d'Isère with no oil in the engine.

In the real world the Skyline is faster than an Aston Vantage and a match for the 355. Mainly because the Nissan's arsenal of driver aids allows you to take diabolical liberties and get away with them. Seriously, you can turn this car into a corner at a preposterous speed, then alarm your passengers by undoing your seatbelt and getting in the back, safe in the knowledge that the unseen silicon will save the day.

Some say that Subaru's Impreza and Mitsubishi's Evo VI can match the Skyline for less than half the price, but then there are those who say marzipan is a foodstuff and that anchovies make an ideal topping for your pizza.

They're wrong. As a driver's car, the Nissan is about as good as it gets.

And would I buy one? No. Not a chance.

Need a winter sun break? Buy a Bora

All is not as it seems. In this month's edition of the Geneva airport in-house magazine, they talk of a city where the people 'add a friendly note to the litany of pretty valleys, castles, cathedrals, abbeys and, of course, the old traditional pubs. A region of unforgettable splendour.'

Would you like to guess what they're talking about? Nope, you're quite wrong. In fact, they're describing Birmingham, which to my knowledge has no pretty valleys, no castles, no abbeys and no unforgettable splendour. Just a lot of cars on bricks.

And this brings me to the television advertisement for the Volkswagen Bora. 'Any excuse' is what it says, and to hammer the point home we see a Dutch architect driving all the way back to an Alpine research institute because he's forgotten his pen.

Well, when I was faced with a trip to Blackpool last weekend, I did indeed choose to use a Bora, rather than any of the other cars lying around in my drive. And why? Because of the new V5 engine? The blue dash or the discreet styling? Because it would offer unflappable reliability and silent running? No, not really. I used it because it was the only car out there that had a full tank of petrol.

And then there was the business of coming home. I was

faced with a simple choice. Take the car or take up the offer of a lift in a helicopter. Ooh, that's a hard one. I'll have to ask the audience.

Obviously, I should have said: 'Look, I know I'm tired after marching round all day with a cannon on my back and a ton of lead shot in each pocket, but what I want now is four hours on the M6. I don't want to fly over Birmingham's pretty valleys and unforgettable splendour. I want to see it all from ground zero. I'm going home in my Bora.'

But, strangely, I was more tempted by the notion of getting home in 50 minutes and leapt into the Squirrel as though it was Saigon in 1975 and Charlie was swarming through the embassy gates. I even thought about filling the seatbelt fastening with Superglue in case someone tried to drag me out again.

So off we went at 113mph in a straight line from Clitheroe to my garden, where we'd touch down in a furious flurry of spinning blades and strobing lamps. My children are going to love this, I thought. Nearly as much as I will.

But with just 13 minutes left to run, snow began to fall, the pilot dived for the deck and dropped me on an industrial estate in Banbury. Naturally, I carry the phone numbers of all Banbury's cab companies in my head. And take it from me: absolutely nobody laughs at you as you tramp around a provincial town on a Saturday evening dressed in tweed plus-fours.

Then we have the children. Be assured that they weren't the slightest bit disappointed that Daddy didn't drop into the garden from a helicopter but came up the drive instead in a Ford Mondeo with a Mr Whippy aerial.

I'm afraid that while helicopters may be man's greatest achievement thus far, they have one big drawback. If the weather goes wrong, you end up miles from home, on an industrial estate, trying to pacify the guard dogs with the pheasants you've shot.

The Bora, on the other hand, can cope with any weather you care to throw at it, even the British winter sun that can't really be arsed to haul itself more than six inches above the horizon. You know what I'm talking about here. It doesn't matter if your car has sun visors the size of barn doors, if they swivel or if they come with illuminated mirrors on the back, the sun will always be in that tiny gap just above the rear-view mirror.

I bet that's what got Q. Over the years he's come up with ejector seats and machine guns in the sidelights, but I bet he was finally and tragically nailed because he never thought to fit his own car with a central sun visor.

The Bora's got one; a bit of plastic six inches wide and an inch deep which, all on its own, justifies the £19,000 price tag. It means you can see where you're going but, unfortunately, you will not necessarily know where you are.

To make the satellite navigation work, you need to slot a CD-ROM into the CD player and, if you want to listen to 'The Best of the Pretenders', you must take it out again. This means you could end up on an industrial estate in Banbury or, worse, one of Birmingham's pretty valleys.

So what of the car itself? Well, bearing in mind that I need to say 'Happy New Year' to everyone, there's only enough space left to say that all is not as it seems. This is not, as we've been told, a driver's car for the thirty-something architect with a lost pen. It's a Golf with a boot, and claims to the contrary are nothing more than 15 feet of warm wind.

Driving fast on borrowed time

Satellite navigation will soon become a standard feature in all new cars, and some of you may be very happy with that. Me? Well, I'm not so sure.

Here's why. Your car will be receiving information from satellites, so how long will it be before it starts to receive instructions? How long before it's restrained from doing more than 70 on a motorway or 40 in the suburbs?

You might think that this is all some kind of pie-in-the-sky dream that could become available, one day, perhaps some time in the new millennium. But I wouldn't be at all surprised to see it squeak into reality before this one is over, 13 months from now.

The impact would be colossal. Think. If you were suddenly unable to break speed limits, there would be absolutely no point, at all, in buying a car with a large engine. And please don't talk to me about track days or big torque making for relaxed driving, because that's non-sense. If you could never go faster than 70, you wouldn't even think about a 1.6, leave alone a supercharged 12. You'd buy a bloody Yaris.

No, worse; you'd buy a hybrid, a half-petrol/half-battery-powered obscenity with smooth rear wheel arches and an electronic Prescott under the rear parcel shelf, charging you £4000 for moving and £4000 every time you stop.

That's coming, too, you know. It doesn't matter how many times the RAC says motorists are up in arms, and it certainly doesn't matter how many pages I manage to fill with pro-car news, Phoney Tony has a 170-seat majority, so he can do whatever he damned well wants. And what he wants is to hang you up and bleed you dry.

He wants empty streets for his new baby to play in, and to get them he's going to impose legislation that'll make the tax disc of today seem about as costly as a penny chew. The technology already exists. Each car will be fitted with a black box, and every time you drive on to a motorway or into a town centre, your credit card will be debited.

There will be automatic debits for lawbreakers too. Obviously, you won't be able to speed, but anyone who jumps a red light will have £50 deducted from their pay at source. We already have this for absentee fathers, and forget the notion that people are innocent until proven guilty. You're a motorist, and that makes you as guilty as hell.

A few classic car magazines will survive, but *Top Gear* will be an early casualty. Along with all the lads' mags. These promote a lifestyle not in accord with the teachings of the Blair Witch Project and, bit by bit, their editors will be made to see the error of their ways.

This is already happening. A government think-tank, made up of no-hoper housewives in ill-fitting trouser suits, decided this month that the time has come to nail some sense into motoring programmes that promote speed. Pretty soon now, James Bond will be on the sparkling mineral water. And she'll not be allowed a car, either.

You probably think that if this were to come to pass,

there would be riots in the streets and burning effigies of Prescott lighting the night sky. But look what's happened already. They've put speed mountains on every back street in the land and no one has done a thing about it. And every time they slide a bus lane down an already congested street, there's a chorus of silence.

They do nothing to bring down car prices, which has only managed to inflame the Consumers' Association – a body with as many teeth as the Padstow Tufty Club. Performance motoring is doomed, and we're all remaining silent.

This is because we don't have a single leg to stand on. They need only to wheel out the bereaved parents of a four-year-old girl who's been killed by someone doing 50 in a 30, and there's not a damned thing you can say. Not a thing. You may say that we'll behave in built-up areas if they leave us alone on derestricted normal roads, but this time, they wheel out the kids of a man who was killed when two nutcases in a brace of 911s ran out of talent at a critical moment. And again, you're stumped.

They have a way of dealing with us, even now. When we turn up in a bespoilered GTR or Evo VI, they smile the smile of someone who has the moral high ground and one day will win.

This is a promise. In 15 years you won't be able to buy a performance car in Britain. Ferrari will survive, making art forms for people's garages, but the days of fire-spitting Subarus and hot Pugs are numbered. Mr Blair is going to win the next election and, with or without European help, he'll make fast driving about as acceptable as rape.

And there is nothing you or I can do to stop it, so I suggest that very early tomorrow morning you head for the Buttertubs Pass in Yorkshire. Drive it hard and fast, concentrating until your back and armpits are flowing like Niagara. Scare yourself, because that thrill, that sense of being over the edge, that moment when you've never felt so alive: soon, it will be a thing of the past.

Welcome to the world of Johnny Cabs. No need to fasten your seatbelts. We'll never be going fast enough.

I've seen the future and it looks a mess

Let me guess. This morning, you did not get dressed in a Bacofoil suit, you did not eat a pill for breakfast and you did not use a robot dog with aerials coming out of its ears to fetch the papers. I'm sure you were given a gadget for Christmas but, let's be honest, it was a lava lamp, and that's about as now as Slade.

I think it's fair to say that pretty well every single prediction about life in the year 2000 was wrong. We weren't hit by a giant meteorite on New Year's Eve. There was no second coming in Bethlehem, and the only millennium bug out there is the one that's making your wife's nose run.

But one thing did change. Over the Christmas holidays, a new type of car crept on to the market and, at a stroke, changed things for ever. Oh, sure, it still uses a series of small explosions to move about, rather than dylithium crystals, but it looks like nothing you've ever seen before.

Rarely do I lament the absence of a picture with this column, but today I could really do with one because using old-fashioned words to describe the new Fiat

Multipla seems almost philistine. We should be tele-pathizing.

The whole back end is square and slopes inwards, like the rear window on a Ford Anglia. The roof is perforated by two sunshine roofs and has a dip in the middle so that, after a rainstorm, you have a lake above your head.

Then there's the front, and that's just insane. It's as though Fiat used two designers. One made a bus, the top half of which has been lowered on to the bottom half of the other's low-slung sports car. Aesthetically, it's a shambles, a jumble of shapes and angles that have no place in the same country, leave alone the same car. It is roast beef and gooseberry fool, served up in a bowl that's part sherry schooner, part fish.

I could tell you that the Multipla is now *the* car in St-Tropez, but it won't make any difference. The first time you see one, your jaw muscles will turn to un-controllable mush. 'Why,' you will wail, 'does it have eyes in its forehead? And why does it have a duck pond on the roof?'

You'll be sucked in for a closer look and then you will be converted because, inside, there are six seats: three in the front and three in the back, each of which does the triple salchow at the touch of a button or the tug of a lever.

So what we have here is a car that's a bit shorter than a hatchback but, because it's wider, can take six people and still leave room for a boot. So who cares if it looks strange?

And we haven't got to the dashboard yet. Obviously, it's carpeted, but, less obviously, all the instruments are in the middle except the satellite navigation screen, which slides out of its box right behind the wheel. This is logical. Who cares how many revs you're doing or what track is on the CD? You want a talking road map in your line of sight because you're in Birmingham and you want to get out.

But people carriers tend to be expensive and thus the preserve of only the more affluent prime ministers. A bottom-of-the-range Ford Galaxy, for instance, costs £18,000; the Renault Espace is even more. This new Fiat, however, is yours for just £13,000.

And that means it appeals on all sorts of levels. On the one hand it's an inexpensive, practical car that will suit the family man in a cardigan. On the other, it's very new Labour. Very Guggenheim. With its truly innovative design, it would fit right in at the Groucho. But it is also ideal for someone who wants to stand out from the crowd and no longer wishes to walk round in a lime-green knitted suit. It even works as a minicab.

You could use it on the moon or to fetch the papers. I dare say you might even be able to eat it as a sort of twenty-first-century food substitute.

And I think we'll be seeing more of its type in the years to come. You see, the days when cars broke down or got punctures are gone, so car companies can now begin to concentrate on being clever rather than worrying about reliability and safety. I mean, the Multipla is available with either a petrol engine or, if you spend more, a diezzzel. Both can get from 0 to 60 and would exceed that on the motorway. Both use some fuel, make a bit of noise and go round corners.

Really, I have no idea what it's like to drive because, while I was there, in the driver's seat, pressing the pedals and things, I wasn't really driving. There was no wrestling with the wheel, no leather helmet, no need for supersonic derring-do.

It was a car, like any other, and yet it just wasn't — and that's why, without any question, it's the best new model we've seen for a long, long while. It crept into the twentieth century by the skin of its teeth, but it's the only car out there that really belongs in the twenty-first.

Nice motor; shame it can't turn corners

When a new car is launched to the motoring press, it is a lavish affair. Hundreds of hacks in Rohan trousers and Christmas jumpers are shepherded into the front of an aeroplane and flown to some exotic hotel, where they spend an evening eating artichokes with butter knives and wrestling with those snail-vice things that have no name. The next day they climb into the new car and drive on a predetermined route back to the airport. Simple, but a complete waste of time.

You see, in order to discover what the car is like, all you need do is ask the manufacturer to fax a copy of that predetermined route. They choose it specifically to suit the car they're launching. So, for instance, if it is made up entirely of twisting mountain roads, the car is obviously noisy on the motorway. Or, if it is short, there's a strong likelihood the car is uncomfortable over a long distance.

When Saab launched the 9-5 Aero, journalists were flown to southern Germany and asked to drive 150 kilometres up a motorway and 150 kilometres back. And what can we deduce from this? Easy. The Saab 9-5 Aero doesn't like corners. More than that, actually: it hates them. I've just spent the Christmas break driving the saloon and the estate, and I'm duty bound to tell you something. In the same way that you would not call a member of the Russian Mafia a big girl's blouse, you should not say to a Saab salesman: 'Yes, I'll take it.' The results will be the same. Great discomfort, followed by lots of bleeding.

It's not the torque steer, the desperate writhing of the wheel under harsh acceleration, and nor is it the astonishing lack of grip. No, it's a combination of the two, made worse by a traction control system that works in geological time. Only after you've left the road, ploughed through a hedge and are halfway to hospital does the silicon brain think 'Oh-oh, something's not right here' and try to cut the power to an engine that, by now, is three fields away.

Saab, of course, is now owned by General Motors, and the 9-5 is basically a Viking version of the Vauxhall Vectra, itself one of the worst-handling cars of the modern age. But the Vectra is never asked to handle more than 200bhp, whereas this Aero is fitted with a 2.3-litre turbo motor that churns out up to 240bhp. It's like fitting a Saturn V rocket to Ben Hur's horse. It's a damned shame, because the engine is wonderful. After 23 years at the forefront of turbo technology, Saab has eliminated 'lag' and come up with a blinder, a strong, immensely torquey, rip-snorting power plant that desperately needs a better home.

I wouldn't mind, but it was an engineer from Saab who once told me that you couldn't possibly put more than

220bhp through the front wheels of a car. 'It would be dangerous,' he said. And it is, matey. It is.

It seems pointless to talk about the rest of the car. I feel like the mother of a murderer, who tells reporters that, apart from his fascination with Nazi memorabilia and axes, Shane was a lovely lad. But it's true. Apart from its allergy to bends, the Saab is a lovely car, the estate version in particular. First of all, despite GM's involvement, it manages to look like a Saab. In fact, with the Aero body kit, it looks fantastic.

And inside it's Saaby too, with a dashboard that seems to have been lifted straight from one of their jets. It even has autopilot in the shape of cruise control and a night panel button that turns everything off except the speedo. And the ignition key is housed in the centre console next to the gear lever. Not better. Just Saaby.

This is key to the appeal of Saab; that and the fact that it's been a long time since the Vikings came up the Humber for a spot of pillage and rape. When we think of Sweden nowadays we think of pine furniture and mobile phones. We like to think that Saabs are made by teams of topless willowy blondes who spend their tea breaks gently beating one another with twigs. And we like this.

Saab people find BMWs a touch too pushy and Mercs way too flash. Saabs tend to be bought by kindly, New Labour souls who don't punch you in the face if you're 0.0001 of a second late leaving the lights. In times of trouble, you'd go to a Saab driver for help because you'd know that Merc-man would draw the curtains and pretend to be out. And that BMW-man would pour boiling oil all over you.

You could have a dinner party for Saab drivers, and it

would be brilliant. They'd be opinionated, interesting and well read. Unless, of course, they'd spent £28,000 on the Aero, in which case they'd be very poor company indeed. Because they'd be dead.

Stop! All this racket is doing my head in

Tiredness can kill, they tell us on motorway warning signs. Well, yes, I'm sure it can, but frankly I'd rather die in a 100mph fireball than pull over for a nap. In fact, nothing is guaranteed to ruin my day quite so effectively as an unplanned pit stop.

I look sometimes at those people mooching around in a motorway service station and I'm overcome with a need to ask them not so much why, but how. How can you have organized your lives so effectively that you have time for lunch in one of Julie's panties? A motorway is quite literally a means to an end. And judging by some of the prices out there, it can also be an end to your means.

When I set off somewhere, I absolutely will not stop until the petrol gauge has broken off the bump stop at the bottom of its range and the car has what sounds like whooping cough. If I need a pee, I will use my left leg on the throttle and my right leg on the clutch. I won't be diverted by those brown signs advising me of an American Adventure ahead and, even if my eyes feel like sandpaper, I'll still keep right on going.

So you can imagine my disappointment this week when, after just 60 miles with Ford's new Racing Puma, and with the gauge still showing full, I had to pull into a filling station.

The problem had started just a few miles down the M40 when, in a rage, my wife had turned the radio off saying that she could only hear the trebly cymbals and it was annoying. Then, a few miles further down the road, she asked me to slow down because, really, the noise was just too much.

And it was. So I eased it down to 80, then 70 and then, in desperation, to 50. But still, the balloon that had begun to inflate in my forehead kept on getting bigger until eventually, just outside High Wycombe, it burst. This was not simply a headache. This was cranial meltdown.

So I broke the cardinal rule and pulled over for a packet of Nurofen. This new Puma is like Ibiza at 3 a.m. It's a Hawaiian barbecue and a plane crash all rolled into a 12-foot package and amplified a thousand times through the Grateful Dead's speaker system.

And it isn't even a nice noise. It's not Supertramp or early Genesis or even the rumble in the jungle that you get from a TVR. It's just noise.

Then there's the ride. On a pockmarked road you can't have a conversation because the ceaseless jiggles add a vibrato warble to your voice. Imagine Lesley Garrett on helium and you're sort of there.

So what the hell is this car, then? Well, it's basically a normal Puma that has been pumped up in every way. The 1.7-litre is beefed up so that it develops 153bhp. The gearbox is beefed up so it can be a bit bolshie from time to time. The seats are beefed up so you can't get in or out easily. And the body is beefed up so that it looks just about as good as any car on the road.

No, really. The wings are flared and filled with massive 17-inch wheels that are smothered in ultra-low-profile 40-section tyres. And in case you were wondering, 40-section means the wheel is painted with nothing more than a thin veneer of rubber. Pull a condom over your head one day and you'll get the picture.

Just a thousand of these Racing Pumas are being made, and all are to be sold in Britain, at a rather steep £22,000 each. To be honest, I'd rather have an Alfa GTV, or a Subaru Impreza. Hell, for that kind of money I'd rather have a new pair of breasts.

But then I'm nearly 40. I no longer find it a hoot to spend the night after a party on one of those fold-up wooden chairs. I'm puzzled by late-night TV. And if you put me in a nightclub where they play white noise through 8m-watt speakers, I'll go home and seek solace in the Yes album.

You, however, are probably different. If you can tell the difference between Westlife and 5ive, then you'll barely notice the Puma's shortcomings. You'll revel in the dash it cuts round town, and in the countryside you'll marvel at its truly electric responses. On the handling front, a normal Puma scores 10. This gets a solid 12.

You won't worry that the back seats are fit only for amputees, and you'll actually be quite glad that it's not really a modern-day Escort Cosworth. With just 153bhp on tap, it takes 8 seconds to get from 0 to 60, and that means cheaper insurance.

Sure, you won't be able to speak to anyone while on the move, but then you don't anyway. I mean, how can you with the stereo making those computerized banging noises all the time?

What I'm trying to say is that the Racing Puma is only for people under 25. Like a good night out, it's deeply uncomfortable and deafeningly loud, but on the way home, when nobody's looking, it'll go like a jack rabbit.

Looks don't matter; it's winning that counts

It's just 20 years since Jaguar was renamed the British Leyland Large Car Division, and its workforce celebrated by going on strike again. Back then, Jaguar didn't have a workforce as such; just a group of men in donkey jackets who stood round a brazier outside the factory gates, throwing things at policemen.

Occasionally, they'd go inside and make a car, in the same way that, occasionally, a dog will go into the bread-bin to make a sandwich, but there was little point, because it wasn't a car in the strictest sense of the word. Oh, it looked like one, and it had wheels, but if anyone tried to go somewhere in it, they'd arrive somewhat later than anticipated, in the back of a tow truck. Jaguar, like the animal after which it was named, was on the verge of extinction.

So it's good to report that, after careful nurturing from American conservationists at Ford, Jaguar's numbers are rising. Indeed, 1999 was its best ever year, with 80 per cent of output going abroad.

Part-time workers have been told they can't go home and, while new lines are built at the Halewood factory, staff have not simply been laid off. They've been told

to go round Liverpool painting schools and helping old
ladies across the road.

And then, of course, there's the new F-type sports
car, which was designed to raise eyebrows at the recent
Detroit Motor Show but found its way instead into every
newspaper, motoring magazine and news bulletin around
the world. Jaguar insiders are saying it's a concept that
could, if people like it, perhaps, be put into production.
To which I say: 'Oh, for crying out loud. Just get on
with it.'

Of course, I know there are difficulties. The Audi TT,
for instance, began life as a concept car but ended up
wrapped round a tree. And Peugeot once made a concept
car that looked great, but there was no space anywhere
for an engine. So if Jaguar ever puts the F-type into full
production, it won't look like that car you saw last week.
But it will be similar, and that's good enough. I mean,
you'd sleep with someone who looked similar to Liz
Hurley, wouldn't you? Furthermore, if it's true Jaguar
could put it on the market for £35,000, you can kiss
goodbye to Porsche's Boxster. And the TT. And the
miserable Z3.

People have forgotten these days that price was the
E-type's biggest selling point. We remember it now for
having that long, long bonnet and for doing 150mph at a
time when most cars wouldn't do 4, but it was the price
tag of just £2000 that mattered most.

And that's what will sell the F-type. Price is what pulls
the punters in. Looks, and the promise of up to 300bhp
from a supercharged V6 engine, will only serve to pull
their trousers down.

I shan't go into the details of this fabulous car, because

Ray Hutton did that last week, but I will say that there is one fly in the Pimm's.

On 12 March Jaguar will field two cars at the Australian Grand Prix. This is like David Batty stepping up to take that penalty against Argentina. He'd never done it before and it would be a very public place in which to miss . . .

I'm desperately glad to see Jaguar moving in on Formula One. I like the idea of a pit crew dressed in tweed helmets and plus-fours. And I'm hoping they'll take that silly drinks thing from Eddie Irvine and give him a pipe instead.

But as a Jag driver I'll find it rather disappointing if I get up in the middle of the night to find a Mercedes-powered McLaren on pole, a BMW-powered Williams in second and the Jaguars down in eleventh and twelfth places.

F1 was fine when autocratic teams ran the show. You can't buy a Minardi, so it didn't really matter that their cars drove round at the back. The team members could be magnanimous in defeat, say they did it only for the thrill anyway and go home. But now the sport is being taken over by manufacturers, failure will be rather more serious.

It's funny, but while we are surprised when a Mercedes or a BMW breaks down, most people are still surprised when a Jaguar doesn't. A large part of F1's audience will remember the days when the XJ6 exploded on the hard shoulder and will nod sagely when the race cars do much the same sort of thing on the track.

Jaguar has to win. We know it has access to Ford's $22 billion bank account, and we know the Ford engine is just about the most powerful unit out there, so there are no excuses. If they are beaten by Mercedes and BMW on the track, they will be beaten by Mercedes and BMW on the road – it's that simple.

F1 isn't a sport any more. With the car makers running the teams, it has become a mobile showcase. And there's no point spending millions to show how brilliant you are if the global TV audience can see full well that you're not.

It's a simple choice: get a life, or get a diesel

I know why people who live in the Scampi Belt buy large, unwieldy off-road cars. And I don't blame them. I have a large, unwieldy off-road car. Lots of my friends have them.

It's because we like the Norman keep driving position. From way up there, among the ozone, we can see the enemy approaching. Only last week, yet another pensioner drove his car the wrong way down the M40 and was eventually killed when he slammed head-on into a BMW. Wouldn't have happened if BMW-man had been in a Range Rover; he'd have seen him coming.

We like the security too. Oh, sure, off-roaders are more prone to turning over, and they can clear a motorway crash barrier with feet to spare, but in the Harvester Zone, where traffic rarely gets above 40, a four-wheel-drive can smash and bash its way through the most vigorous accidents, causing nothing more than light bruising to those inside.

So, on the suburban school run, the simple fact of the matter is this: your children are safer in a heavyweight off-roader than in a normal car.

Unfortunately, words like 'big' and 'heavy' and 'high' mean that off-road cars cleave the air like wardrobes.

Which means the fuel injectors on their large engines have to operate with the ferocity of that fountain in Geneva.

Let me put it this way. My daughter faces an 18-mile trip to school each morning, so that's 36 miles, twice a day . . . at 12 miles to the gallon. This equates to £93 a week, or nearly £5000 a year. For petrol. To do the school run. And that makes me wonder, for the first time in my life, whether maybe it's sensible to think a little bit more seriously about switching to Satanism. And that's why I chose to spend the whole of last week tooling around in a diesel-powered Jeep Grand Cherokee: £31,000-worth of carcinogenic soot and evil.

You may think that this was the wrong place to start, because the Americans don't understand diesel engines, but the Cherokee is built in Austria and uses a 3.1-litre turbo unit designed in Italy. So it should have been OK. I'd only gone five yards before I knew it wasn't. My foot was welded to the floor, and there was enough noise to cause an earthquake, but the speedometer was climbing with the verve of continental drift: 0 to 60 takes 14 seconds.

Aware of this shortfall, I planned my overtaking manoeuvres with great care. But time and again I'd pull out and sit on the wrong side of the road, going nowhere, until a flurry of flashing lights coming the other way forced me to get back in line.

So it's all very well saying I got all the way from Oxfordshire to a shoot in Yorkshire, and back, on one 17-gallon tankful, but you're bound to do 23mpg if you spend the entire time stuck behind old people in Rovers doing 40.

Obviously, I eased it up a bit on the motorway but, at 7 p.m., Johnnie Walker handed over to Bob Harris, and suddenly the radio fell silent. No, really, at 70mph in a diesel-powered Jeep, Whispering Bob is completely inaudible. With no chitchat to while away the hours, I reached into the back and found a pair of headphones. They say you shouldn't drive while wearing cans, but in a car of this type it really doesn't matter. You can't hear anything anyway, and what does it matter if you're killed? The damned thing is so slow you'd never get where you were going, so you may as well be dead.

Buying a diesel-powered Jeep rather than the 4.7-litre petrol-driven V8 would save perhaps £1500 a year in fuel, but it will add half an hour to every journey. And half an hour twice every working day equates to five hours a week. And that, in a lifetime, is 9000 hours – 375 wasted days. Just to save a few quid. It's like cutting your hands off to save money on gloves.

What's really annoying is that the diesel engine spoilt what I suspect is rather a good car. Oh, it's too expensive and luxurious for the gamekeeper and it's way too small on the inside for the school run. Furthermore, its jiggly ride and plasticky switches, allied to some truly disgusting World of Leather seats, means it's no match for the Range Rover. But then it doesn't cost fifty grand, and you get a lot of toys as standard, from he-man stuff like permanent four-wheel drive to light-in-the-loaf features like a CD player. It also looks good and, with a V8, it will do 0 to 60 in 8 seconds.

Unlike the Japanese competition, it's not a utilitarian box. It can cope with the rigours of a grouse moor and works, too, in the executive bathroom.

I'm therefore such a fan that I decided to leave the pheasants I shot in the boot, as a sort of present to the public relations man who lent it to me. Shame it was a diesel, though, in the same way that it's a shame he's a vegetarian.

Insecure server?

When I first began to write for a living, I used a manual typewriter that provided very little in the way of distractions. You could type in black ink, and when that became boring you could type in red ink. And that was about it.

But now, I'm simply staggered that I'm sitting here writing anything at all, because my new computer can do so much more. When I turn it on in a morning, knowing that I must write something before lunchtime or I'll be killed and eaten, I still get waylaid by the promise of a quick game of FreeCell to get me in the mood.

And what's this? Heavens, it seems I can also sit here all day watching DVD movies with CD-quality sound. So now I face a choice. Write, or spend half an hour or so on board *Das Boot*.

Das Boot won, but now I'm back and the deadline is getting awfully close. But I fancy looking for pictures of naked girls on the web, so I'll just do that for a while, if you'll excuse me.

Right. Now the thrust. I heard a chap on the radio saying he'd just bought a car on the Internet. He'd found a dealership, negotiated a price, chosen a colour and had the whole transaction done and dusted within seven days. Well, I bet he's fun on a night out. You can't buy a car

over the Net, you idiot. You'll never know whether the
seat gives you backache, whether the salesman's a git, or
if you're talking to a silicone Maxwell who'll take your
credit card number and fall into the sea with it.

And what about second-hand cars? Even if you could
find something for sale that isn't in Minnesota, how could
you possibly know what it's like without taking at least a
tiny test drive? Something that's impossible online.

I know Ford has built a hologram car for Tony's
Dome, but this won't give an accurate picture at all.
In fact, it is the most useless invention I've ever heard of.
What is the point of a car that doesn't exist? Sure, you
could make it go into town, but why, if you can't go
with it?

Now, where was I? Ah, yes. What if you decided to do
all your shopping via computer? Think. You could work
from home, watch the latest movies and have everything
you need brought to your door. You'd never need to go
out. So then you'd lose your social skills, become covered
in boils and, eventually, you'd die. No one would know
until goo started to seep into the flat below. And your
holo-car began to pixelate.

For me, though, the biggest risk with the Net is fraud. I
have been asked many times for my credit card number
and, occasionally, I've felt tempted to tap it in. But I never
will, because, for all I know, the vendor is a Colombian
drug lord who will not be willing to uphold any money-
back guarantee.

So, if I'm not going to buy anything on the Net, why
is every Internet-based company worth £2000 billion?
If they can't sell anything, they'll go bust. It'd be like
opening a restaurant and refusing to unlock the doors. Or,

more accurately, like hiring Ronnie Biggs to take people's credit cards after dinner.

As I see it, the Net has two purposes. First of all, it's a giant library that can tell you anything at any time of day or night. But none of the information contained in the silicone-nerve centre can be trusted. So far as I can tell, there's nothing to stop me setting up a web site that says that Tara Palmer-Tomkinson is 47 and has a degree in robotics from Cambridge University.

Try it. Go into the Net tonight and ask for biographical details of, say, James Garner. You'll find that every site contradicts the next, whereas if you look in a book you know every fact has been checked and then checked again. And most books are not written by 14-year-old boys with apple-sized zits.

So this leaves us with the Net's only real purpose: pornography. If you want to see what can be inserted in whom by what, then there is a bewildering array of photographic evidence. Every star has been disrobed for your pleasure, and every act, no matter how deranged, is reproduced in full grisly detail.

Which brings me back to the original point. Why is every Internet company worth £2000 billion? Why, if I paid a visit to a venture capitalist this afternoon with some half-baked Internet-based idea, would he be willing to give me his house and all its contents?

I suspect we are looking at the emperor's new clothes here, and that no one has yet stepped forward to say, Hang on a minute. This is all b******s. And breasts, bosoms and pubic hair.

Vauxhall recently offered a thousand-pound discount to anyone who bought one of their cars over the Internet, and I'm absolutely dying to see just how many people take them up on it. And more than that, how many people meant to but were distracted en route by the promise of some Hot Asian Babes. Or even a game of FreeCell.

The sooner we all remember that a computer is a tool, like an electric drill, a hammer or a washing-up bowl, the better it will be for everyone. And the sooner we remember that cars need to be tried before you buy, the better it will be for your peace of mind.

Ahoy, shipmates, that's a cheap car ahead

We've never needed an excuse to go to France. The food, the wine and, in the south at least, the sumptuous climate are enough. But now the allure is even stronger, because you can go over there and, between mouthfuls of foie gras, wave wads of sterling at their unemployed youths. Then, after lunch, you can go into one of their job centres and stand in the middle of the room laughing.

Do not, however, go there through the Channel Tunnel, because they will think you've been eating beef and are an idiot. If time is tight, you are better off flying, and if it is not, why start your trip in a box? If you were a veal calf, they wouldn't allow it.

But, you will now wail, the ferries are so much worse. They are full of French schoolchildren sent to England to steal from the rich, or sarf London darts teams who bought 30,000 cans of extra-strong lager on the outward trip and drank it all on the way home.

Not any more. Duty-free shopping was abolished in

June, and afterwards ferries became little islands of Victorian calm between hurly-burly Britain and the poverty and despair of France.

On a ferry, your umbrella becomes a parasol and you are filled with an overwhelming urge to take up bee-keeping. You park your car and pop up on deck to wave goodbye to the white cliffs with a Dunkirk spirit of adventure in your heart. Seagulls floating on the salty breeze, a quick promenade to the front to make sure the doors are shut, and then inside for lunch at one of Langan's brasseries.

Sadly, however, I fear that this Jane Austenesque idyll may be short-lived, because this week P&O decided to become a car dealer. It will buy cars on the Continent, which it will sell in Britain at Continental prices. So, you ring them up, choose the car you want, the colour, the spec, and 12 weeks later it will be delivered to your door. You can even get part exchange.

Some of the discounts are breathtaking. A Mercedes CL500, which would cost £83,000 in Britain, is available from P&O for just £69,900. You can save £7300 on a Range Rover, £6500 on a Jaguar XK8 and £2000 on a Golf. And remember, they come with right-hand drive and warranties that British dealers are bound to honour.

Now you're probably thinking that this is nothing new, that hundreds of companies have been doing the same sort of thing ever since the rip-off Britain stories began. Well, yes, but we all watch *Watchdog*, and we know that some of these guys will get our deposit cheques and enjoy dinner that night in Rio with Ronnie Biggs. You could, of course, remove the risk element by going over there and ordering the car yourself, but let's be honest. Hand on heart, do you speak Flemish?

I have spoken to hundreds of people about buying cars on the Continent. I've told them about the savings and the ease with which it can be done. I've explained that Ford in Britain will even give you a factsheet to facilitate the buying of a car in Belgium. It lists not only individual dealers but also an English-speaking contact. But the response is always the same: 'Oh, I can't be bothered.'

We've been told that personally imported cars from the Continent have hit new-car sales hard and that soon the car makers will have to lower their prices. But they won't, because it's not true. Car sales are not really down at all, while it's easier and less risky to buy a new car here.

Well, this P&O deal stops all that. It is a blue-chip company, up there with Marks & Spencer and Tommy Cooper as a name you can trust. Instead of taking your money to South America, it will deliver a brand-new car, of your choosing, to your door, with huge savings. It says it'll bring in 10,000 cars a year, but I really don't think so. It'll be more like 1.9 million, because anyone who buys a new car now from a main agent is not simply daft. He's a fully certified window-licker. A loony. Madder than the result of a liaison between a March hare and Mad Jack McMad, who as you all know was winner of last year's Mr Mad competition.

A P&O spokeswoman told me that the company had been hit by the abolition of duty free and that it needed something to fill the hole. Well, you've got it now, love. Because this car thing is a little bit bigger than a darts team buying booze for the Christmas party.

The trouble is that all the ferries coming back to Dover will be jammed solid with new cars, so ordinary people will have to make like moles and come home through the tunnel. Pity, really, but it's a small price to pay for making a worthwhile point to our poverty-stricken next-door neighbours. Les rosbifs. Not so mad after all, vieux haricot.

So modern it's been left behind already

When we think of the French, we think of Breton jerseys, an onion necklace and a sit-up-and-beg bicycle. And when they think of us, they think of le pub. Morning, John.

The pub. A gentle murmur of Sunday morning, corduroy bonhomie. The ceaseless winking of a fruit machine in the corner and a tray of yellowed dominoes left out from the night before. Brown beer, plaice in breadcrumbs with a lemon wedge. Horse brasses. The usual, John.

That's not the Britain I know. The Britain I know has fired-earth walls and Macy Gray on the stereo. Linen tableclothes, low-voltage lighting and a glass of Chablis. It's not nicotine yellow. It's ice white and garnished with a brushed-aluminium handrail. Not a clump of cress.

I was in what the French would call a traditional pub last week, and I couldn't believe how backward it felt. The patterned carpet. The cheese and onion crisps. And that infernal winking machine.

If I ran Rover, I'd ban all my design staff from pubs. I'd make them go to Gary Rhodes's place in the City, and I'd tell them to have one of his lamb sausages. Then I'd take them to Pharmacy in Notting Hill and I'd say: 'Look, fools. This is Britain now. So stop trying to make our cars look like the Coach and bloody Horses.'

They don't put an onion holder in Renaults, so why

do British car designers feel they need all that leather and walnut? This has never occurred to me before, but then I'd never driven the Audi S6 Quattro before. Now I have. And I can tell you this. It's the most 'now' car you can buy.

They've taken a normal A6 and flared the wheel arches, not subtly, but with a swath of Healeyesque eyebrows. These shroud massive alloy wheels that sit well with the enlarged radiator grille, chromed door mirrors and brace of superfat exhaust pipes.

Of course, you don't do all this to a car unless there is some meaningful meat under the bonnet, and there is. They've taken the 300bhp 4.2-litre V8 out of the bigger A8, fitted it with afterburners and dropped it into the A6. So now there's 340bhp, and that's enough to get you from rest to 60mph in around six seconds. Top speed, to keep the German Greens happy, is limited to 155mph.

Good car, then? Oh, yes, and inside it just keeps getting better. The seats are finished in mock suede, and the dash is carpeted. Sounds ghastly, especially when I tell you the carpet in question is fronted with polythene. But it works. Then night falls and you have to turn on the lights. And at this point you'll want to pull over and invite perfect strangers to come and have a look. The dash becomes a teeming mass of small red lights. It's like looking down on Los Angeles from the Hollywood hills, only in the Audi you're doing 75mph and just moving into third.

Climbing out of the S6 and back into my Jag felt as though I'd moved back two centuries. From Conran to Wren. From Tony Blair's New Labour to Harold Wilson's bottle of HP sauce.

But now it's time for that 'however' moment. Ready? OK, then, here goes.

However, while the Audi may well be the soup of the day, there's a fly in it.

It costs £52,250, which makes it a direct rival to the BMW M5 and Jaguar's recently tweaked supercharged XJR. Now I know that, in terms of ambience, the Jag is a pub and the BMW is a Harvester, and that in a traffic jam I'd much rather be in the Audi. I also know that the S6 is available as an estate, which makes it useful at the gymkhana. And when it comes to pulling water-skiers, the Audi is in a class of one. But what about those times when you're not in New Zealand and the boat's broken. Then what?

Well, sorry, but on normal roads the S6 is outgunned: 340bhp may be a lot, but the Jag serves up 370 and the BMW a massive 400.

Time and again, I'd pull out to overtake a truck and be left there, on the wrong side of the road, wishing that a) I'd stayed where I was or that b) I was in an XJR. Put simply, the S6 is fast, but not fast enough.

And while the four-wheel-drive system makes it tidy on a wet roundabout, it does not have the poise of its rivals everywhere else. From time to time it feels leaden, ponderous even. While the M5 and XJR iron out bumps in the road, the Audi transmits each one with faithful accuracy. It's sad this. It's like sitting down at the finest-looking restaurant only to find the chairs are uncomfortable and the food is better at your local.

The funny thing is, though, that I'd go back. Food is only one bit of a restaurant's make-up, in the same way that high-speed poise is only one part of a car's. And in so many other areas, the S6 is absolutely bang-on.

Something to shout about

More news from Rover on the 75 front. With 8000 sold, it's outperforming the Alfa 156 here in the UK, while over in Italy it's been voted the 'most beautiful car in the world'. Furthermore, a bunch of Middle Eastern motoring observers have voted it their car of the year. So, there we are then. It's brilliant.

Well, sorry to be the one who relieves himself all over the bonfire, but I'm not convinced. I don't care how many LCD readouts they put on the dashboard or whether the K-series power plant is an engineering masterpiece, the Rover name still smacks of postwar austerity; as a result, the 75 is a sort of wheeled Werther's Original.

And then there's that advertisement where the new 25 is seen driving round a roulette wheel. What's that all about? It should have Dr Finlay behind the wheel, not some bird in a silk nightie.

And I don't see how the situation will ever get better, not so long as BMW remains at the helm. It's a bit like Manchester United buying Liverpool FC and telling them: 'Be good . . . but not as good as us.' The best Rover can hope to achieve is second place, and that's why they are about to post losses of around £600 million. A sum described in City circles as 'a lot'.

Then there's Marks & Spencer who, like Rover, have a middle-aged, middle-England appeal and who are also about to announce some catastrophic results. And meanwhile we have a £758 million Dome that no one wants to visit, a river of fire that didn't happen, a big wheel that broke and a flame of hope – which was designed to burn all year in Birmingham but fizzled out after five days.

In Brazil, some of our football players lost an important game of football, and I understand that our cricketers, too, failed to do well in South Africa. So, all in all, it's not been a good start to the third millennium for the Mr Smiths and Mr Robinsons of the world.

Some, of course, would say that this is predictable, that we should accept the fact that these days England is just a 44 dial code, .uk on the Web, the fifty-first state of America and the thirteenth member of the EU. They would argue that the empire is gone, along with Scotland, Wales and Northern Ireland, and that we are nothing more than a two-bit island race in a global village.

I, however, am proud of being English, in a passive, now-that-you-mention-it sort of way. I like the fact it's always 57 degrees and drizzling, because this means we spend more time at work and less on the beach. And this, in turn, makes us richer.

I mean, look at France. Yes, they won the World Cup and, yes, they came damned close to taking the ultimate rugby crown, too, but so what? Their idea of a luxury car is a Peugeot 406, and their students have to get jobs in London since there are none in Paris.

And Germany? Think how delighted they must have been when they bought Rover, how they'd put one over on Tommy. But now it turns out their longest-serving chancellor was corrupt and their little acquisition is costing them £600 million a year.

Sure, I'm no great fan of Phoney Tony, but then he's Scottish. As is his Chancellor, his Lord Chancellor, the Chief Secretary of the Treasury, his Foreign Secretary and the new bloke at Transport. Then there's Prescott, who's Welsh, and most of the rest are homosexual. England's contribution to the Cabinet is Mo Mowlem, and she's the best of the lot, by far.

And then there's Richard Curtis, Marco Pierre White and Tara Palmer-Tomkinson. There's *Notting Hill* and *The Full Monty*. I even had some British wine the other night, and it was bloody good.

But best of all, there's Jaguar. My old XJR has just gone back after two years and 20,000 totally trouble-free miles. No, really, in all that time not a single thing went wrong, whereas life with my Toyota Landcruiser is a nonstop return trip to the dealers.

I've looked at all the alternatives. There's a Jeep Grand Cherokee outside my house right now, but it's too jiggly. The Mitsubishi Shogun is too brash, and the Merc M-class is just too Guildford. Which means that, some time this year, we shall get either a Discovery or a Range Rover, because they're still the best 4x4s by far.

And what about sports cars? I know the new Boxster is a fine-handling machine that now goes as quickly as its badge would suggest, and I'm aware that six-cylinder SLKs are about to burst out of the pipeline. But, come on, neither of these is a match for the sheer brutality you get from a TVR. These things are so aggressive that they could almost be Scottish.

But if they're out of your price range, then it's off to Mazda for an MX-5, a car that wouldn't be half as good if it were not for the Lotus Elan.

And anyway, we do still have an empire. It is a small island in the Pacific Ocean, and last time I looked the population was 8000. And all of them, curiously, have Rover 75s.

Appendix

A taste of what Postman Pat has pushed through the Clarkson letterbox over the years.

Dear Jeremy ...

'If Clarkson found Norfolk flat and featureless he is in a minority. Norwich has a shopping centre that is as good as any in the country ...'

P.G.

'I think most Norfolk people wish that Jeremy Clarkson would revert back to his previous job selling Paddington Bears. I do not care for his road testing attitude and even less his patronising and sanctimonious views of Norfolk.'

C.M.

'I was shocked to learn that the French Gendarmerie is using your photo for training purposes of how an English hooligan looks when he is full of britpiss. You should complain.'

T.V.

'Clarkson, you are a freak. You scare the children the way you look on television. And it gets worse when you open your mouth. Unbelievable.'

T.V.

'I am a squaddie on top of a hill near the border of Kosovo and recently saw an article calling you a fashion freak. I don't agree with what they say and I think people from Norfolk still point at cars as well. But getting to the original point, I think you are the coolest dude to put his foot on planet earth . . . keep up the good work.'

M.S.

'I am 83 years old and I've been driving every day for a living since 1930. The modern cars you write about today, I wouldn't have one as a gift. They are rubbish. Who wants to do over 50mph anyway?'

J.J.

'Jeremy, wonderful how you sorted out those navish foreigners and those poofters, and German ones at that. Your friends urge you to consult a doctor and your enemies hope you don't.'

T.V.

'Just fill the magazine with lots of pictures of Jeremy and lots of articles written by him. He's so gorgeous and sexy I'd like to cover him with chocolate and lick it all off . . .'

S.H.

'As part of an English project, we are allowed to write about our favourite celebrity. I chose you because I think you're funny and get to drive ace vehicles. My friend Max is writing to Tiff Nodel, the one who helps to present Top Gear with you. I think you're better than him though.'

G.F.

'Congratulations on your new talk show on the BBC. This is an absolute breakthrough. For the first time a baboon will have his own talk show.'

T.V.

'I have a large collection of toy cars and trucks. The fact that you said collectors of toy cars are child molesters I found not only highly offensive to thousands of ordinary people, but of such you should be sent to a shrink to see what makes you tick ... I wish upon you an eternity stuck in an old car in a convoy of trucks and caravans ...'

J.F.

'If the VC were awarded for stupidity and ignorance you would be one of the first to receive it. Nature seems to have given you a large body but a very small brain ...'

B.C.

'People who commit crimes are dysfunctional. They are alienated, bitter and resentful. So they attack symbols of success, like JC's Cosworth and he wants to flog them within an inch of their lives, which will make them even more resentful. JC is intelligent, gifted and graced by success. He should not insult our intelligence by uttering such bollocks.'

A.D.

'Jeremy Clarkson is without doubt the most appallingly sexist person to strut across planet earth but he has a valid, if slightly liberal point of view regarding the treatment of the vehicle villain ... I have just had the misfortune of being the victim, for the fourth time, of car crime. These bastards should be staked out naked in the desert . . . etc.'

G.M.

'We are out there, the Supertramp music fans. I have all the music and if you would like anything taped please drop me a line.'

P.S. Did you see them at the Albert Hall in 1997?

M.O.

'Dear Mr Clarkson, You're a prick.'

Jeremy Clarkson

CLARKSON ON CARS

Jeremy Clarkson is the second best motoring writer in Britain. For twenty years he's been driving cars, writing about them and occasionally voicing his opinions on *Top Gear*.

No one on in the business is taller.

Here, he has collected his best car columns and stories in which he waxes lyrical on topics as useful and diverse as:

The perils of bicycle ownership

Why Australians – not Brits – need bull bars

Why soon only geriatrics will be driving BMWs

The difficulty of deciding on the best car for your wedding

Why Jesus's dad would have owned a Nissan Bluebird

… And why it is that bus lanes cause traffic jams

Irreverent, damn funny and offensive to almost everyone, this is writing with its foot to the floor, the brake lines cut and the speed limit smashed to smithereens. Sit back and enjoy the ride.

JEREMY CLARKSON

I KNOW YOU GOT SOUL

Some machines have it and others don't: Soul. They take your breath away, and your heart beats a little faster just knowing that they exist. They may not be the fastest, most efficient, even the best in their class – but they were designed and built by people who loved them, and we can't help but love them back.

For instance,

Zeppelin airships, whilst disastrously explosive in almost every case, were elegant and beautiful bubbles in the air.

The battleships were some of the least effective weapons of war ever built, but made the people who paid for them feel good.

Despite two tragic crashes, the *Space Shuttle* still leaves you with a rocket in your pocket.

Some might dismiss this list as simply being for boys and their toys, but, as Jeremy Clarkson shows, that is to miss the point of what makes the sweep of the Hoover Dam sexier than a supermodel's curves; why the *Princess* flying boat could give white elephants a good name; and why the *Flying Scotsman* beats the Bullet Train every time.

In *I Know You Got Soul*, Jeremy Clarkson celebrates, in his own inimitable style, the machines that matter to us, and tells the stories of the geniuses, boffins and crackpots who put the ghost in the machine.

JEREMY CLARKSON

THE WORLD ACCORDING TO CLARKSON

Jeremy Clarkson has seen rather more of the world than most. He has, as they say, been around a bit. And as a result, he's got one or two things to tell us about how it all works – and being Jeremy Clarkson he's not about to voice them quietly, humbly and without great dollops of humour.

With a strong dose of common sense that is rarely, if ever, found inside the M25, Clarkson hilariously attacks the pompous, the ridiculous, the absurd and the downright idiotic ideas, people and institutions that we all have to put up with at home and abroad, whilst also celebrating the eccentric, the clever and the sheer bloody brilliant.

'Hilarious ... it'll make you appreciate the ludicrousness of modern life and have you in stitches' *Sun*

PENGUIN BOOKS

Born to be Riled

Jeremy Clarkson began his writing career on the *Rotherham Advertiser*. Since then he has written for the *Sun*, the *Sunday Times*, the *Rochdale Observer*, the *Wolverhampton Express and Star*, all of the Associated Kent Newspapers, and *Lincolnshire Life*. Today he is the tallest person working in British television.

Jeremy Clarkson's other books are *Clarkson's Hot 100*, *Clarkson on Cars*, *Motorworld*, *Planet Dagenham*, *The World According to Clarkson*, *I Know You Got Soul* and *And Another Thing: The World According to Clarkson Volume 2*